# A TEXTBOOK OF TOPOLOGY

# A TEXTBOOK OF TOPOLOGY

**B.C. Chatterjee**
*Formerly, Department of Pure Mathematics*
*Calcutta University*

**S. Ganguly**
*Department of Pure Mathematics*
*Calcutta University*

**M.R. Adhikari**
*Department of Pure Mathematics*
*Calcutta University;*
*Formerly, Department of Mathematics*
*Burdwan University*

AB **Asian Books Private Limited**
7/28, Mahavir Lane, Vardan House
Ansari Road, Darya Ganj, New Delhi-110 002

**AB**

# Asian Books Private Limited

*Registered and Editorial Office* 7/28, Mahavir Lane, Vardan House, Ansari Road, Darya Ganj, New Delhi-110 002.
E-Mail : asian@nda.vsnl.net.in
World Wide Web : http://www.asianbooksindia.com
Phones : 3287577, 3282098, 3271887, 3259151-71-81
Fax : 91 11 3262021

*Sales Offices*

*Bangalore* 103, Swiss Complex No. 33, Race Course Road, Bangalore-560 001
Ph. : 2200438 Fax : 91 80 2256583 Email : asianblr@blr.vsnl.net.in

*Kolkata* 10 A, Hospital Street, Calcutta-700 072
Ph. : 2363040 Fax : 91 33 2369899 Email : calasian@cal.vsnl.net.in

*Chennai* 2, Aziz Mulk, 9th St. Thousand Lights, Chennai-600 006
Ph. : 8295093, 8295094 Email : asianmds@vsnl.net

*Guwahati* Anandaram Barua Road, Panbazar (Opp. Don Bosco Church)
Guwahati, Assam-781 001
Ph. : 0361 513020, 635729 Email : asianghy@sancharnet.in

*Hyderabad* 3-5-1101/1/B IInd Floor, Opp. Blood Bank,
Narayanguda Hyderabad-500 029
Ph. : 4754941, 4750951 Fax : 91 40 4751152
Email : hydasian@hd2.vsnl.net.in

*Mumbai* 5, Himalaya House, 79 Palton Road Mumbai-400 001
Ph. : 2619322, 2623572 Fax : 91-22-2623137
Email : asianbk@bom3.vsnl.net.in



1st Edition 2002

ISBN 81-86299-23-8



*Published by* Kamal Jagasia for Asian Books Pvt. Ltd., 7/28, Mahavir Lane, Vardan House, Darya Ganj, New Delhi-110 002.

*Typeset at* DataVision, D-2, MCD Flat, R Block, Greater Kailash Part-I, New Delhi-48

*Printed at* Rajkamal Electric Press, Delhi-33

# PREFACE

The book is written as a text book of Topology and provides the readers with a foundation in general topology. This book will facilitate further study in a broad variety of mathematical disciplines. The materials in the book have been mainly drawn from the book 'General Topology and Elements of Set Theory' (1978), written by Professor B.C. Chatterjee, the first author of the present book. He incorporated the fruits of his long experience of teaching topology in the Department of Pure Mathematics, Calcutta University during the period 1948-1978. Professor Chatterjee took help of French and German books and made his topics arranged with an eye to the benefit of the students.

Professor Chatterjee had a mind to include some more topics to make the book more organised but unfortunately, this could not be done by him due to his sudden demise. We were his students and were blessed enough to have his love and affection. We want to pay homage to our revered teacher by adding more materials to complete the book with our limited knowledge and ability. The book is an outcome of this endeavour. The book is designed to develop the fundamental concepts of general topology which are the basic tools of working mathematicians in variety of mathematical disciplines. The book also provides results of general topology which are necessary for the study of algebraic topology. The book has nine chapters. Chapter 1 provides 'Elements of Set Theory', Chapter 2 provides 'Topological Spaces'. We study 'Separation Axioms' in Chapter 3, 'Real Functions' in Chapter 4, 'Countability' in Chapter 5, 'Compactness' in Chapter 6, 'Connectedness' in Chapter 7, 'Metric Spaces' in Chapter 8, 'Homotopy and Fundamental Groups' in Chapter 9. Another book is designed to serve as an introduction to the basic materials of 'Algebraic Topology'

Any suggestion for improvements of the book is welcome.

S. Ganguly

M.R. Adhikari

January 2, 2002

# CONTENTS

*Chapter 1*

# ELEMENTS OF SET THEORY

The theory of sets as a mathematical discipline was created by the German mathematician G. Cantor (1845-1918). The set theory, as given by Cantor in the intuitive form, is now called the ''naive theory of sets''. It underwent radical changes and has been supplanted by the ''axiomatic theory of sets''.

## Art. 1 SETS AND SUBSETS

According to Cantor ''a *set* is any collection of definite, distinguishable objects of our intuition or of our intellect to be conceived as a whole''. The essential feature of the definition is that a collection of objects is to be regarded as a single entity in the form of a set. The objects are called the *elements* or *members* of the set. If $A$ denotes a set, and $x$ is an element of the set $A$, then we express this symbolically by $x \in A$. If on the other hand $y$ does not belong to the set $A$, we express this by $y \notin A$.

Let $A$ and $B$ denote two sets. If every element of $B$ is also an element of $A$, then $B$ is said to be a *subset* of $A$, and this is expressed symbolically by $B \subset A$ or $A \supset B$. $\subset$ (or $\supset$) is called the *inclusion relation* for sets.

Let $B$ be a subset of a set $A$. Then the elements of $B$ are distinguished from other elements of $A$ by some property, say $\pi$. We denote this symbolically in the form

$$B = \{x \in A : x \text{ has the property } \pi\}$$

which means that $B$ consists of all those elements $x$ of $A$ which have the property $\pi$. For instance, if $\mathbf{Z}$ denotes the set of all integers, then the (sub)-set $Y$ of all even integers is denoted by

$$Y = \{x \in \mathbf{Z} : x \text{ is even}\}.$$

We conceive of a set, which contains no element at all, and call it the *empty set*, the *void set*, or the *null set*, and denote it by $\Phi$. The null set is considered to be a subset of every set.

It now readily follows that :

(*i*) Two sets $A$ and $B$ are equal, *i.e.*, $A = B$, iff $A \subset B$ and $B \subset A$.

(*ii*) For three sets $A, B, C$, $A \subset B$ and $B \subset C$ imply $A \subset C$.

(*iii*) Every set $A$ is a subset of itself, *i.e.*, $A \subset A$.

A set $A$ is called the *improper subset* of itself; any other non-null subset (if it exists) is called a *proper subset.*

## 1.1 Algebra of Sets

Let $\{A_v\}$ be a given family of sets.

The *intersection* (or *meet*) of the family of sets $\{A_v\}$ is a set $M$, determined by the property that an element belongs to $M$, iff it belongs to every member of the family $\{A_v\}$; and this is denoted by $M = \cap \{A_v\}$.

The *union* (or *join*) of the family of sets $\{A_v\}$ is a set $K$, determined by the property that an element belongs to $K$ iff it belongs to at least one member of the given family $\{A_v\}$; and this is denoted by $K = \cup \{A_v\}$.

It is evident from the above definitions that, if $G$ is any set contained in all members of the family $\{A_v\}$, and $H$ is any set which contains all members of the family $\{A_v\}$, then

$$G \subset \cap \{A_v\} \subset \text{each } A_v \subset \cup \{A_v\} \subset H.$$

The intersection of two sets $A$ and $B$ is denoted by $A \cap B$. $A$ and $B$ are said to be disjoint iff $A \cap B = \Phi$.

The union of two sets $A$ and $B$ is denoted by $A \cup B$. The union of a family of sets is empty iff each member of the family is empty.

One can also readily verify (verification left as exercise) the following properties for intersection, union and inclusion relations for sets :

(*i*) $A \cap A = A$ and $A \cup A = A$; (*Idempotent properties*)

(*ii*) $A \cap B = B \cap A$, and $A \cup B = B \cup A$; (*Commutative properties*)

(*iii*) $(A \cap B) \cap C = A \cap (B \cap C)$, and $(A \cup B) \cup C = A \cup (B \cup C)$; (*Associative properties*)

(*iv*) $A \cap (B \cup C) = (A \cap B) \cup (A \cap C)$, and $A \cup (B \cap C) = (A \cup B) \cap (A \cup C)$; (*Distributive properties*)

(*v*) $A \cap (A \cup B) = A$, $A \cup (A \cap B) = A$; (*Absorptive properties*)

(*vi*) Each of the three properties :

$$A \subset B,\ A \cap B = A \text{ and } A \cup B = B,$$

implies the other two. (*Consistency property*)

As $\Phi \subset A$, it follows by (*vi*) that $A \cap \Phi = \Phi$ and $A \cup \Phi = A$, for every set $A$.

Let $A$ and $B$ be two sets, then their *difference* $A - B$ (or $A \setminus B$) is defined to be the set, consisting of all those elements of $A$ which are not elements of $B$. The definition may be put in the symbolic form,

$$A - B = \{x \in A : x \notin B\}.$$

In particular, $A - B = \Phi$ iff $A \subset B$, and $A - B = A$, iff $A \cap B = \Phi$.

If $A$ is a subset of a set $X$, then $X - A$ is called the *complement* of $A$ in $X$; it is also a subset of $X$. A subset $B$ of $X$ is the complement of $A$ in $X$ iff $A \cup B = X$ and $A \cap B = \Phi$. Also $A$ is the complement of $B$ in $X$; that is $X - (X - A) = A$.

Let $X$ be a fixed set, and $A, B, C, \ldots$ be subsets of $X$. In considering the intersection, union, complements in $X$ of these subsets, the set $X$ is usually called the *universal set* (and is sometimes denoted by $I$). The complement of a subset $A$ (in $X$) is then denoted by $X \setminus A$ or $X - A$ (or $I - A$).

Let $\{A_\alpha\}$ be a family of subsets of a fixed set $X$, then the following *dualisation properties* can readily be verified :

$$X - \cap \{A_\alpha\} = \cup \{X - A_\alpha\}, \text{ and } X - \cup \{A_\alpha\} = \cap \{X - A_\alpha\}.$$

The above results are known as *DeMorgan's rules.*

For two subsets $A$, $B$ of $X$, the above are reduced to

$$X - (A \cap B) = (X - A) \cup (X - B)$$

and $$X - (A \cup B) = (X - A) \cap (X - B);$$

and these can be proved as follows.

Let $p \in X - (A \cup B)$. Then $p \in X$ and $p \notin A \cup B$. $p \notin A \cup B$ means that $p \notin A$ and $p \notin B$. Thus $p \in X$ and $p \notin A$, so that $p \in X - A$, and also $p \in X$ and $p \notin B$, imply $p \in X - B$. Hence $p \in (X - A) \cap (X - B)$. Thus $X - (A \cup B) \subset (X - A) \cap (X - B)$.

Next, let $u \in (X - A) \cap (X - B)$. Then $u \in X - A$ and $u \in X - B$, so that $u \in X$ and $u \notin A$ and $u \notin B$. That is, $u \in X$ and $u \notin A \cup B$, so that $u \in X - (A \cup B)$. Thus $(X - A) \cup (X - B) \subset X - (A \cup B)$.

Combining the above two results, we get

$$X - (A \cup B) = (X - A) \cap (X - B).$$

The other part can also be proved in a similar way.

## EXERCISES

**1.** Verify the properties (*i*) – (*vi*) in page 2.

**2.** Prove that for two subsets $A, B$ of $X$,

(*i*) $A \subset B$ iff $X - B \subset X - A$,

(*ii*) $A \subset X - B$ iff $A \cap B = \Phi$,

(*iii*) $X - A \subset B$ iff $A \cup B = X$,

(*iv*) $A - B = A \cap (X - B)$.

3. Verify the dualisation properties in page 3.

4. For two sets $A$ and $B$, $(A - B) \cup (B - A) = (A \cup B) - (A \cap B)$, and it is called the *symmetric difference* of $A$ and $B$, and is denoted by $A \Delta B$. If $A, B$ are subsets of a (universal) set $X$, show that
$$A \cap B = A \cap [X - \{A \cap (X - B)\}],$$
and
$$A \Delta B = (A \cup B) \cap [X - (A \cap B)].$$

## 1.2 Cartesian Product of Sets

Let $P_1, P_2, ..., P_n$ be a finite collection of non-empty sets. Then their *Cartesian product* (or *combinatorial product*), denoted by $P_1 \times P_2 \times \ldots \times P_n$, is the set defined by $P_1 \times P_2 \times \ldots \times P_n = \{(x_1, x_2, ..., x_n) : x_k \in P_k, k = 1, 2, ..., n\}$; thus it is the set whose elements are all the ordered $n$-tuples $(x_1, x_2, ..., x_n)$, whose $k$th co-ordinate $x_k$ runs over the set $P_k$, for $k = 1, 2, ..., n$.

If, in particular, $P_1 = P_2 = \ldots = P_n = P$, then the Cartesian product $P_1 \times P_2 \times \ldots \times P_n = P \times P \times \ldots \times P$ is also denoted by $P^n$, and called the $n$th *Cartesian power* of the set $P$. $P^n$ consists of all $n$-tuples with the co-ordinates from the set $P$. [Thus $P^2$ consists of all pairs $(x, y)$, where $x$ and $y$ run independently over $P$]. The $n$-tuples $(x, x, ..., x)$, for all $x \in P$, are called the *diagonal elements* of $P^n$.

### EXERCISES

Verify the following properties :

1. $P \times Q = A \times B$ iff $P = A$ and $Q = B$.

2. $P \subset A$ and $Q \subset B$ imply $P \times Q \subset A \times B$, provided that $P$ and $Q$ are non-empty.

3. $(P - Q) \times S = (P \times S) - (Q \times S)$.

4. $(P \cap Q) \times (A \cap B) = (P \times A) \cap (Q \times B)$.

5. $(P \cup Q) \times (A \cup B) = (P \times A) \cup (P \times B) \cup (Q \times A) \cup (Q \times B)$.

## Art. 2 RELATIONS

### 2.1 Binary Relations

Let $X$ be a non-empty set. A subset $R$ of $X \times X$ is called a *binary relation* or simply, a *relation* on $X$. When $(a, b) \in R$, one expresses this by saying that "$a$ has the relation $R$ with $b$", and denotes this symbolically by $aRb$. If $(a, b) \notin R$, one writes $a\bar{R}b$, which means that "$a$ has not the relation $R$ with $b$".

When $R = X \times X$, $R$ is called the *universal relation* on $X$, and $aRb$ holds for every pair of elements $a, b \in X$. If, on the other hand, $R = \Phi$, then $R$ is called the *null relation*; and $a\bar{R}b$ holds for every pair of elements $a$, $b \in X$.

A relation $R$, defined on a set $X$, is said to be

(*i*) *reflexive*, iff $aRa$ holds for every $a \in X$; that is all the diagonal elements of $X \times X$ belong to $R$;

(*ii*) *transitive*, iff $aRb$ and $bRc$ imply $aRc$;

(*iii*) *symmetric*, iff $aRb$ implies $bRa$;

(*iv*) *asymmetric*, iff it is not symmetric;

(*v*) *anti-symmetric*, iff $aRb$ and $bRa$ imply $a = b$.

The *inverse of a relation* $R$ is a relation $R^{-1}$, defined by the property that $xR^{-1}y$ holds iff $yRx$ holds. That is, a pair $(x, y) \in R^{-1}$ iff the pair $(y, x) \in R$. Thus a relation $R$ is symmetric iff $R = R^{-1}$.

### 2.2 Equivalence Relation

A relation $R$ on a set $X$ is said to be an *equivalence relation* if it is reflexive, symmetric and transitive. The importance of an equivalence relation lies in the fact that :

**Theorem 1 :** *If an equivalence relation $R$ is defined on a set $X$, then $X$ is partitioned into disjoint classes in such a manner that two elements $a$, $b$ of $X$ belong to the same class iff $aRb$ holds.*

*Conversely, given any partition of a set $X$ (into disjoint classes), there exists correspondingly an equivalence relation $R$ on $X$, such that the partition of $X$ determined by $R$ is identical with the given partition of $X$.*

**Proof :** Let $x \in X$, and let $C_x$ denote the class of all those elements $x' \in X$, such that $xRx'$ holds. Let $C_x$ and $C_y$ be two classes such that $C_x \cap C_y \neq \Phi$. Then there exists an element $z \in X$, such that $z \in C_x$ and $z \in C_y$, so that $xRz$ and $yRz$ both hold. Now, $yRz$ implies $zRy$, by the symmetric property. Then, by the transitive property, $xRz$ and $zRy$ imply $xRy$, which implies $yRx$ (by symmetric property). Hence, for any $y' \in C_y$, $xRy$ and $yRy'$ imply $xRy'$, so that $y' \in C_x$; and for any $x' \in C_x$, $yRx$ and $xRx'$ imply $yRx'$, so that $x' \in C_y$. Thus $C_x = C_y$. Hence, for any two classes $C_x$ and $C_y$, either $C_x \cap C_y = \Phi$ or $C_x = C_y$. This proves the first part of the theorem.

Conversely, let a partition of $X$ into disjoint classes be given. Let $R$ be a binary relation on $X$, defined by $aRb$ iff $a$ and $b$ belong to the same class of the given partition. Then it follows readily that $R$ is an equivalence relation, and that the partition determined by $R$ is identical with the given partition.//

The family of the classes into which a set $X$ is partitioned with respect to an equivalence relation $R$, defined on $X$, is denoted by $X/R$, and called the *quotient set of X modulo R*.

## 2.3 Order Relations

An asymmetric, transitive relation, defined on a set $X$, is called on *order relation* on $X$; and $X$ is said to be ordered with respect to the particular order relation. If $R$ is an order relation on $X$, then the *ordered set X* is denoted by $(X, R)$.

A reflexive, antisymmetric and transitive relation $R$ on a set $X$ is said to be a *relation of partial order*; and $(X, R)$ is then called a *partially ordered set* or, simply, a *poset*.

The set of all subsets of a given set $X$ is called the *power set* of $X$, and is denoted by $P(X)$. Then the inclusion relation (among subsets of $X$) is a relation of partial order in $P(X)$; thus $(P(X), \subset)$ is a poset.

In a poset, the relation of partial order is usually denoted by "$\subset$", and called the *inclusion relation*; and $a$ is said to be included in $b$, or equivalently, $b$ is said to include $a$, if $a \subset b$.

A relation of partial order $R$, defined on a set $X$, is said to be a relation of *linear order*, or *total order*, or *complete order*, if the following *property of trichotomy* holds : "For any pair of elements $a, b \in X$, either $aRb$ or $bRa$ holds." Accordingly, $(X, R)$ is called a *linearly ordered*, or *totally ordered*, or *completely ordered* set or a chain.

**Example 1. (i) : *The (weak) inequality relation "$\leq$" is a relation of linear order on the set of real numbers.***

**(ii) *The (strong) inequality relation "$<$" is an order relation on the set of real numbers; but it is not a relation of partial order.***

A relation of linear order is usually denoted by "$\leqq$".

Let $(X, \leq)$ be a linearly ordered set, and $A$ be a non-empty subset of $X$. If $v$ is an element of $A$, such that $v \leq x$ for all $x \in A$, then $v$ is called the *smallest element* or the *least element* of $A$. If every non-empty subset of $X$ possesses a least element, then "$\leq$" is called a *well ordering relation* on $X$, and $(X, \leq)$ is correspondingly called a *well-ordered set.*

## EXERCISES

**1.** Let $R$ be a binary relation on a set $X$. Find :

(*i*) the smallest symmetric relation on $X$ including $R$,

(*ii*) the largest symmetric relation on $X$ included in $R$.

**2.** Let $R$ be a reflexive and transitive relation on a set $X$.

(*i*) Let a relation $S$ be defined on $X$ by $(x, y) \in S$ iff $(x, y) \in R$ and $(y, x) \in R$. Then $S$ is an equivalence relation on $R$.

(*ii*) Let $T$ be a relation defined on the quotient set $X/S$ by $(C_x, C_y) \in T$ iff $(x, y) \in R$. Then $T$ is a relation of partial order on $X/S$.

## 2.4 Lattices

Let $(X, \subset)$ be a poset, and $a$, $b$ be two elements of $X$. An element $c \in X$, such that $c \subset a$, and $c \subset b$ is called a *lower bound* of the pair $a$, $b$. A lower bound, which includes every lower bound, is called the *infimum* or *greatest lower bound* or *g.l.b.* (in abbreviation) of the given pair of elements $a$, $b$. Any two *g.l.b.*'s of $a$ and $b$ would include one another, and would be equal by the antisymmetric property.

Likewise, an element which includes both $a$ and $b$ is called an *upper bound* of the pair $a$, $b$. An upper bound which is included in every upper bound is called the *supremum* or *least upper bound* or, *l.u.b.* (in abbreviation) of the pair $a$, $b$; and it is unique, by the anti-symmetric property, if it exists at all.

The concepts of upper bound, lower, bound, *l.u.b.* and *g.l.b.* can likewise be defined for any given subset of $X$.

A poset, in which every pair of elements has a *g.l.b.* is called a lower semi-lattice. The *g.l.b.* of two elements $a$ and $b$ is then denoted by $a \cap b$ or $a \,.\, b$. Likewise, a poset, in which every pair of elements has an *l.u.b.*, is called an upper semi-lattice. The *l.u.b.* of two elements $a$ and $b$ is then denoted by $a \cup b$ or $a + b$. A poset, in which every pair of elements has both a *g.l.b.* and an *l.u.b.*, is called a lattice. It follow readily from the definitions that the *g.l.b.* and *l.u.b.* of pairs of elements in a lattice satisfy the properties as follows:

L (1) : $a \cap a = a$, and $a \cup a = a$. (*Idempotent properties*)

L (2) : $a \cap b = b \cap a$, and $a \cup b = b \cup a$.

(*Commutative properties*)

L (3) : $(a \cap b) \cap c = a \cap (b \cap c)$, and $(a \cup b) \cup c = a \cup (b \cup c)$.

(*Associative properties*)

L (4) : $a \cap (a \cup b) = a$, and $a \cup (a \cap b) = a$.

(*Absorptive properties*)

The idempotent properties $L$ (1) can be deduced from the other three properties $L$ (2 – 4). Also, it follows directly from the definitions of *g.l.b.* and *l.u.b.* that :

Any one of the three properties $a \subset b$, $a \cap b = a$ and $a \cup b = b$, implies the other two. (*Consistency property*)

[A lattice can also be defined as a system of double compositions, where the two compositions $\cap$ and $\cup$ satisfy the properties $L$ (2), $L$ (3) and $L$ (4). A binary relation "$\subset$" is then defined in the set, such that $a \subset b$ iff $a \cap b = a$. Then the consistency property follows from $L$ (4). Also it can readily be shown that $\subset$ is a relation of partial order, and that $a \cap b$ and $a \cup b$ are the *g.l.b.* and *l.u.b.* of the two elements $a$ and $b$. Thus the set forms a lattice with respect to the inclusion relation].

A lattice is said to be

(*i*) *complete*, iff every non-empty subset of $X$ has a *g.l.b.* and an *l.u.b.*

(*ii*) *modular*, if the following modular property is satisfied in it :

$c \subset a$ implies $(a \cap b) \cup c = a \cap (b \cup c)$; (*Modular property*)

(*iii*) *distributive*, if the following distributive property holds in it :

$a \cap (b \cup c) = (a \cap b) \cup (a \cap c)$; (*Distributive property*)

or, equivalently,

$$a \cup (b \cap c) = (a \cup b) \cap (a \cup c),$$

which follows by dualising the above property.

It may be noticed that the modular property is self-dual.

If $c \subset a$, we have $a \cup c = a$; hence the modular property follows from the distributive property. Consequently, *every distributive lattice is modular.*

Let $(X, \subset)$ be a poset. An element 0, such that $0 \subset a$, for every $a \in X$, is called the *null element* of $X$. Dually, an element 1, such that $a \subset 1$, for every $a \in X$, is called the *universal element* of $X$. A poset may not contain the null or the universal element.

Let $(X, \subset)$ be a lattice which contains both the null and the universal element; then two elements $a, b$ are said to be complements of each other, if $a \cap b = 0$ and $a \cup b = 1$. The lattice is said to be complemented, iff every element of $X$ has at least one complement in $X$.

**Example 2.** ***For a non-empty set X, the power set P (X) forms a lattice with respect to the set theoretic inclusion relation. This lattice has $\phi$ as its null element, and X as the universal element. Also this lattice is distributive (hence also modular) and complemented (as can be seen from Art. 1.1). It is also a complete lattice, the g.l.b. and the l.u.b. of a given collection of non-empty subsets of X being given by their intersection and union respectively.***

**Example 3.** ***The set of positive integers forms a lattice with respect to the divisibility relation, the g.l.b. and the l.u.b. of two positive integers being their h.c.f. and l.c.m. respectively. This lattice has the integer 1 as its null element, but it has no universal element.***

## Art. 3 FUNCTIONS

Let $X$ and $Y$ be two non-empty sets. Then a subset $f \subset X \times Y$, such that for each $x \in X$, there exists at least one $y \in Y$, for which $(x, y) \in f$, is called a *function on X into Y*, or a *mapping of X into Y*. The set $X$ is called the *domain* of $f$, and is denoted by *dom* $(f)$ or $D_f$; and the set $Y$ is called the *codomain* of $f$.

Let $A$ be a non-empty subset of $X$, then we define the *image of A under f* to be the set, consisting of all those elements $y \in Y$, such that $(x, y) \in f$, for some $x \in A$. This is denoted symbolically by :

$$f(A) = \{ y \in Y : (x, y) \in f, \text{ for some } x \in A \}.$$

The image of $X$ under $f$, *i.e.*, the set $f(X)$ is called the *range* of $f$, and is denoted by *Rng* $(f)$ or Im $(f)$.

For a given point $x \in X$, the set $f(x)$ is a non-empty sub-set of $Y$, and it may consist of more than one element. $f(x)$ is called the *image(s) of x under f*, or the *value(s) of f at the point x*.

**Single-valued Functions :** A function $f$ on $X$ into $Y$ is said to be *single-valued* if, for each $x \in X$, the set $f(x)$ consists of a single element only; otherwise the function $f$ is said to be *many-valued, or multiple-valued.*

It now follows immediately from the above definitions that if $f$ is a single-valued function on $X$ into $Y$, then $(x, y) \in f$ and $(x, z) \in f$ imply $y = z = f(x)$. It also follows readily that :

1. $A \subset B \subset X$ implies $f(A) \subset f(B) \subset Y$.

2. If $\{A_\alpha\}$ is any collection of subsets of $X$, then

$$f(\cup\{A_\alpha\}) = \cup\{f(A_\alpha)\} \text{ and } f(\cap\{A_\alpha\}) \subset \cap\{f(A_\alpha)\}.$$

A single-valued function $f$ on $X$ into $Y$ is said to be

(*i*) *injective* if $x \neq x'$ in $X$ implies $f(x) \neq f(x')$ in $Y$;

(*ii*) *surjective* if $f(X) = Y$;

(*iii*) *bijective* if it is both injective and surjective.

When $f(X) = Y$, $f$ is called a function or mapping of $X$ *onto* $Y$. Injective, surjective and bijective functions are also called *injections, surjections* and *bijections* respectively.

**Inverse functions :** Let $f$ be a single-valued function on $X$ into $Y$ and let $M$ be a subset of $Y$. Then the inverse image of $M$ under $f$, denoted by $f^{-1}(M)$, is defined to be the subset of $X$ given by

$$f^{-1}(M) = \{x \in X : f(x) \in M\}.$$

The following properties can now be readily verified (proof left as exercise) :

(*a*) $f^{-1}(\Phi) = \Phi$, and $f^{-1}(Y) = X$.

(*b*) $f^{-1}(Y - M) = X - f^{-1}(M)$, where $M \subset Y$.

(*c*) $M \subset N \subset Y$ implies $f^{-1}(M) \subset f^{-1}(N) \subset X$.

(*d*) For any collection of subsets $\{M_\alpha\}$ of $Y$,

$$f^{-1}(\cup\{M_\alpha\}) = \cup\{f^{-1}(M_\alpha)\} \text{ and } f^{-1}(\cap\{M_\alpha\}) = \cap\{f^{-1}(M_\alpha)\}.$$

(*e*) $M \cap N = \Phi$ in $Y$ implies $f^{-1}(M) \cap f^{-1}(N) = \Phi$ in $X$.

(*f*) $A \subset f^{-1}(M)$ implies $f(A) \subset M$, where $A \subset X$ and $M \subset Y$.

It should be noted that although, for every subset $M$ of $Y$, $f^{-1}(M)$ is a uniquely determined subset of $X$, even when $M$ is a singleton (*i.e.*, consists of one element only), $f^{-1}$ cannot be considered as a function on $Y$ into $X$. In fact, for an element $y \in Y$, $f^{-1}(y)$ may as well be the null subset of $X$, that is, there may not exist any element $x \in X$ such that $f(x) = y$, unless $f$ is, in particular, surjective.

Let $f$ be a surjection of $X$ onto $Y$. Then the function $g$ on $Y$ onto $X$, such that $g(y) = \{x \in X : f(x) = y\}$, is said to be the *inverse* of $f$, and is denoted by $f^{-1}$. The function $f^{-1}$ may not however be single-valued, unless $f$ is injective. If $f$ is also injective, *i.e.*, if $f$ is bijective, then $f^{-1}$ is also bijective (*i.e.*, it is single-valued, surjective and injective). In this case, every $x \in X$ has a unique image $y \in Y$, and every $y \in Y$ is the image of a unique element $x \in X$. Thus $f$ establishes a *biunique* or a (1, 1)-*correspondence* between the elements of $X$ and $Y$.

**Theorem 2 :** *A single-valued function f on a set X into a set Y induces an equivalence relation R on X, such that $xRx'$ holds iff $f(x) = f(x')$; and one obtains a bijective function g on the quotient set X/R onto the subset $f(X) \subset Y$, defined by $g(C_x) = f(x)$.*

**Proof :** Let $f$ be a single-valued function on a set $X$ into a set $Y$, then $f$ is a surjection on $X$ onto $f(X) \subset Y$. Then $f$ induces a binary relation $R$ among the elements of $X$, such that $xRx'$ holds iff $f(x) = f(x')$. Also $R$ is obviously an equivalence relation (verification left as an exercise). The quotient set $X/R$ consists of classes of elements of $X$, such that all the elements of the same class have one and the same image under $f$, and two elements from two different classes have distinct images in $Y$. Thus there is a (1, 1)-correspondence between the different members (*i.e.*, classes) of $X/R$ and the different elements of $f(X)$. That is, if $C_x$ denotes the particular class (belonging to $X/R$) to which the element $x$ belongs, then the mapping $g$, defined by $g(C_x) = f(x)$, is a bijection on the quotient set $X/R$ onto the subset $f(X) \subset Y$.//

**Restrictions and extensions :** Let $f$ be a single-valued function on $X$ into $Y$. Then $f(X)$ is a subset of $Y$. Let $K$ be a subset of $X$, then $f(K) \subset f(X) \subset Y$. Let $h$ be the function on $K$ onto $f(K)$, defined by $h(x) = f(x)$ for all $x \in K$. Then $h$ is called the *restriction* of $f$ to $K$, and is denoted by $f1_K$.

Conversely let $h$ be a single-valued function on $K \subset X$ into $Y$. And let $f$ be a single-valued function on $X$ into $Y$, such that $f(x) = h(x)$ for every $x \in K$. Then $f$ is called *an extension of h* to $X$. An extension $f$ of $h$ is not necessarily unique : since the values of $f(x)$, for $x \in X - K$, may be arbitrarily assigned.

**Composition of functions :** Let $f$ be a single-valued function on a set $X$ into a set $Y$, and $g$ be a single-valued function on $Y$ into a set $Z$. Let $h$ be the function on $X$ into $Z$, defined by $h(x) = g(f(x))$, for every $x \in X$. Then $h$ is a single-valued function on $X$ into $Z$; it is called the *composite* or the *product* of the functions $f$ and $g$, and is denoted by $g.f$ or $gof$.

The following results can readily by proved :

1. *Composition of single-valued mappings is associative.*

   In fact, if $f: X \to Y$, $g: Y \to Z$ and $k: Z \to W$ be three single-valued mappings, then both $(k \cdot g) \cdot f$ and $k \cdot (g \cdot f)$ are single valued mappings of $X$ into $W$. Also, for any element $x \in X$,

$$[(k \cdot g \cdot) \cdot f]\,(x) = [(k \cdot g)]\ (f(x)) = k\,(g\,(f(x)))$$

and $$[k \cdot (g \cdot f)]\ (x) = k\,([g \cdot f]\ (x)) = k\,(g\,(f(x))).$$

Thus $(k \cdot g) \cdot f = k \cdot (g \cdot f)$; hence the composition of mappings is associative.

2. *Let f be a single-valued mapping of a set X into a set Y. Then f can be expressed as a product of some mappings of particular types, as shown below :*

(*i*) *$f = i.g.$, where g is surjective and i is an injective mapping of $f(X)$ into Y.*

Let $g(x) = f(x)$ for all $x \in X$; then the range of $g$ is $\text{Im}(f) \subset Y$. Next, let $i(u) = u$ for all $u \in \text{Im}(f)$; then $f = i.g$.

[Let $K$ be a non-empty subset of a set $X$. Then the mapping $i$, defined by $i(x) = x$, for every element $x \in K$, is an injective mapping of $K$ into $X$. It is called the *inclusion mapping* of $K$ into $X$, and is usually denoted by $i_K$].

(*ii*) *$f = i.h.q.$, where q is the surjective mapping of X onto X/R, R being the equivalence relation induced by f (vide Theorem 2, p.11), h is the bijective mapping of X/R onto Im (f), defined by $h(C_x) = f(x)$, and i is the inclusion mapping of Im (f) into Y.*

If $f$ is in particular, surjective, then $\text{Im}(f) = Y$, and one obtains $f = h \cdot q$.

The decompositions of $f$, described above, are all unique.

**Operators :** A single-valued mapping on a set $X$ into itself is called an *operator* or an *endomapping* or a *transformation* of the set $X$. The set of all operators on a set $X$ is denoted by $X^X$. A bijective mapping of $X$ into itself is called a *non-singular transformation* or a *permutation* of $X$ (the term permutation being generally used for finite sets). A transformation which is not non-singular is said to be *singular*.

The *identity operator* I is defined by $I(x) = x$, for every $x \in X$. It is evidently a non-singular transformation of $X$.

## 3.1 Indexing of Sets

Let $f$ be a surjective function on a set $A$ onto a set $X$. Let $a \in A$ and let $f(a)$, *i.e.*, the element of $X$ which is the image of $a$ (under $f$), be denoted by $x_a$. As $f$ is surjective, every element of $X$ is represented as $x_a$ for some element $a \in A$. Thus, the elements of $X$ are *indexed*, each index being an element of $A$. $A$ is said to be the *indexing set* (or *index set*), and $f$ is said to determine the particular indexing. As there may exist different surjective functions on $A$ onto $X$, the set $X$ may be indexed, in different ways, having $A$ as the indexing set.

## 3.2 Cartesian Product of Sets

The Cartesian product of a finite collection of sets $X_1, X_1, ..., X_n$ has been defined to be the set consisting of all ordered $n$-tuples $(x_1, x_2, ..., x_n)$, with $x_k \in X_k$, for $k = 1, 2, ..., n$. Each $n$-tuple $(x_1, x_2, ..., x_n)$ may be considered as a mapping of the indexing set $\{1, 2, ..., n\}$ into the union of the sets $X_1, X_2, ..., X_n$ in such a manner that the image of $i$, denoted by $x_i$, lies in the set $X_i$, for $i = 1, 2, ..., n$.

We may now define the *Cartesian product of any given collection of sets* $\{X_a : a \in A\}$, $A$ being the indexing set, as the collection of all mappings of $A$ into $\cup \{X_a : a \in A\}$, where the image of the index $a \in A$ lies in the set $X_a$. The product is denoted by $\pi \{X_a : a \in A\}$.

The Cartesian product $\pi \{X_a : a \in A\}$ is thus the set of all functions $x$, having the domain $A$ such that $x_a = x(a) \in X_a$ for each $a \in A$. Each such function $x$ is called a *choice function* for the family of sets $\{X_a : a \in A\}$. For $x \in \{X_a : a \in A\}$ and $i \in A$, the value $x_i \in X_i$ is called the $i$th *coordinate of* $x$.

One *cannot*, however, *prove* the existence of a choice function for an arbitrarily given family of sets. To avoid this difficulty, we take recourse to an axiom, known as the axiom of choice (discussed in Art 4-1). The axiom of choice asserts that, for any family of sets $\{X_a : a \in A\}$, such that $A \neq \Phi$ and $X_a \neq \Phi$ for each $a \in A$, there always exists at least one choice function.

If the indexing set $A = \Phi$, then the null function is the only possible choice function. On the other hand, if one of the sets $X_a$ is empty, then the Cartesian product of the sets $\{X_a\}$ is empty.

### EXERCISES

**1.** Let $X$ and $Y$ be non-empty sets, and let $f : X \to Y$ and $g : Y \to X$ be two mappings. Then prove that :

(*i*) If $g.f$ is the identity mapping on $X$, then $f$ is an injection and $g$ is a surjection.

(*ii*) If $g.f$ is the identity mapping on $X$ and $f.g$ is the identity mapping on $Y$, then both $f$ and $g$ are bijections and $g = f^{-1}$.

**2.** Let $X$ be a non-empty set, and let $f \in X^X$. Prove that :

(*i*) if $f.g = f.h$, then $g = h$ for all $g, h \in X^X$ iff $f$ is injective;

(*ii*) if $g.f = h.f$, then $g = h$ for all $g, h \in X^X$ iff $f$ is a surjection.

**3.** Prove that $X^Y = \Phi$ iff $X = \Phi$ and $Y \neq \Phi$.

**4.** Prove that for two non-empty sets $X, Y, X^Y = Y^X$ implies $X = Y$.

## Art. 4 ON AXIOMATIC SET THEORY

Theory of sets developed so far lacks strict axiomatic foundation. Formulation of definitions and propositions and methods used in the proofs are more or less intuitive, and are found sometime to lead to contradictions or paradoxes. The naive theory of sets suffers from the presence of several well-known paradoxes. These paradoxes are avoided in the axiomatic set theory by eliminating sets that are "too large".

In the axiomatic set theory, one takes as primitive concepts, the concepts of set, element, and the relation of an element belonging to a set. These are subject to the following axioms :

(*i*) **Axiom of uniqueness :** *If the sets A and B have the same elements then A and B are identical* (and we denote this by $A = B$).

(*ii*) **Axiom of union :** *Given two arbitrary sets A and B, there exists a set C whose elements are all the elements of the set A and all the elements of B, and C does not contain any other elements.* (We call $C$ the union of $A$ and $B$, and denote this symbolically by $C = A \cup B$).

(*iii*) **Axiom of the difference :** *For any two arbitrary sets A and B, there exists a set D, whose elements are those and only those elements of set A which are not elements of the set B.* (We call $D$ the *difference* of the sets $A$ and $B$, and denote this symbolically by $D = A - B$ or $A \setminus B$).

(*iv*) **Axiom of existence :** There exists at least one set.

*Intersection* of two sets $A$ and $B$ can be defined in terms of difference, by $A \cap B = A - (A - B)$.

All the properties, relating to union, intersection, difference, symmetric difference, complements, obtained so far, can be deduced from the above four axioms.

We require a few other axioms in terms of which (taken together with the above four axioms) it is possible to express all those properties of set theory which suffice for the application of the theory (of sets) to other branches of mathematics. These new axioms (three in number) are the following :

(*v*) *For every propositional function* $p(x)$ *and for every set* $A$ *there exists a set consisting of those and only those elements of the set* $A$ *which satisfy the propositional function* $p(x)$.

[Let $p(x)$ be an expression which becomes a proposition when one substitutes for $x$ an arbitrary value of $x$ belonging to some fixed set. Then the expression $p(x)$ is called a propositional function (with the fixed set as the domain of the argument). We sometimes consider propositional functions for which the domain of the argument $x$ is not restricted to any particular set].

(*vi*) **The axiom for power set :** *For every set A there exists a set whose elements are all the subsets of the set A.*

The set formed by all subsets of a set $A$ is called *the power set of* $A$, and is denoted by $\mathbf{P}(A)$.

(*vii*) **The axiom of choice :** *(or any equivalent form of it), described below:*

The seven axioms (*i*) – (*vii*) are not independent of each other. In fact, the axioms (*iii*) and (*iv*) can readily be deduced from the remaining ones. The axiom (*ii*), relating to the formation of the union of two sets $A$ and $B$, can be deduced from axiom (*v*), provided that we assume that $A$ and $B$ are subsets of some fixed set $X$.

## 4.1 Axiom of Choice

The axiom of choice was formulated by Ernst Zermelo (in 1904). This axiom, which is also known as *Zermelo's axiom*, excited vigorous controversy and criticism. Although validity of this axiom has been questioned by many eminent mathematicians, it has proved most useful, if not indispensible,in much of analysis.

1. **The axiom of choice (or Zermelo's axiom) :** *The Cartesian product of a non-empty family of non-empty sets is non-empty.*

   This axiom had been introduced by Peano in 1889, in a special form, known as the *principle* of *finite induction*, in his axioms for natural numbers.

   It has been proved (in 1963-64) by P.J. Cohen that this axiom is independent of the other axioms of set theory.

In its application to different branches of mathematics, e.g. set theory, algebra, analysis, in different contexts, the axiom of choice, in the form as given above, is not found appropriate in many cases. Some equivalent forms of the axiom of choice, formulated by different mathematicians, are found more convenient in many such cases; these are given below :

2. **Tukey's lemma :** *Every non-empty family of sets of finite character has a maximal member.*

   [A family of sets **F** is said to be of *finite character*, if $\Phi \in$ **F**, and a non-empty set $A \in$ **F** if and only if each finite subset of $A$ belongs to **F**].

3. **Hausdorff's maximality principle :** *Every non-empty partially ordered set containts a maximal chain.*

4. **Kuratowski-Zorn lemma (or Zorn's lemma) :** *If every linearly ordered subset of a non-empty partially ordered set $X$ has an upper bound, then there exists a maximal element in $X$.*

5. **Well-ordering theorem (Zermelo) :** *Every set can be well-ordered.*

   It can be shown that the five propositions (1) – (5), stated above are pairwise equivalent. We do not propose to prove the equivalence. One may refer to any book on axiomatic set theory for this purpose.

## EXERCISES

**1.** Deduce axioms (*iii*) and (*iv*) from the remaining axioms of the set theory.

**2.** Prove that the following proposition (sometimes called Zermelo's postulate) is equivalent to the axiom of choice :

Given a family $\Omega$ of disjoint non-empty sets, there exists a set $A$, such that $X \cap A$ is singleton for each $X \in \Omega$.

## Art. 5 CARDINAL NUMBERS

A set $X$ is said to be *equipotent* (or *equipollent*, or *equivalent*) with a set $Y$, denoted symbolically by $X \sim Y$, if there exists a bijective mapping of $X$ onto $Y$.

Since (*i*) the indentity operator is bijective; (*ii*) the inverse of a bijective mapping is bijective, and (*iii*) the product of two bijective mappings is a bijective mapping (proof left as an exercise), it follows that *the relation of being equipotent is an equivalence relation.* Hence

Any given collection of sets is partitioned into disjoint classes, in such a way that any two sets belonging to the same class are equipotent with one another, while two sets from two different classes are not equipotent with one another. Each such class is called a *cardinal number*. So two equipotent sets belong to the same cardinal number, and they are said to have the same *cardinal number* (or *cardinality* or *power* or *potency*).

Thus, a cardinal number is assigned to every set. The cardinal number of a set $X$ is denoted by card ($X$); and for two sets $X$ and $Y$, card ($X$) = card ($Y$) if and only if $X \sim Y$.

A cardinal number signifies a property which is common to all sets which are equipotent with each other. More precise definitions for the concept of a cardinal number have been attempted, but they have been found to be unsatisfactory. Relations between cardinal numbers are merely convenient ways of expressing relations between sets. "We must leave the determination of the essence of the cardinal numbers to philosophy."

Two finite sets are evidently equipotent if and only if they have the same number of elements. A finite set, consisting of $n$ elements, *i.e.*, a set which is equipotent with the set $\{1, 2, ..., n\}$, is assigned the number $n$ as its cardinal number (for $n = 1, 2, \ldots$). The cardinal number of the null set $\Phi$ is taken to be the number 0.

The cardinal number of an infinite set is sometimes called a *transfinite cardinal number*. In particular, the cardinal number assigned to the set of natural numbers **N** is denoted by the Hebrew letter aleph with suffix 0 or, more commonly, by $d$. The cardinal number assigned to the set of real numbers **R** is denoted by aleph with suffix 1 or, more commonly, by $c$; it is also called the *power* (or *potency*) of the continuum.

## 5.1 Countable Sets

A set which is equipotent with the set of natural numbers **N**, *i.e.*, a set whose cardinal number is $d$, is called a *denumerable* (or *enumerable*) set. A set, which is either finite or denumerable, is said to be *countable*.

Let $X$ be a countable set. If it is finite, with $n$ elements, then, using $\{1, 2, ..., n\}$ as the indexing set, the elements of $X$ can be indexed as $x_1, x_2, ..., x_n$. If, on the other hand, $X$ is infinite, it is a denumerable set. Then using the set of natural numbers $\mathbf{N} = \{1, 2, \ldots\}$ as the indexing set, the elements of $X$ can be indexed, and thereby the elements of $X$ are arranged in a *sequence*, $x_1, x_2, ..., x_n, \ldots$; it is known as an *enumeration* of the elements of the set $X$ The sequence will be denoted by $\{x_n : n = 1, 2, \ldots\}$ or $\{x_n\}$. Conversely, any sequence is either a finite set or a denumerable set.

**Theorem 3 :** *Every infinite set has a denumerable subset.*

**Proof :** Let $X$ be an infinite set. We shall show that, for each natural number $n$, there exists a subset $A_n \subset X$, such that card $(A_n) = n$. As $X \neq \Phi$, there exists a one-pointic subset $A_1 \subset X$. Let $A_k \subset X$, where card $(A_k) = k$. As $X$ is an infinite set, $X - A_k \neq \Phi$; and let $x \in X - A_k$. Then $A_{k+1} = A_k \cup \{x\} \subset X$, and card $(A_{k+1}) = k + 1$. Hence, by the principle of mathematical induction, there exists a subset $A_n \subset X$, with card $(A_n) = n$, for each natural number $n$.

Now, we define a family of sets $B_i$ by

$$B_n = A_{2^n} - (A_1 \cup A_2 \cup A_{2^2} \cup \ldots \cup A_{2^{n-1}}), \text{ for } n = 1, 2, 3, \ldots,$$

(so that $B_1 = A_2 - A_1$, $B_2 = A_4 - (A_1 \cup A_2)$, ...).

Then $\{B_n : n \in \mathbf{N}\}$ is a family of pairwise disjoint non-empty subsets of $X$. (In fact, card $(B_n) \geq 2^n - (2^n - 1) = 1$).

Let $f$ be a choice function for the non-empty family of non-empty sets $\{B_n : n \in \mathbf{N}\}$. Then $f$ is an injective mapping of the set of natural numbers $\mathbf{N}$ into $X$; and so *rng* $(f)$ is a denumerable subset of $X$.//

**Theorem 4 :** *Any subset of a countable set is countable.*

**Proof :** Let $X$ be a countable set, and $A$ be a subset of $X$. If $A$ is finite; then it is countable, and we have nothing more to prove. Next, let $A$ be an infinite set. $X$ is then denumerable, and let $\{x_n : n = 1, 2, \ldots\}$ be an enumeration of $X$. Let $f$ be a function on $\mathbf{N}$ into $A$, defined recursively by

$$f(1) = x_{n_1}, \text{ where } n_1 \text{ is the smallest } n \in \mathbf{N}, \text{ such that } a_n \in A,$$

$$f(k+1) = x_{n_{k+1}}, \text{ where } n_{k+1} \text{ is the smallest } n \in \mathbf{N}, \text{ such that}$$
$$a_n \in A - \{a_1, a_2, \ldots, a_{n_k}\}$$

Then $f$ is a bijection of $\mathbf{N}$ onto $A$; hence $A$ is denumerable.//

**Theorem 5 :** *The Cartesian product of two countable sets is countable.*

**Proof :** Let $X$ and $Y$ be two countable sets, and let $P$ denote their Cartesian product. Now, if both the sets $X$ and $Y$ are finite, then $P$ is also a finite set, and hence also countable. Next, let $X = \{x_1, \ldots, x_r\}$ be a finite set, and $Y$ a denumerable set; and let $Y = \{y_1, y_2, \ldots, y_n, \ldots\}$ be an enumeration of $Y$. Then the elements of the Cartesian product $P$ can be arranged as a sequence $(x_1, y_1), \ldots, (x_r, y_1)$; $(x_1, y_2), \ldots, (x_r, y_2)$; ..., $(x_1, y_n), \ldots, (x_r, y_n)$; ...; and $P$ is therefore a denumerable set (and hence also countable). Finally, let both $X$ and $Y$ be denumerable, and $X = \{x_1, x_2, \ldots\}$ and $Y = \{y_1, y_2, \ldots\}$ be their enumerations. Then the elements of their Cartesian product $P$ can be arranged as a sequence,

$$(x_1, y_1);\ (x_1, y_2),\ (x_2, y_1);\ \ldots;\ (x_1, y_n),\ (x_2, y_{n-1}),\ \ldots,\ (x_n, y_1);\ \ldots,$$

where $(x_i, y_j)$ precedes $(x_m, y_n)$ iff either $i + j < m + n$, or $i < m$ when $i + j = m + n$.

Explicitly, a bijection $f$ of $P$ onto **N**, is given by

$$f(x_i, y_j) = [1 + 2 + \ldots + (i + j - 2)] + i. //$$

The above result can readily be generalised for the Cartesian product of any finite collection of countable sets. Thus

*The Cartesian product of any finite collection of countable sets is countable.*

**Theorem 6 :** *The union of any countable family of countable sets is countable.*

**Proof :** Firstly, let $X_1, X_2, \ldots, X_n$ be a finite collection of denumerable sets, and let $\{x_{r1}, x_{r2}, \ldots, x_{rn}, \ldots\}$ be an enumeration of the set $X_r$, for $r = 1, 2, \ldots, n$. Then their union consists of the elements $\{x_{ij}\}$, for $i = 1, 2, \ldots, n$ and $j = 1, 2, \ldots, n, \ldots$ (or a subset of this collection, in case the given sets are not pair-wise disjoint). The elements $x_{ij}$ can be arranged in a sequence,

$$x_{11}, x_{21}, \ldots, x_{n1};\ x_{12}, x_{22}, \ldots, x_{n2};\ \ldots;\ x_{1m}, x_{2m}, \ldots, x_{nm};\ \ldots,$$

where $x_{ij}$ precedes $x_{rs}$ if, either $j < s$, or $i < r$ when $j = s$.

In fact, $f(x_{ij}) = n(j - 1) + i$ is a bijection of the collection $\{x_{ij}\}$ onto **N**.

Next, let $X_1, X_2, \ldots, X_n, \ldots$ be a denumerable family of denumerable sets. Let $\{x_{r1}, x_{r2}, \ldots, x_{rn}, \ldots\}$ be an enumeration of the elements of the set $X_r$, for $r = 1, 2, \ldots$ Then the elements $\{x_{ij}\}$, for $i = 1, 2, \ldots,$ and $j = 1, 2, \ldots,$ can be arranged as a sequence,

$$x_{11};\ x_{12}, x_{21};\ x_{13}, x_{22}, x_{31};\ \ldots;\ x_{1n}, x_{2,n-1}, \ldots, x_{n1};\ \ldots,$$

where $x_{ij}$ precedes $x_{mn}$ iff either $i + j < m + n$, or $i < m$ when $i + j = m + n$.

$f(x_{ij}) = [1 + 2 + \ldots + (i + j - 2)] + i$ is a bijection of the set of elements $\{x_{ij} : i = 1, 2, \ldots, j = 1, 2, \ldots\}$ onto **N**.

The union of the given family of sets is given by the denumerable set of elements $\{x_{ij} : i = 1, 2, \ldots, j = 1, 2, \ldots\}$, or a subset of this set; hence the union is countable.//

**Theorem 7 :** *Each of the following sets is countable :*

(*i*) *the set of all integers* **Z**;

(*ii*) *the set of all rational numbers* $\boldsymbol{Q}$;

(*iii*) *the set of all polynomials with rational coefficients;*

(*iv*) *the set of all (real) algebraic numbers.*

**Proof :** (*i*) $\mathbf{Z} = \mathbf{N} \cup \{0\} \cup \{-n : n \in \mathbf{N}\}$; hence **Z** is countable.

(*ii*) The set $Q_r^+ = \{1/r, 2/r, ..., n/r, \ldots\}$ is denumerable, for $r = 1, 2, ...$ So, $Q_r = Q_r^+ \cup \{0\} \cup \{-n/r : n \in N\}$ is also a denumerable set, for $r = 1, 2, \ldots$ Hence, $Q_1 \cup Q_2 \cup \ldots \cup Q_n \cup \ldots$ is also denumerable, and is therefore countable. The set of rational numbers **Q** is a subset of the set $Q_1 \cup Q_2 \cup \ldots$; hence the set **Q** is also countable.

(*iii*) Let $P_n$ denote the set of all polynomials of degree $n$ and with rational coefficients. Every such polynomial is determined by its $n+1$ coefficients (of which the leading coefficient $\neq 0$). Hence $P_n$ is equipotent with the Cartesian product $Q_1 \times Q_2 \times \ldots \times Q_n \times [\mathbf{Q} - \{0\}]$, where $Q_1 = Q_2 = \ldots = Q_n = \mathbf{Q}$ denotes the set of all rational numbers. $P_n$ is therefore denumerable (by generalisation of Theorem 5). Hence the union, $P_0 \cup P_1 \cup P_2 \cup \ldots \cup P_n \cup \ldots$ is also denumerable; in other words, the set of all polynomials with rational coefficients is countable.

(*iv*) [A real number is said to be a (*real*) *algebraic number*, if it is a root of some (non-null) polynomial with rational coefficients (or, equivalently, with integral coefficients)].

Let $P$ denote the set of all polynomials with rational coefficients. Then $P$ is a countable set, by (*iii*). Again every polynomial belonging to $P$, has only a finite number of real roots, not exceeding the degree of the polynomial. Hence, the set of all real algebraic numbers, being a subset of the denumerable set formed by the union of a denumerable collection of finite sets, is necessarily a countable set.//

An infinite set, which is not denumerable, is called an *uncountable set.* As examples of uncountable sets, one has :

**Theorem 8 :** (*i*) *The set of all real numbers* **R** *is uncountable.*

(*ii*) *The set of all (real) transcendental numbers is uncountable.*

**Proof :** (*i*) If possible, let **R** be denumerable. Then its elements have an enumeration, $\{a_1, a_2, ..., a_n, \ldots\}$, say. Now, any real number $a_n$ can be expressed uniquely as a non-terminating decimal expression

$$a_n = *** \cdot a_{n1}\, a_{n2}\, a_{n3} \cdots$$

in a non-trivial manner (*i.e.*, where all $a_{ni}$, for all $i$ greater than some $m$, are not zero. In fact, one can always replace a terminating decimal expression by a non-terminating one, viz. $5.62 = 5.619999\ldots$). Let the sequence $\{a_n\}$ of all the real numbers, with their corresponding decimal expressions, be arranged in a vertical sequence :

$$\left.\begin{array}{l} a_1 = *** \cdot a_{11}\ a_{12}\ a_{13} \cdots \\ a_2 = *** \cdot a_{21}\ a_{22}\ a_{23} \cdots \\ \cdots\ \ \cdots\ \ \ \cdots\ \ \ \cdots \\ a_n = *** \cdot a_{n1}\ a_{n2}\ a_{n3} \cdots \\ \cdots\ \ \cdots\ \ \ \cdots\ \ \ \cdots \end{array}\right\} \cdots \qquad \cdots \qquad \ldots(a)$$

Then a non-terminating decimal expression $\cdot\, b_1\, b_2\, b_3 \ldots$, where $b_n$ is taken to be a digit different from $a_{nn}$ and 0, for $n = 1, 2, \ldots$, does not occur in the vertical sequence ($a$), since it differs from the decimal expression for $a_n$ in the $n$th (decimal) digit, $n = 1, 2, \ldots$ But the non-terminating decimal expression $\cdot\, b_1\, b_2\, b_3 \ldots$ represents a real number, and should therefore occur in the vertical sequence ($a$). Thus there is a contradiction. Hence, **R** must be uncountable.

(*ii*) [A (real) number, which is not algebraic, *i.e.*, which is not a root of any (non-null) polynomial with rational coefficients, in called a (real) *transcendental number*].

Let $T$ denote the set of all real transcendental numbers. Then $T$ is necessarily uncountable, otherwise, $\mathbf{R} = A \cup T$ would be countable, where $A$ denotes the countable set (vide Theorem 7 (*iv*)) of all real algebraic numbers; but **R** is uncountable (as seen above in Theorem 8 (*i*)).//

### 5.2 Sets with the Cardinal Number *c*

(*i.e., sets which have the power of the continuum*):

Any set, which is equipotent with **R** (the set of all real numbers), is said to have the cardinal number $c$ (or that it has the power of the continuum).

**Theorem 9 :** (*i*) *The open ray* $(a, +\infty) = \{x \in \mathbf{R} : a < x\}$ *has the cardinal number c.*

(*ii*) *Any open interval* $(a, b) = \{x \in \mathbf{R} : a < x < b\}$ *has the cardinal number c.*

**Proof :** (*i*) $x \to a + e^x = y$, with the inverse $y \to \log_e (y - a) = x$, is a bijective mapping of **R** onto the open ray $(a, +\infty)$.

(*ii*) $x \to e^x = y$ is a bijection on **R** onto the open ray $(0, +\infty)$, $y \to y/(1+y) = z$ is a bijection on $(0, +\infty)$ onto the open interval $(0, 1)$, and finally, $z \to z(b-a) + a = w$ is a bijection on $(0, 1)$ onto the open interval $(a, b)$. The composite of these three mappings,

$$x \to (be^x + a)/(e^x + 1) = w,$$

with the inverse $w \to \log_e \dfrac{w-a}{b-w} = x$, is a bijection on **R** onto the open interval $(a, b)$.//

Since half-open intervals and closed intervals and closed rays are obtained from the corresponding open intervals or open rays by adjoining one or two (end) points only, it follows immediately from the above theorem that:

**Corollary :** *Any closed ray* $[a, +\infty)$, *any half open interval* $(a, b]$, $[a, b)$, *or any closed interval* $[a, b]$ *has the cardinal number c.*

It also follows from above that any open ray $(-\infty, b)$ or closed ray $(-\infty, b]$ has the cardinal number $c$.

**Theorem 10 :** *Cantor's ternary set has the cardinal number c.*

Cantor's ternary set $T$ is defined to be the set of all those real numbers $a$, which are expressible in the form,

$$a = a_1/3 + a_2/3^2 + \ldots + a_n/3^n + \ldots, \qquad \ldots(1)$$

where $a_1, a_2, \ldots, a_n, \ldots$ assume the values 0 and 2 only.

[Cantor's ternary set may be obtained by an interesting geometrical method: The closed unit interval [0, 1] is divided into three equal parts, and the middle open segment (1/3, 2/3) is removed. The remaining closed subsegments [0, 1/3] and [2/3, 1] are again trisected, and their middle open segments, viz. $(1/3^2, 2/3^2)$ and $(2/3 + 1/3^2, 2/3 + 2/3^2)$, are removed. The remaining 4 closed subsegments are then trisected, and their middle open segments are removed. This process is continued ad infinitum, and the points, which survive, constitute *Cantor's ternary set*].

**Proof :** (of Theorem 10) : Any real number $b$ in [0, 1] can be expressed as a dyadic fraction,

$$b = b_1/2 + b_2/2^2 + \ldots, \quad \ldots \quad \ldots \qquad \ldots(2)$$

where $b_1, b_2, \ldots$ assume independently the values 0 and 1 only. Now the mapping $b \to a$, defined by $a_i = 0$ iff $b_i = 0$, and $a_i = 2$ iff $b_i = 1$, is a bijection on the closed unit interval [0, 1] onto the Cantor's ternary set $T$. Hence $T$ has the cardinal number $c$.//

### 5.3 Order Relation for Cardinal Numbers

Let $\alpha$ and $\beta$ be two cardinal numbers, and $A$ and $B$ be two sets, such that card $(A) = \alpha$ and card $(B) = \beta$. We define a binary relation $\alpha \leq \beta$ (or equivalently, $\beta \geq \alpha$), iff the set $A$ is equipotent with a subset of the set $B$. Also, we write $\alpha < \beta$ (or, equivalently $\beta > \alpha$), iff $\alpha \leq \beta$ and $\alpha \neq \beta$.

**Theorem 11 :** *Let $\alpha$, $\beta$ and $\gamma$ be cardinal numbers. Then*

(*i*) $\alpha \leq \alpha$.

(*ii*) $\alpha \leq \beta$ *and* $\beta \leq \gamma$ *imply* $\alpha \leq \gamma$.

**Proof :** Let $A$, $B$ and $C$ be three sets, such that card $(A) = \alpha$, card $(B) = \beta$ and card $(C) = \gamma$.

(*i*) Since the identity mapping of $A$ onto $A$ is a bijection (and $A$ is a subset of itself), it follows that $\alpha \leq \alpha$.

(*ii*) By hypothesis, there exists a bijection $f$ of $A$ onto a subset $B_1 \subset B$, and also a bijection $g$ of $B$ onto a subset $C_1 \subset C$. Let $g(B_1) = C_2$, then $C_2 \subset C_1 \subset C$, and the composite $gf$ is a bijection of $A$ onto the subset $C_2 \subset C$. Hence $\alpha \leq \gamma$.//

**Corollary :** *"$\leq$" is an order relation on the set of cardinal numbers.*

**Proof :** The relation is transitive (by Theorem 11 (*ii*)). It is also asymmetric, since if $A$ and $B$ are two finite sets with 3 and 5 elements respectively, then card $(A) = 3$ and card $(B) = 5$, and whereas $3 \leq 5$ holds, but $5 \leq 3$ does not hold.//

We shall show that the relation "$\leq$" is, in particular, antisymmetric. This follows from the following theorem :

**Theorem 12 :** *If $A$ and $B$ are two sets, such that $A \sim B_1 \subset B$ and $B \sim A_1 \subset A$, then $A \sim B$.*

*(The above is known as the equivalence theorem of Schroeder and Bernstein).*

We shall use the following lemma in the proof of this theorem :

**Lemma :** *Let $A_2 \subset A_1 \subset A$, and $A \sim A_2$, then $A \sim A_1$.*

**Proof of the lemma :** Let $f$ be a bijective mapping of $A$ onto $A_2$.

Then $f$ induces a bijective mapping on $A_1$ onto a subset $A_3 \subset A_2$, a bijective mapping on $A_2$ onto a subset $A_4 \subset A_3$, and so on. And thus one gets

$$A \sim A_2 \sim A_4 \sim ..., \text{ and } A_1 \sim A_3 \sim A_5 \sim ..., \quad ...(1)$$

where $A \supset A_1 \supset A_2 \supset A_3 \supset A_4 \supset A_5 \supset \ldots$

Since $f$ is bijective, one obtains the following equiptoent sets, under $f$, by (1) :

$$(A - A_1) \sim (A_2 - A_3) \sim (A_4 - A_5) \sim \ldots \quad ...(2)$$

Let $D = A \cap A_1 \cap A_2 \cap A_3 \cap \ldots$ Then one obtains the following decompositions of $A$ and $A_1$ in terms of disjoint components :

$$A = D \cup (A - A_1) \cup (A_1 - A_2) \cup (A_2 - A_3) \cup \ldots$$

and $$A_1 = D \cup (A_1 - A_2) \cup (A_2 - A_3) \cup (A_3 - A_4) \cup ...,$$

since any element $a \in A$ belongs either to every $A_i$, and therefore to $D$, or there exists a suffix $n$ such that $a \in A_{n-1} - A_n$ (one takes $A_0 = A$).

By (2), the subset $(A - A_1) \cup (A_2 - A_3) \cup (A_4 - A_5) \cup \ldots$ is mapped by $f$ onto the subset $(A_2 - A_3) \cup (A_4 - A_5) \cup (A_6 - A_7) \cup \ldots$. Let the residual subset $D \cup (A_1 - A_2) \cup (A_3 - A_4) \cup \ldots$ be mapped onto itself by the identity mapping $I$. Then the mapping $g$, defined by

$$g(x) = f(x) \text{ if } x \in (A - A_1) \cup (A_2 - A_3) \cup (A_4 - A_5) \cup \ldots$$

and $$= x \text{ if } x \in D \cup (A_1 - A_2) \cup (A_3 - A_4) \cup (A_5 - A_6) \cup \ldots$$

is a bijective mapping of

$$[(A - A_1) \cup (A_2 - A_3) \cup \ldots] \cup [D \cup (A_1 - A_2) \cup (A_3 - A_4) \cup ...] = A$$

onto

$$[(A_2 - A_3) \cup (A_4 - A_5) \cup \ldots] \cup [D \cup (A_1 - A_2) \cup (A_3 - A_4) \cup ...] = A_1.$$

Thus $A \sim A_1$. This completes the proof of the lemma.

**Proof of theorem 12 :** The relation $A \sim B_1$ induces a relation $A_1 \sim B_2$, where $B_2 \subset B_1$. Thus $B_2 \subset B_1 \subset B$. Also, $B \sim A_1$ and $A_1 \sim B_2$ imply $B \sim B_2$. Hence, by the above lemma, one obtains $B \sim B_1$. Now, $A \sim B_1$ and $B_1 \sim B$ imply $A \sim B$.//

As an immediate consequence of Theorem 12 we obtain :

**Theorem 13 :** *If $\alpha$ and $\beta$ are cardinal numbers, such that $\alpha \leq \beta$ and $\beta \leq \alpha$, then $\alpha = \beta$.*

It now follows from Theorems 11 and 13 that "$\leq$" *is a relation of partial order on the set of cardinal numbers*. We shall now show that the law of trichotomy also holds in this poset.

**Theorem 14 :** *If $\alpha$ and $\beta$ are any two cardinal numbers, then either $\alpha \leq \beta$ or $\beta \leq \alpha$.*

**Proof :** Let $A$ and $B$ be two sets, such that card $(A) = \alpha$ and card $(B) = \beta$. Let **F** denote the family of all bijective functions $f$, such that dom $(f) \subset A$ and rng $(f) \subset B$. The family **F** is non-empty (iff the sets $A$ and $B$ are non-empty). Now, we introduce an order relation "$\subset$" in **F**, such that, for $f_1, f_2 \in \mathbf{F}, f_1 \subset f_2$ iff $f_2$ is an extension of $f_1$. Then "$\subset$" is a relation of partial order on **F** (proof left as an exercise). Let **C** be a chain (*i.e.*, a linearly ordered subset) in **F**, then $\cup \{f : f \in \mathbf{C}\}$ is an upper bound of **C** and it lies in **F**. Thus the hypothesis of Kuratowski-Zorn lemma is satisfied for the partially ordered set **F**; hence there exists a maximal element $g$ in **F**.

We shall show that either dom $(g) = A$ or rang $(g) = B$. If possible let both be false. Then $A$ – dom $(g) \neq \Phi$ and $B$ – rang $(g) \neq \Phi$; so let $x \in A$ – dom $(g)$ and $y \in B$ – rang $(g)$. Let now $h$ be a function defined by $h(t) = g(t)$, for $t \in$ dom $(g)$ and $h(x) = y$. Then $h \in \mathbf{F}$ and $h$ is an extension of $g$. This contradicts the maximality of $g$. Hence, either dom $(g) = A$ or rng $(g) = B$ must hold. Now, if dom $(g) = A$, then rng $(g) \subset B$ implies that $\alpha \leq \beta$. On the other hand, if rng $(g) = B$, then $g^{-1}$ is a bijective mapping of $B$ onto dom$(g) \subset A$, and therefore $\beta \leq \alpha$. Thus either $\alpha \leq \beta$ or $\beta \leq \alpha$ holds.//

It now follows from Theorems 11, 13 and 14 that :

**Theorem 15 :** *The set of cardinal numbers is linearly ordered by the relation "$\leq$".*

As the null set $\Phi$ is a subset of every set, it follows that card $(\Phi) = 0$ is the smallest cardinal number. But, by virtue of the following theorem, there is no largest cardinal number.

**Theorem 16 :** *For any set A, card $(\mathbf{P}(A)) >$ card $(A)$, where $\mathbf{P}(A)$ denotes the power set of A.*

**Proof :** If $A = \Phi$, then $\mathbf{P}(\Phi)$ consists of only one subset, *i.e.*, $\Phi$, so that card $(\mathbf{P}(\Phi)) = 1 > 0 =$ card $(\Phi)$. Next, let $A \neq \Phi$, and let card $(A) = \alpha$ and card $(\mathbf{P}(A)) = \beta$. Now, the function $f$, defined by $f(x) = \{x\}$, for every $x \in A$, is an injective function on $A$ into $\mathbf{P}(A)$, so that it is a bijection on $A$ onto a subset rng $(f) \subset \mathbf{P}(A)$. Hence $\alpha \leq \beta$. Next we show that $\alpha \neq \beta$.

If possible, let $\alpha = \beta$. Then there exists a bijection $h$ on $A$ onto $\mathbf{P}(A)$, *i.e.*, dom $(h) = A$ and rng $(h) = \mathbf{P}(A)$. Let $C$ be a subset of $A$, defined by $C = \{x \in A : x \notin h(x)\}$, where $C$ may as well be the null set $\Phi$. Now, $C \in \mathbf{P}(A)$, since $C \subset A$. And, as $h$ is bijection on $A$ onto $\mathbf{P}(A)$, there exists an element $p \in A$, such that $h(p) = C$. Now, two cases are possible; either

$p \in C$ or $p \notin C$. If we assume that $p \in C$, we are lead to a contradiction, as by the definition of $C$, $p \notin h(p) = C$. Hence, $p \notin C$. But this also leads to a contradiction, since $C = h(p)$ implies that $p \notin h(p)$, and consequently $p$ must belong to $C$. Thus there is a contradiction in either case. Hence our assumption $\alpha = \beta$ is not valid; in other words $\alpha \neq \beta$.

Now combining the results $\alpha \leq \beta$ and $\alpha \neq \beta$, we get $\alpha < \beta$.//

**Theorem 17 :** (*i*) *$d$ is the smallest transfinite cardinal number; i.e., if $\alpha$ is any transfinite cardinal number, then $d \leq \alpha$.*

(*ii*) $d < c$.

**Proof :** (*i*) This result follows directly from Theorem 3.

(*ii*) By (*i*), $d \leq c$; and by Theorem 8 (*i*), $d \neq c$. Hence $d < c$.//

Schroder-Bernstein Theorem finds an important application in the following example (one of the famous *theorems of Cantor*) :

**Example 1.** ***The n-dimensional unit cube :***

$$I^n = \{(x_1, x_2, ..., x_n) : 0 \leq x_i \leq 1, \text{ for } i = 1, 2, ..., n\}$$

***has the cardinal number c.***

**Solution :** Let $I$ denote the unit line segment $I = \{x \,\varepsilon\, \mathbf{R} : 0 \leq x \leq 1\}$. Let $T = \{(x, 0, 0, ..., 0),\ 0 \leq x \leq 1\}$, then $T \subset I^n$; and $I \sim T$ under the bijective mapping, $x \longleftrightarrow (x, 0, 0, ..., 0)$.

Next, let $(x_1, x_2, ..., x_n) \,\varepsilon\, I^n$, then $x_i$ can be expressed uniquely as a non-terminating decimal expression, $x_i = .\, x_{i1}\, x_{i2}$ ..., for $i = 1, 2, .., n$. Now $(x_{11}\, x_{12}\, ..., x_{21}\, x_{22}\, ..., ..., .\, x_{n1}\, x_{n2}\, ...) \to .\, x_{11}\, x_{21} \dots x_{n1}\, x_{12}\, x_{22} \dots x_{n2} \dots$ is an injective mapping of $I^n$ into $I$; that is, it is a bijective mapping of $I^n$ onto a subset $U$ of $I$. Thus $I^n \sim U \subset I$.

By Schroder-Bernstein theorem, $I \sim T \subset I^n$ and $I^n \sim U \subset I$ imply $I^n \sim I$. But $I$ has the cardinal number $c$ (by corollary to Theorem 9). Consequently, $I^n$ has also the cardinal number $c$.//

## 5.4 Sum, Product and Powers of Cardinal Numbers

Let $\alpha$ and $\beta$ be two cardinal numbers, and let $A$ and $B$ be two sets such that card $(A) = \alpha$ and card $(B) = \beta$. Then we define

$$\alpha + \beta = \text{card}\,(A \cup B), \text{ provided that } A \cap B = \Phi;$$

$$\alpha \cdot \beta = \text{card}\,(A \times B), \text{ provided that } A \neq \Phi \text{ and } B \neq \Phi;$$

$$\alpha^\beta = \text{card}\,(A^B), \text{ where } A^B \text{ denotes the set of all functions } f, \text{ such that dom}\,(f) = B \text{ and rng}\,(f) = A.$$

In case, $A \cap B \neq \Phi$, we substitute two sets $P$, $Q$ for $A$ and $B$ such that $P \sim A$ and $Q \sim B$, and $P \cap Q = \Phi$, and then $\alpha + \beta =$ card $(P \cup Q)$ is well defined. In fact, we can take

$$P = \{(a, x) : a \in A\} \text{ and } Q = \{ y, b) : b \in B\},$$

where $x$ and $y$ are any two elements, different from $b$ and $a$ respectively.

For $A = \Phi$ and/or $B = \Phi$, we have $\alpha = 0$ and/or $\beta = 0$. And we define $\alpha\beta = 0$, in case either (or both) of $\alpha$ and $\beta$ be 0.

Thus the sum, product and powers of cardinal numbers are defined in all cases. And also we have :

**Theorem 18 :** *If $\alpha$, $\beta$, $\gamma$ and $\nu$ are any cardinal numbers, then*

(*i*) $\alpha + (\beta + \gamma) = (\alpha + \beta) + \gamma$; (*ii*) $\alpha + \beta = \beta + \alpha$;

(*iii*) $\alpha\,(\beta\gamma) = (\alpha\beta)\,\gamma$; (*iv*) $\alpha\beta = \beta\alpha$;

(*v*) $\alpha\,(\beta + \gamma) = \alpha\beta + \alpha\gamma$ (*vi*) $\alpha^{\beta+\nu} = \alpha^\beta\,\alpha^\nu$;

(*vii*) $(\alpha\beta)^\nu = \alpha^\nu\,\beta^\nu$; (*viii*) $(\alpha^\beta)^\nu = \alpha^{\beta\nu}$;

(*ix*) $\alpha \leq \beta$ *implies* $\alpha + \gamma \leq \beta + \gamma$; (*x*) $\alpha \leq \beta$ *implies* $\alpha\gamma \leq \beta\gamma$;

(*xi*) $\alpha \leq \beta$ *implies* $\alpha^\nu \leq \beta^\nu$; (*xi*) $\alpha \leq \beta$ *implies* $\gamma^\alpha \leq \gamma^\beta$.

**Proof :** The above propositions are proved by defining appropriate bijective mappings. The proofs are left as exercises.//

**Theorem 19 :** (*i*) *For any two cardinal numbers $\alpha$, $\beta$, $\alpha \leq \alpha + \beta$ and $\alpha \leq \alpha\beta$.*

(*ii*) *For any transfinite cardinal number $\alpha$ and any non-negative integer (i.e., a finite cardinal number) $n$, we have $\alpha + n = \alpha$.*

**Proof :** (*i*) Let $A$ and $B$ be two disjoint sets, such that card $(A) = \alpha$ and card $(B) = \beta$. Then $\alpha + \beta =$ card $(A \cup B)$ and $\alpha\beta =$ card $(A \times B)$. Now $f(x) = x$, for every $x \in A$, is a bijective mapping of the set $A$ onto the subset $A \subset A \cup B$; hence $\alpha \leq \alpha + \beta$. Also $g(x) = (x, t)$, for every $x \in A$ and for a fixed element $t \in B$, is a bijective mapping of the set $A$ onto the subset $A \times \{t\} \subset A \times B$; hence $\alpha \leq \alpha\beta$.

(*ii*) Let $A$ and $B$ be two disjoint sets, such that card $(A) = \alpha$ and card $(B) = n$. Let $B = \{y_1, y_2, ..., y_n\}$, where $y_i \neq y_j$ for $i \neq j$. The infinite set $A$ has a denumerable subset $C$ (say), and let $C = \{x_r : r \in \mathbf{N}\}$ be an enumeration of the elements of $C$, where $x_i \neq x_j$ for $i \neq j$. Let a function $f$ be defined by

$$f(x) = y_j \text{ for } x = x_j,\ 1 \le j \le n,$$
$$= x_{j-n} \text{ for } x = x_j,\ j > n,$$
$$= x \text{ for } x \in A - C.$$

Then $f$ is a bijective mapping of $A$ onto $A \cup B$. Hence $\alpha + n = \alpha$.//

In particular, for $\alpha = \beta = d$, we have the following :

**Theorem 20 :** *(i)* $d + d = d$, *and* $d \,.\, d = d$;

*(ii)* $d + d + \ldots + d$ *(n terms)* $= d$, *and* $d \,.\, d \ldots d$ *(n terms)* $= d$, *for any natural number n.*

**Proof :** (*ii*) The two results follow from Theorem 6 and a deduction from Theorem 5. And then (*i*) follows by taking $n = 2$.//

**Theorem 21 :** *For any cardinal number* $\alpha$, *we have* $\alpha < 2^{\alpha}$.

**Proof :** Let $A$ be a set, such that card $(A) = \alpha$. Then $2^{\alpha} =$ card $(B^A)$, where $B$ is any set with two distinct elements only, say $B = \{0, 1\}$. Also, by Theorem 16, we have $\alpha <$ card $(\mathbf{P}(A))$. So, it suffices to show $\mathbf{P}(A) \sim \{0, 1\}^A$.

Let a function $f$ be defined on $\mathbf{P}(A)$, by

$$f(E) = \psi_E \in \{0, 1\}^E, \text{ for } E \subset A,$$

where $\psi_E(x) = 1$ for $x \in E$, and $\psi_E(x) = 0$ for $x \in A - E$.

Then $f$ is a bijection on $\mathbf{P}(A)$ onto the set $\{0, 1\}^A$. Hence the result follows by Theorem 16.//

In particular, when $\alpha = d$, we have the following :

**Theorem 22 :** $2^d = c$.

**Proof :** Every real number $x$ in the unit interval $[0, 1]$ can be expressed uniquely, in the dyadic scale, as a non-terminating expression :

$$x = x_1/2 + x_2/2^2 + \ldots + x_n/2^n + \ldots,$$

where $x_i = 0$ or $1$, for $i = 1, 2, \ldots$. The number is thus associated with an infinite sequence $\{x_1, x_2, \ldots, x_n, \ldots\}$, whose elements assume only two values 0 and 1. The set of all such sequences has the cardinal number $2^d$. And the unit interval $[0, 1]$ has the cardinal number $c$. Hence $c \le 2^d$.

On the other hand, the set of all non-terminating decimal expressions formed with only two digits, 4 and 7 say, has the cardinal number $2^d$. Such decimal expressions form a proper subset of the unit interval $[0, 1]$. Hence $2^d \le c$.

Now, combining the results $c \le 2^d$ and $2^d \le c$, one gets $2^d = c$.//

The results, proved for $d$ in Theorem 20, can be generalised to any transfinite cardinal number. In fact :

**Theorem 23 :** *For any transfinite cardinal number* $\alpha$, $\alpha + \alpha = \alpha$.

**Proof :** Let $A$ be a set, such that card $(A) = \alpha$. Then the set $B$, defined by $B = A \times \{0, 1\} = \{(a, 0) : a \in A\} \cup \{(a, 1) : a \in A\}$, has the cardinal number $\alpha + \alpha$; *i.e.*, card $(B) = \alpha + \alpha$. Let **F** denote the set of all injective functions $f$, such that dom $(f) \subset A$ and rng $(f) =$ dom $(f) \times \{0, 1\}$. The infinite set $A$ contains a denumerable subset $C$ (say). Then the set $B \times \{0, 1\}$ is also denumerable (by Theorem 5). Hence there exists an injective function $f$, such that dom $(f) = C$ and rng $(f) = C \times \{0, 1\}$. This shows that the set **F** is non-empty. The set **F** is partially ordered by the order relation "$\subset$", where $f_1 \subset f_2$ means that $f_2$ is an extension of $f_1$ (proof left as an exercise). Hence, by the Hausdorff's maximality principle, **F** contains a maximal chain **C** (say). Let $g = \cup$ **C**, then $g \in$ **F**. Let dom $(g) = D$. Then $g$ is a bijective mapping of $D$ onto $D \times \{0, 1\}$, and therefore

$$\text{card } (D) = \text{card } (D) + \text{ card } (D).$$

The proof will be completed by showing that card $(D) = \alpha$.

Let $E = A - D$. If $E$ is a finite set, say with card $(E) = n$. Then card $(D)$ = card $(D \cup E)$ (by Theorem 19 (*ii*)) = card $(A) = \alpha$. Next, let $E$ be an infinite set; then it contains a denumerable subset $G$ (say). Then $G \times \{0, 1\}$ is also a denumerable set. Hence there exists a bijective mapping $f$ of $G$ onto $G \times \{0, 1\}$. Then $h = f \cup g \in$ **F**, and $g \subset h$ and $g \ne h$. This contradicts the maximality of **C**. Therefore $E$ must be a finite set, and consequently card $(D) = \alpha$.//

**Corollary :** *If* $\alpha$ *is any transfinite cardinal number, and* $\beta$ *is any cardinal number, such that* $\beta \le \alpha$, *then* $\alpha + \beta = \alpha$.

**Proof :** By Theorems 19 (*i*), 18 (*ix*) and 23, we have successively $\alpha \le \alpha + \beta$, $\alpha + \beta \le \alpha + \alpha$ (since $\beta \le \alpha$, by hypothesis), and $\alpha + \alpha = \alpha$. And combining these results, we get $\alpha \le \alpha + \beta \le \alpha + \alpha = \alpha$, which implies $\alpha + \beta = \alpha$.//

**Theorem 24 :** *For any transfinite cardinal number* $\alpha$,

$$\alpha^2 = \alpha\ \alpha = \alpha.$$

**Proof :** Let $A$ be a set, such that card $(A) = \alpha$. Let **F** denote the set of all injective functions $f$, such that

$$\text{dom } (f) \subset A \text{ and rng } (f) = [\text{dom}(f)] \times [\text{dom } (f)].$$

The infinite set $A$ contains a denumerable subset $C$ (say). Then $C \times C$ is also denumerable (by Theorem 5); hence there exists a bijective mapping $f$ of $C$ onto $C \times C$. And, as $f \in \mathbf{F}$, it follows that $\mathbf{F}$ is non-empty. The set $\mathbf{F}$ is partially ordered by the relation "$\subset$", where $f_1 \subset f_2$ means that $f_2$ is an extension of $f_1$ (verify). Hence, by Hausdorff's maximality principle, there exists a maximal chain $\mathbf{C}$ (say) in $\mathbf{F}$. Let $g = \cup\, C$, then $g \in \mathbf{F}$ and is a maximal element of $\mathbf{F}$. Let dom $(g) = D$. Then $g$ is a bijective mapping of $D$ onto $D \times D$. Hence card $(D) =$ card $(D \times D) =$ card $(D)$ . card $(D)$. The proof will be completed by showing that card $(D) = \alpha$.

Let card $(D) = \beta$, and let $E = A - D$. If card $(E) \leq \beta$, then (by Corollary to Theorem 23) $\beta = \beta +$ card $(E) =$ card $(D \cup E) =$ card $(A) = \alpha$, so that card $(D) = \alpha$. There is only another possibility, viz. $\beta <$ card $(E)$; and let us assume that $\beta <$ card $(E)$. Then there exists a subset $G \subset E$, such that card $(G) = \beta$. Also, from $\beta =$ card $(D) =$ card $(D) \times$ card $(D) = \beta \cdot \beta = \beta^2$ and card $(D) = \beta =$ card $(G)$, we get card $(D \times G) =$ card $(G \times D) =$ card $(G \times G) = \beta$. Hence card $[(D \times G) \cup (G \times D) \cup (G \times G)] = \beta + \beta + \beta = \beta$ (by Theorem 23). Consequently, there exists a bijective mapping $f$ of $G$ onto the set $(D \times G) \cup (G \times D) \cup (G \times G)$. Let $h = f \cup g$, then $h$ is a bijective mapping of $D \cup G$ onto $(D \cup G) \times (D \cup G)$. Hence $h \in \mathbf{F}$, also $g \subset h$ and $g \neq h$. This contradicts the maximality of $g$ in $\mathbf{F}$. Hence $\beta <$ card $(E)$ is not possible, so that card $(E) \leq \beta$ must hold and therefore $\beta =$ card $(D) = \alpha$ (as shown above).//

**Corollary :** *If $\alpha$ is any transfinite cardinal number, and $\beta$ is a cardinal number, such that $0 \leq \beta \leq \alpha$, then $\alpha\beta = \alpha$.*

**Proof :** By Theorems 19 (*i*), 18 (*x*) and 24, we have successively $\alpha \leq \alpha\beta$, $\alpha\beta \leq \alpha\alpha$ (since $0 < \beta \leq \alpha$, by hypothesis), and $\alpha\alpha = \alpha$. And combining these, we get $\alpha \leq \alpha\beta \leq \alpha\alpha = \alpha$, from which it follows that $\alpha\beta = \alpha$.//

We can now deduce a number of interesting results for the cardinal numbers $c$ and $d$ :

**Theorem 25 :** (*i*) $c^d = c$; (*ii*) $c \,.\, d = c$;

(*iii*) $c^c = 2^c$ (*iv*) $n^d = d^d = c^d = c$;

(*v*) $c \,.\, d = c \,.\, c = c^n = c$; (*vi*) $n^c = d^c = c^c = 2^c$.

**Proof :** (*i*) $c^d = (2^d)^d = 2^{d \cdot d} = 2^d = c$.

(*ii*) The result follows directly from the Corollary to Theorem 24, since $0 < d < c$.

(*iii*) $c^c = (2^d)^c = 2^{cd} = 2^c$. (*iv*) $c = 2^d \leq n^d \leq d^d \leq c^d = c$.

(*v*) $c \leq cd \leq cc = c^2 \leq c^n \leq c^d = c$.

(*vi*) $2^c \leq n^c \leq d^c \leq c^c = 2^c$.//

**Note :** 1. It follows from $d^d = c$ that the set of all infinite sequences of positive integers (or rational numbers) has the power of the continuum (*i.e.*, has the cardinal number $c$).

2. It follows from $c^d = c$ that *the set of all infinite sequences of real numbers has the power of the continuum.*

3. The result $c^n = c$ is a direct consequence of the Cantor's theorem given in Example 1, (Page 26).

**Example 2.** ***The set $C[a, b]$ formed by all real-valued continuous functions on a closed interval $[a, b]$ has the cardinal number c.***

**Solution :** Let $f \varepsilon C[a, b]$, then $f$ is determined uniquely by its values $f(r)$ at the rational points $r$ in $[a, b]$. The set of *all* functions $f$, with the set of rational numbers in $[a, b]$ as their domain and having real values, has the cardinal number $c^d = c$ (since the set of rational points in $[a, b]$ has the cardinal number $d$ and the set of real numbers has the cardinal number $c$). Hence the cardinal number of $C[a, b] \leq c$.

Again, the constant functions $f(x) = k$, where $k$ is any real number, form a subset of $C[a, b]$ having the cardinal number $c$. Hence the cardinal number of $C[a, b] \geq c$.

Now, combining the results, card $(C[a, b]) \leq c$ and card $(C[a, b]) \geq c$, we get card $[C[a, b]) = c$.

## 5.5 Continuum Hypothesis

We are, so far, familiar with three distinct transfinite cardinal numbers, $d < c < 2^c$. Also, $d$, $c$, and $2^c$ are the cardinal numbers of the set of natural numbers, the set of real numbers, and the set of all subsets of the set real numbers respectively. Theorem 16 provides us with a definite method of obtaining a set, whose cardinal number is greater than the cardinal number of any *given* (*i.e.*, *known*) set.

It is, as yet, an unsolved problem in Set Theory as to whether there exists any cardinal number $\alpha$ lying strictly between $d$ and $c$, *i.e.*, satisfying $d < \alpha < c$. The *continuum hypothesis* asserts that there is no cardinal number $\alpha$ satisfying $d < \alpha < c$. A similar hypothesis that there is no cardinal number $\beta$, satisfying $\alpha < \beta < 2^\alpha$, for any transfinite cardinal number $\alpha$, is known as the *generalised continuum hypothesis.*

Godel proved that if the axioms of Set Theory are consistent, then the introduction of the continuum hypothesis as an additional axiom does not cause any inconsistency.

Later, it was proved by Cohen that the generalised continuum hypothesis is also independent of the other axioms of the Set Theory (e.g., Zermelo-Frankel axioms of Set Theory).

## EXERCISES

**1.** Prove that if $\alpha, \beta, \nu$ are cardinal numbers, such that $1 < \alpha \leq \nu$ and $1 < \beta \leq \nu$, then $\alpha^{\nu} = \beta^{\nu}$.

**2.** If $B \subset A$ and $\omega \leq \text{card}(B) < \text{card}(A)$, then $\text{card}(A - B) = \text{card}(A)$.

## Art. 6 ORDINAL NUMBERS

The principal difference between the concepts of cardinal numbers and ordinal numbers lies in the fact that every set has a cardinal number, while only well-ordered sets have ordinal numbers.

A set may be well-ordered in essentially different ways, and correspondingly different ordinal numbers are associated with the set; but a set has a unique cardinal number.

### 6.1 Order Types

Let $(X, \leq)$ be a linearly ordered set. For two elements $a, b \in X$, we define a relation "$<$" such that $a < b$ (or, equivalently $b > a$) holds iff $a \leq b$ and $a \neq b$, and we say that "*a precedes b*" or that "*b succeeds a*". Then "$<$" is an order relation, since it is asymmetric and transitive. Let it be also irreflexive, and *connected* (*i.e.*, for any two elements $a$, $b$, one and only one of the following holds : $a < b$, $a = b$ and $b < a$). Such an irreflexive, connected, order relation $<$ will, simply, be called an order relation in the whole of Art. 6. The relation $>$ which is also an order relation, is called the *inverse* of the relation $<$.

An element $a$ is called *the first element* in $(X, <)$, if $a < b$ holds for every element $b$ $(\neq a)$ of $X$; the first element, if it exists, is evidently unique (by the property of connectedness). An element $c$ is called the *last element* in $(X, <)$, iff $b < c$ holds for every element $b$ (other than $c$) of $X$; and the last element, if it exists, is also unique.

Let $(X, <_X)$ and $(Y, <_Y)$ be two ordered sets with their respective order relations $<_X$ and $<_Y$. A bijective mapping $f$ of $X$ onto $Y$ is called a *similarity* or an *order-isomorphism*, iff $x_1 <_X x_2$ (in $X$) implies $f(x_1) <_Y f(x_2)$ (in $Y$). And, as the inverse of a similarity is also a similarity, the two ordered sets $X$ and $Y$ are said to be *similar* (or *order-isomorphic*) to each other.

As the identity mapping is a similarity (for any ordered set), the product of two similarities is a similarity, and the inverse of a similarity is a similarity (verification left as an exercise), it follows that *the relation of being similar is an equivalence relation in any collection of ordered sets.* The collection is accordingly partitioned into disjoint classes, and each such class is called an *order type.* Thus each ordered set belongs to an order type or, as is usually said, has an order type. And, two sets have the same order type if, and only if, they are similar. The order type of an ordered set $(X, <)$ will be denoted by ord $(X, <)$.

Every similarity is a bijective mapping, but *not* conversely (a bijective mapping does not take into account the order relation). Hence two ordered sets, having the same order type, are equipotent. But two equipotent sets may not be similar, and therefore may not have the same order type.

It is evident that if an ordered set has the first (and/or the last) element, then any ordered set, similar to it, has also the first (and/or the last) element. Violation of this necessary property is a sufficient condition for two ordered sets to be non-similar.

**Example 1.** ***Let N denote the set of natural numbers. Then (N, <) is an ordered set*** $\{1 < 2 < 3 \ldots\}$***. Let N * denote the same set N, but with the inverse order relation >; i.e., (N *, >) is the ordered set*** $\{\ldots > 3 > 2 > 1\}$***. The two sets are obviously equipotent. But they are not similar, since (N, <) has 1 as its first element, but (N *, >) has no first element.***

If $\alpha$ be the order type of an ordered set $(X, <)$, then the same set with the inverse order relation will be denoted by $(X^*, >)$, and its order type will be denoted by $\alpha^*$.

The order type of the ordered set $(\mathbf{N}, <)$ of natural numbers is denoted by $\omega$, whereas that of $(\mathbf{N}^*, >)$ is denoted by $\omega^*$.

The order type of the set of rational numbers, ordered by the usual inequality relation $<$, *i.e.*, ord $(\mathbf{Q}, <)$, is denoted by $\eta$.

The order type of the set of real numbers, ordered by the usual inequality relation $<$, *i.e.*, ord $(\mathbf{R}, <)$, is denoted by $\lambda$.

Two finite sets $X$ and $Y$, ordered by their respective order relations $<_X$ and $<_Y$, are similar, iff they are equipotent. Hence, the order type of a finite set with $n$ (distinct) elements will be denoted by the natural number $n$. In particular, we define ord $(\Phi) = 0$.

Let $\alpha$ and $\beta$ be the order types of two ordered sets $(X, <_X)$ and $(Y, <_Y)$ respectively. Then we define $\alpha \leq \beta$, if $(X, <_X)$ is similar to $(Z, <_Y)$, where $Z$ is a subset of $Y$. It can readily be seen that *inequality among order types is a relation of partial order, but not a relation of linear* *order,* since the law of trichtomy does not hold. In fact, (**N**, <) is never similar to any subset of ($\mathbf{N}^*$, <), and ($\mathbf{N}^*$, <) is not similar to any subset of (**N**, <), since **N** and any subset of it have always a first element, but $\mathbf{N}^*$ or any infinite subset of it never possesses a first element.

## 6.2 Well-ordered Sets

An ordered set is said to be *well-ordered,* if every non-empty subset of it has a first element.

Any finite set or any set of the order type $\omega$ is well-ordered, whereas any set of the order type $\omega^*$, $\eta$ or $\lambda$ is *not* well-ordered.

It is generally possible to define different order relations on the same set, and a particular order relation may not be a well-ordering for the set. The well-ordering theorem (Zermelo's axiom) asserts that for each set there exists an order relation, which is a well-ordering relation on the set. [In fact, it can be shown that *"an order relation on a set is a well-ordering relation if, and only if, the set thus ordered does not contain any subset of the order type* $\omega^*$*"*].

**Theorem 26 :** *If f is an order-isomorphism of a well-ordered set* $(X, <)$ *onto a subset* $Y \subset X$*, then* $x < f(x)$*, for every* $x \in X$.

**Proof :** If possible, let $x$ be a element of $X$, such that $f(x) < x$. The set of all such elements $x$ forms a non-empty subset of $X$, and has therefore a first element, say $u$. Let $f(u) = y$, then $y < u$. As $f$ is an order-isomorphism, $y < u$ implies $f(y) < f(u)$, *i.e.*, $f(y) < y$. But this contradicts the defining property of $u$. Hence $x < f(x)$ must hold for every element $x \in X$.//

Let $x$ be an element of a well-ordered set $(X, <)$, and let $X_x$ be a subset of $X$, defined by $X_x = \{y \in X : y < x\}$. Then $X_x$ is called the *initial segment of X determined by x.*

**Theorem 27 :** (*i*) *A well-ordered set is never order-isomorphic to any of its initial segments.*

(*ii*) *Initial segments determined by two distinct elements are never order-isomorphic.*

**Proof :** (*i*) If possible, let a well-ordered set $(X, <)$ be order-isomorphic to an initial segment $X_x$, under an order-isomorphism $f$. Then $f(x) \in X_x$, and therefore $f(x) < x$, which is impossible, by Theorem 26.

(*ii*) Let $x$, $y$ be two distinct elements of $X$, and let $y < x$. Then the initial segment $X_x$ is a well-ordered set, and $y \in X_x$. Hence the initial segment

$X_y$ is also an initial segment of the well-ordered set $(X_x, <)$. Consequently, $X_x$ cannot be order-isomorphic to $X_y$, by (*i*).//

**Theorem 28 :** (*i*) *The only order-isomorphism of a well-ordered set onto itself is the identity mapping.*

(*ii*) *Two order-isomorphic well-ordered sets can be mapped each onto the other, order-isomorphically, in only one way.*

**Proof :** (*i*) Let $f$ be an order-isomorphism of a well-ordered set $(X, <)$ onto itself. Then $f^{-1}$ is also an order-isomorphism of $(X, <)$ onto itself. If $f(x) \neq x$, for an element $x \in X$, then $f^{-1}(x) \neq x$. Also, by Theorem 26, $x < f(x)$ and $x < f^{-1}(x)$. But $x < f^{-1}(x)$ implies $f(x) < x$. Thus both $x < f(x)$ and $f(x) < x$ must hold, which is impossible. Hence $f(x) = x$ holds for all $x \in X$; in other words, $f$ is the identity mapping.

(*ii*) Let $f$ and $g$ be two order-isomorphisms of a well-ordered set $(X, <_X)$ onto a well-ordered set $(Y, <_Y)$. Then $g^{-1}f$ is an order-isomorphism of $(X, <_X)$ onto itself. Hence, by (*i*), $g^{-1}f$ is the identity mapping $I_X$, and therefore $g = f$.

**Theorem 29 :** *Principle of transfinite induction : If $(X, <_X)$ is a well-ordered set, and $Y$ be a subset of $X$, such that $X_x \subset Y$ implies $x \in Y$, then $Y = X$.*

**Proof :** If possible, let $Y \neq X$. Then $X - Y$ is a non-empty subset of $X$, and it has a first element $x$ (say), since $X$ is well-ordered. Any element of $X_x$ precedes $x$ and cannot therefore belong to $X - Y$; hence it must belong to $Y$. Thus $X_x \subset Y$. Hence, by hypothesis, $x \in Y$, which is impossible. Consequently, $Y = X$.//

[It can be shown that *the principle of transfinite induction is applicable to an ordered set if, and only if, the set is well-ordered*].

## 6.3 Ordinal Numbers

The order type of a well-ordered set is called an *ordinal number.* $\omega$ is an ordinal number, since $(\mathbf{N}, <)$ is a well-ordered set. Every finite set, for any order relation on it, is always well-ordered; hence every natural number is also an ordinal number. On the other hand, the order types $\omega^*$, $\eta$ and $\lambda$ are not ordinal numbers, since the ordered sets $(\mathbf{N}^*, >)$, $(\mathbf{Q}, <)$ and $(\mathbf{R}, <)$ are not well-ordered.

Inequality among order types has already been defined. It now follows, by Theorem 27 (*i*), that if $\alpha$ and $\beta$ are two ordinal numbers, then

$\alpha < \beta$ iff $A$ is similar to an initial segment of $B$, where $A$ and $B$ are two well-ordered sets with order types $\alpha$ and $\beta$ respectively.

Inequality "$\leq$" of order types is a relation of partial order, for which the law of trichotomy does not hold in general (vide page 33). But, for the ordinal numbers, we have the following :

**Theorem 30 :** *For any two ordinal numbers $\alpha$ and $\beta$, one and only one of the following relations must hold :*

$$\alpha < \beta,\ \alpha = \beta,\ \textit{and}\ \beta < \alpha.$$

**Proof :** Let $A$ and $B$ be two well-ordered sets, having $\alpha$ and $\beta$ as their respective order types. If $A$ and $B$ are both empty, then $\alpha = \beta = 0$. If $A = \Phi$ and $B \neq \Phi$, then $\alpha = 0 < \beta$ (and if $A \neq \Phi$ and $B = \Phi$, then $\beta = 0 < \alpha$). Next, let both $A$ and $B$ be non-empty. One can then proceed to establish a correspondence between the elements of $A$ and $B$, by associating the first element of $B$ with the first element of $A$, and then the next element of $B$ is associated with the next element of $A$, and so on. Let $a \in A$, and let each element of the initial segment $A_a$ be thus associated with some element of $B$. Then, either every element of $B$ corresponds to some element of $A_a$, so that $B$ is order-isomorphic to a subset of $A_a$, and accordingly $\beta < \alpha$, or there exist some elements of $B$ which do not correspond to any element of $A_a$, and then the first such element (of $B$) is associated with $a$, so that $A_a$ is order-isomorphic to a subset of $B$. For the second alternative, it follows by the principle of transfinite induction that either $A$ is order-isomorphic to $B$, so that $\alpha = \beta$, or $A$ is order-isomorphic to a proper subset of $B$, so that $\alpha < \beta$. Thus one of the relations $\alpha < \beta$, $\alpha = \beta$ and $\beta < \alpha$ must hold.

It also follows, by Theorem 27, that more than one of the above three relations can never hold. Thus one and only one of the above three relations always holds.//

**Theorem 31 :** *If $\alpha$ is an ordinal number different from 0, and $A$ is a well-ordered set of order type $\alpha$, then $A$ is order-isomorphic to the set of all ordinal numbers $\geq 0$ and $< \alpha$, ordered according to their magnitudes.*

**Proof :** Let $x$ be an arbitrary element of $A$, and let $\psi(x)$ denote the order type of the initial segment $A_x$, where $\psi(x) = 0$ if $x$ is the first element of $A$. Then $\psi(x) < \alpha$ for any $x \in A$, and $\psi(x_1) < \psi(x_2)$ if $x_1, x_2 \in A$ and $x_1$ precedes $x_2$. Thus to each element $x \in A$, there corresponds an ordinal number $\psi(x) < \alpha$, such that $\psi(x_1) < \psi(x_2)$ iff $x_1$ precedes $x_2$.

On the other hand every ordinal number $\beta < \alpha$ corresponds to some element of the set $A$. In fact, if $B$ is a well-ordered set of the order type

$\beta$, then $B$ is order-isomorphic to an initial segment $A_y$ of the set $A$, where $A_y \neq A$; and then $\beta = \psi(y)$.//

**Theorem 32 :** *Every set of ordinal numbers is well-ordered according to their magnitudes.*

**Proof :** Let $E$ denote a non-empty set of distinct ordinal numbers. Let $\alpha$ be any ordinal number belonging to the set $E$. Then, by Theorem 31, the set of all ordinal numbers $\geq 0$ and $< \alpha$ is of the order type $\alpha$, and is well-ordered according to their magnitudes. It then follows that the set of all ordinal numbers $\geq 0$ and $\leq \alpha$ is also well-ordered; and therefore there exists a smallest number, say $\beta$, in this set. Then $\beta$ belongs to $E$, since $\alpha \in E$. Also $\beta$ must be the smallest number in $E$. Otherwise, any number $\gamma$, smaller than $\beta$, must also be smaller than $\alpha$; and this is impossible, since $\beta$ is the smallest among the numbers in $E$ which are smaller than $\alpha$. Thus there exists a smallest number in $E$. Consequently, any set of ordinal numbers is well-ordered according to their magnitudes.//

Let $W_\alpha$ denote the set of all ordinal numbers less than the ordinal number $\alpha$, where $\alpha$ is any given ordinal number. Then it follows, by Theorem 31, that $\alpha$ is the order type of the (well-ordered) set $W_\alpha$.

Let us associate with the ordinal number $\alpha$ the cardinal number of the set $W_\alpha$. Since two order-isomorphic sets are equipotent, but two equipotent sets may not be order-isomorphic, it follows that *two different ordinal numbers may correspond to the same cardinal number.* By using the well-ordering theorem, we can associate with each cardinal number a non-empty collection of ordinal numbers, viz. those which are associated with the given cardinal number. This non-empty collection of ordinal numbers has a unique first element (by Theorem 32), which is called the *initial ordinal* of the given cardinal number.

The initial ordinal number of the cardinal number $d$ is denoted by $\omega$. It is characterised by the property that $W_\omega$ is an infinite well-ordered set, every initial segment of which is finite. The order type of the (well-ordered) set of all ordinal numbers associated with all finite and denumerable cardinal numbers is an ordinal number denoted by $\Omega$. It is characterised by the property that $W_\Omega$ is an uncountable well-ordered set, every initial segment of which is countable. $\Omega$ is called the *first uncountable ordinal.*

## 6.4 Ordinal Arithmetic

We shall now give a brief outline of the ordinal arithmetic, mentioning some properties of the sum, product and powers of ordinal numbers (pointing out a few cases where the property differs from the corresponding property for cardinal numbers). For a fuller treatment, one may refer to the classic "Cardinal and Ordinal Numbers by W. Sierpinski".

**Sum :** Let $\alpha$ and $\beta$ be two ordinal numbers, and let $A$ and $B$ two disjoint well-ordered sets with ord $A = \alpha$ and ord $B = \beta$, and let $C = A \cup B$. We define an order relation $<$ in $C$ such that for two elements $a, b \in C$, $a < b$ holds, if either $a \in A$ and $b \in B$, or $a, b$ both belong to $A$ or $B$ and $a$ precedes $b$ in the order relation, as defined in that set ($A$ or $B$, as the case may be). Then $C$ is necessarily a well ordered set; it is called the *ordinal sum* of $A$ and $B$, and is denoted by $A + B$. We define $\alpha + \beta = \text{ord } C = \text{ord } (A + B)$.

If $A_1$ and $B_1$ are also two disjoint well-ordered sets, with ord $A_1 = \alpha$ and ord $B_1 = \beta$, then their ordinal sum $A_1 + B_1$ is order-isomorphic to $A + B$; and therefore ord $(A_1 + B_1) = \alpha + \beta$. Thus the definition of the sum is independent of the choice of the particular sets $A$ and $B$.

We can readily extend the above definition to the sum of any collection of ordinal numbers.

A few properties of the sum of ordinal numbers are given below:

1. *For any ordinal number* $\alpha$, $\alpha + 0 = \alpha$, $0 + \alpha = \alpha$.
2. *For any three ordinal numbers* $\alpha, \beta, \gamma$, $(\alpha + \beta) + \gamma = \alpha + (\beta + \gamma)$.
(Associative property)
3. *If* $\alpha$ *and* $\beta$ *are ordinal numbers, and* $\beta \neq 0$, *then* $\alpha + \beta > \alpha$.

**Proof :** Let $A$ and $B$ be two disjoint well-ordered sets with ordinal numbers $\alpha$ and $\beta$ respectively; and let $C$ be their ordinal sum, *i.e.*, $C = A + B$. Then $\alpha + \beta = \text{ord } C$. As $\beta \neq 0$, $B \neq \Phi$, and so $A$ is an initial segment of the well-ordered set $C$, and different from $C$. Hence ord $C >$ ord $A$, that is, $\alpha + \beta > \alpha$.//

4. *For ordinal numbers* $\alpha, \beta$, *the inequality* $\alpha + \beta > \beta$ *is, in general, not true, e.g.* $1 + \omega = \text{ord } \{a, 1, 2, 3, \ldots\} = \omega$.
5. *If* $\alpha$ *and* $\beta$ *are ordinal numbers and* $\alpha > \beta$, *then there exists a unique ordinal number* $\gamma \neq 0$, *such that* $\alpha = \beta + \gamma$.

**Proof :** Let $A$ be a well-ordered set with ord $A = \alpha$. Since $\alpha > \beta$, there exists a well-ordered set $B$ with ord $B = \beta$, such that $B$ is an initial segment of $A$ and different from $A$. Let the set $C$ be obtained from $A$ by removing the initial segment $B$. Then $C$ is a non-empty, well-ordered set, disjoint with $B$; and $A = B + C$ (ordinal sum). Let ord $C = \gamma$, then $\gamma \neq 0$ and $\alpha = \beta + \gamma$.

Suppose that $\alpha = \beta + \gamma$ and $\alpha = \beta + \gamma_1$, where $\gamma \neq \gamma_1$. Let $\gamma > \gamma_1$, then $\gamma = \gamma_1 + \delta$, for some ordinal number $\delta \neq 0$ (by the first part of the proof).

Then $\alpha = \beta + \gamma = \beta + (\gamma_1 + \delta) = (\beta + \gamma_1) + \delta = \alpha + \delta$, which is impossible, by property (3). This proves the uniqueness of $\gamma$.//

The number $\gamma$ in (5), is often denoted by $\alpha - \beta$.

6. *For ordinal numbers* $\beta, \gamma, \gamma_1$, $\beta + \gamma = \beta + \gamma_1$ *implies* $\gamma = \gamma_1$.

**Proof :** It is an immediate consequence of the property (5).

7. *For ordinal numbers* $\beta, \beta_1, \gamma$, $\beta + \gamma = \beta_1 + \gamma$ *does not (in general) imply* $\beta = \beta_1$. *e.g.* $1 + \omega = \omega = 0 + \omega$.

8. *For any ordinal number* $\alpha$, *there does not exist any ordinal number lying between* $\alpha$ *and* $\alpha + 1$.

**Proof :** It follows from property (3) that $\alpha + 1 > \alpha$. Let $A$ be a well-ordered set with ord $A = \alpha$. Let $D$ be the set obtained by adding to the set $A$ an element $x$, not belonging to $A$, and $x$ should be regarded as succeeding each element of the set $A$. Then $D$ is an well-ordered set, and ord $D = \alpha + 1$. Let $\gamma$ denote an ordinal number $< \alpha + 1$; and let $E$ denote a well-ordered set with ord $E = \gamma$. Then $E$ is order-isomorphic to an initial segment $D_\alpha$ of the well-ordered set $D$. As $x$ is the last element of $D$, we have either $a = x$ or the element $a$ precedes $x$. Correspondingly, we have either $\gamma = \alpha$ or $\gamma < \alpha$, so that $\gamma \leq \alpha$. Thus there is no ordinal number $\gamma$ such that $\alpha < \gamma < \alpha + 1$.//

It follows from the property (8) that for every ordinal number, there exists a *next* ordinal number; namely, the number $\alpha$ has $\alpha + 1$ as its next number. But a given ordinal number may not be next to any ordinal number; that is, an ordinal number may not have any predecessor. e.g. the number $\omega$ has no predecessor.

An ordinal number, which has a predecessor, is called a number of the *first kind*; and an ordinal number, which has no predecessor, is called a number of the *second kind*. For instance, 5, $\omega + 1$, $\omega + 4$ are numbers of the first kind, and $\omega$, $\omega 3$ are numbers of the second kind (where $\omega 3 = \omega + \omega + \omega$).

**Product :** As in the case of natural numbers, we may define product of ordinal numbers as a repeated sum. To define the product $\alpha\beta$ of two ordinal numbers $\alpha$, $\beta$, we consider two well-ordered sets $A$, $B$ (not necessarily disjoint), with ord $A = \alpha$ and ord $B = \beta$. We then form a set $D$ by collecting $\beta$ copies of the set $A$; that is $D = \{A \times b : b \in B\} = A \times B$. We next define an order relation $<$ in $D$, such that if $(a, b), (c, d) \in D$, then $(a, b) < (c, d)$ means that either $b <_B d$ or else $b = d$ and $a <_A c$. $D$ is then a well-ordered set (it is called the *ordinal product* of $A$ and $B$); and we define $\alpha\beta = \text{ord } D$.

As in the case of sum, the above definition of product is also independent of the choice of the particular sets $A$ and $B$.

The above definition can readily be extended to any given collection of ordinal numbers.

Since the multiplication has been considered as a repeated ordered addition of equal addends, and in view of the fact that multiplication is non-commutative (as shown below), it is particularly necessary to distinguish between the *multiplicand* (the ordinal number that is repeatedly added) and the *multiplier* (the ordinal number that prescribes how often the repeated addition is to be carried out). In the product expression $\alpha \cdot \beta$, $\alpha$ is the multiplicand and $\beta$ the multiplier.

It can readily be shown, from the definitions, that :

9. *For any ordinal number* $\alpha$,

   (*i*) $\alpha . 0 = 0,\ 0 . \alpha = 0;$ (*ii*) $\alpha . 1 = \alpha,\ 1 . \alpha = \alpha.$

10. *For any three ordinal numbers* $\alpha, \beta, \gamma$

    (*i*) $(\alpha . \beta) . \gamma = \alpha . (\beta . \gamma)$ (Associative property)

    (*ii*) $\alpha . (\beta + \gamma) = \alpha . \beta + \alpha . \gamma.$ (Left distributive property)

11. *Multiplication of ordinal numbers is non-commutative. e.g.*

$$2 . \omega = 2 + 2 + \dots = \text{ord } \{a_1, b_1\} + \text{ord } \{a_2, b_2\} + \dots$$
$$= \text{ord } \{a_1, b_1, a_2, b_2, \dots\} = \omega;$$

and
$$\omega . 2 = \omega + \omega = \text{ord } \{a_1, a_2, \dots\} + \text{ord } \{b_1, b_2, \dots\}$$
$$= \text{ord } \{a_1, a_2, \dots, b_1 . b_2, \dots\} \neq \omega.$$

Thus $2 . \omega \neq \omega . 2.$

12. *The right distributive property does not hold, in general;*

    that is, for three arbitrary ordinal numbers $\alpha, \beta, \gamma$,
$$(\alpha + \beta) . \gamma \neq \alpha . \gamma + \beta . \gamma.$$
    e.g. $(1 + 1)\, \omega = 2 . \omega$, and $1 . \omega + 1\omega = \omega + \omega = \omega . 2$; but $2 . \omega \neq \omega . 2$.

**Powers :** Just as multiplication has been defined as a repeated addition, we can define a power of an ordinal number by repeated multiplication. Let $\alpha, \beta$ be two ordinal numbers, where $\beta$ is an ordinal number of the first kind; then we define the power $\alpha^\beta$ recursively as follows :

$$\alpha^0 = 1,$$

and
$$\alpha^\beta = \alpha^\nu . \alpha,$$

where $\nu$ is the predecessor of $\beta$; *i.e.*, $\beta = \nu + 1$.

Since an ordinal number $\beta$ of the second kind has no predecessor, the power $\alpha^{\beta}$ cannot, in this case, be defined recursively, as above. To overcome this difficulty, we show that every ordinal number of the second kind can be obtained from a collection of numbers of the first kind as follows.

By a fundamental sequence of ordinal numbers, we mean an increasing sequence of ordinal numbers containing no greatest number. It can be shown that for a given fundamental sequence of ordinal numbers, there exists a *next larger ordinal number* $\alpha$ (that is, the number $\alpha$ is greater than every member of the given sequence, and no ordinal number less than $\alpha$ has this property). This next larger ordinal number $\alpha$ has this property). This next larger ordinal number $\alpha$ cannot be a number of the first kind, since in that case its predecessor would be the greatest member of the given sequence, which is not possible. Thus the next ordinal number of any fundamental sequence of ordinal numbers is necessarily an ordinal number of the second kind. The number $\alpha$ is called the limit of this fundamental sequence, and is denoted by $\alpha = \lim\limits_{\nu < \alpha} \nu$.

Let $\beta$ be an ordinal number of the second kind, then we can choose a fundamental sequence of ordinal numbers $\nu$ of the first kind, such that $\beta = \lim\limits_{\nu < \beta} \nu$. We have already defined the power $\alpha^{\nu}$, for every ordinal number $\nu$ of the first kind. We now define the power $\alpha^{\beta}$, for an ordinal number $\beta$ of the second kind, by prescribing $\alpha^{\beta} = \lim\limits_{\nu < \beta} \alpha^{\nu}$.

As for laws of indices, it can be proved that :

13. *For arbitrary ordinal numbers* $\alpha, \beta, \gamma$

(*i*) $\gamma^{\alpha} \gamma^{\beta} = \gamma^{\alpha+\beta}$; (*ii*) $(\gamma^{\alpha})^{\beta} = \gamma^{\alpha\beta}$.

But $(\beta\gamma)^{\alpha} \neq \beta^{\alpha}\gamma^{\alpha}$, as can be seen by taking $\gamma = \alpha = 2,\ \beta = \omega$.

## EXERCISES

[$\alpha, \beta, \gamma, \delta, \nu$, etc. denote ordinal numbers].

**1.** Prove that $1 + \alpha = \alpha$ holds iff $\alpha \geq \omega$.

**2.** If $\alpha < \beta$, then

(*i*) $\alpha + \gamma < \beta + \gamma$; (*ii*) $\gamma + \alpha < \gamma + \beta$;

(*iii*) $\gamma\alpha < \gamma\beta$, for $\gamma > 0$

(*iv*) $\alpha\gamma \leq \beta\gamma$, and $\alpha\gamma < \beta\gamma$ if $\gamma$ is an ordinal number of the first kind.

**3.** (*i*) $\gamma + \alpha < \gamma + \beta$ implies $\alpha < \beta$.

(*ii*) $\alpha+\gamma<\beta+\gamma$ implies $\alpha<\beta$.

(*iii*) $\gamma\alpha=\gamma\beta$ and $\gamma>0$ imply $\alpha=\beta$.

(*iv*) $\alpha\gamma=\beta\gamma$ and $\gamma$ is a number of the first kind imply $\alpha=\beta$.

(*v*) $\gamma\alpha<\gamma\beta$ implies $\alpha<\beta$; and

(*vi*) $\alpha\gamma<\beta\gamma$ implies $\alpha<\beta$.

**4.** (*i*) $\alpha<\beta$ and $\gamma<\delta$ imply $\alpha+\gamma<\beta+\delta$.

(*ii*) $0<\alpha<\beta$ and $0<\gamma<\delta$ imply $\alpha\gamma<\beta\delta$.

**5.** Find $\alpha, \beta, \gamma, \delta$, such that $\alpha<\beta$ and $\gamma\leq\delta$ hold, but $\alpha+\gamma<\beta+\delta$ does not hold.

[**Ans :** $\alpha=2,\ \beta=1,\ \gamma=\delta=\omega$].

**6.** $\alpha>0,\ \beta>0,\ \alpha+\beta=\omega$ imply $\alpha\beta=\omega$.

**7.** (*i*) $\alpha\leq\beta$ and $\nu\geq 0$ imply $\alpha^{\nu}\leq\beta^{\nu}$.

(*ii*) $\alpha<\beta$ and $\nu>1$ imply $\nu^{\alpha}<\nu^{\beta}$.

(*iii*) $\alpha\geq 1$ and $\nu\geq 1$ imply $\nu^{\alpha}\geq\alpha$.

*Chapter 2*

# TOPOLOGICAL SPACES

## Art. 1 TOPOLOGICAL STRUCTURES

### 1.1 Topology

Let $X$ be a non-empty set. Let $\tau$ be a family of some subsets of $X$, satisfying the following conditions :

0 (1) : *The null set $\Phi$ belongs to $\tau$.*

0 (2) : *The (universal) set $X$ belongs to $\tau$.*

0 (3) : *The union of any collection of subsets belonging to $\tau$ also belongs to $\tau$.*

0 (4) : *The intersection of any two subsets belonging to $\tau$ also belongs to $\tau$.*

Then $\tau$ is called a *topology* of $X$ (or a *topology on* $X$); and $\tau$ is said to determine a *topological structure* on $X$. The subsets belonging to $\tau$ are called *open sets* (or open subsets) of $X$, relative to the topology $\tau$; and the conditions 0 (1)-0 (4) are called the (defining) *axioms for open sets.*

A non-empty set $X$ with a topology $\tau$ is called a *topological space* $(X, \tau)$; the elements of $X$ are called points of the topological space.

It is generally possible to define different topologies on any given set $X$, and thereby obtain different topological spaces on the same set $X$. e.g.

**Example 1.** ***Trivial (or indiscrete) topology : The two subsets $\phi$ and $X$ constitute a topology of $X$, called the trivial topology or the indiscrete topology (or, sometimes, the chaotic topology) of $X$.***

**Example 2.** ***Discrete topology : The family of all subsets of $X$ constitutes a topology of $X$, called the discrete topology of $X$. This topology is distinct from the trivial topology, if $X$ has more than one element.***

These two topologies are the two extreme ones. The trivial topology contains the smallest number of open sets, and the discrete topology the largest number. Usually there exist other topologies of $X$, which are intermediate between these two.

**Example 3.** ***Confinite topology (or the topology of finite complements): Let X be an infinite set, and let $\tau$ consist of the null set $\phi$ and all those subsets of X whose complements in X are finite subsets. Then $\tau$ constitutes a topology on X. This is known as the confinite topology or the topology of finite complements of X.***

Evidently 0 (1) and 0 (2) hold for $\tau$. If $G_\alpha \in \tau$ for $\alpha \in I$ (where $I$ is an indexing set), so that each $X - G_\alpha$ is a finite set, then

$$X - \cup \{G_\alpha\} = \cap \{X - G_\alpha\}$$

is also a finite set; thus $\cup \{G_\alpha : \alpha \in I\} \in \tau$. This proves 0 (3). Next, let $G$, $H \in \tau$, so that $X - G$ and $X - H$ are finite sets. Then $X - (G \cap H) = (X - G) \cup (X - H)$ is also a finite set. Hence $G \cap H \in \tau$; and this proves 0 (4).

**Note :** The set $G_\alpha$, $G$ and $H$ have been taken above to be different from $\phi$. If some $G_\alpha$ or $G$ or $H$ be the null set, then also 0 (3) and 0 (4) hold, as the null set can be ignored while forming union whereas the intersection becomes the null set.

**Example 4.** ***Topology of countable complements : Let X be a non-countable set, and let $\tau$ consist of $\phi$ and all those subsets of X whose complements in X are countable subsets. Then $\tau$ constitutes a topology of X, known as the topology of countable complements. The proof is similar to that in Example 3.***

**Comparison of topologies :** Any topology $\tau$ of a given set $X$ is a subset of the power set **P** $(X)$, *i.e.*, $\tau \subset \mathbf{P}(X)$. Hence it is possible to establish an order relation among the topologies that can be defined on $X$, in terms of the inclusion relation for subsets. Thus, for two topologies $\tau_1$ and $\tau_2$, defined on a set $X$, $\tau_1$ is said to be *coarser* or *weaker* than $\tau_2$ iff $\tau_1 \subset \tau_2$; in that case $\tau_2$ is said to be *finer* or *stronger* than $\tau_1$. If, moreover, the inclusion relation be proper (*i.e.*, $\tau_1 \neq \tau_2$), then $\tau_1$ is said to be *strictly coarser* than $\tau_2$, and $\tau_2$ is *strictly finer* than $\tau_1$.

Of all topologies on a set $X$, the trivial topology is the coarsest, and the discrete topology is the finest. For an infinite set $X$, the cofinite topology is strictly finer than the trivial topology and strictly coarser than the discrete topology; and for a non-countable set $X$, the topology of countable complements is strictly finer than the cofinite topology and strictly coarser than the discrete topology.

## 1.2 Base

Let $(X, \tau)$ be a given topological space. A family of open sets **B** is said to form *an open base* (or *basis*) *of the topology* $\tau$, iff every open set (relative to the topology $\tau$) is expressible as the union of some sets belonging to **B**. Hence the unions of all sub-collections of **B** constitute the topology $\tau$. Thus the topology $\tau$ is completely determined by any open base **B**. But there may exist different open bases for any particular topology. e.g.

(*i*) any topology $\tau$ always forms an open base of itself;

(*ii*) for the discrete topology, the singletons (*i.e.*, subsets consisting of one point only) also form an open base.

A characteristic property of an open base is given by :

**Theorem 1 :** *A given collection of open sets **B** forms an open base of the topology $\tau$ of a topological space $(X, \tau)$ if, and only if, for any open set G and any point $p \in G$ there exists a set $V \in$ **B**, such that $p \in V \subset G$.*

**Proof :** Let **B** be an open base of $\tau$. Then any open set $G$ is a union of some sets belonging to **B**. Hence, for any point $p \in G$, there exists a set $V \in$ **B**, such that $p \in V \subset G$.

Conversely, let the given condition be satisfied for **B**, and let $G$ be any open set. Then, for any point $p \in G$, there exists a set $V(p) \in$ **B**, such that $p \in V(p) \subset G$. Let $p$ run over $G$, then one obtains, by forming union:

$$\cup \{p\} \subset \cup \{V(p)\} \subset G, \text{ i.e., } G \subset \cup \{V(p)\} \subset G.$$

Thus $G = \cup \{V(p)\}$ is a union of some members of **B**. Hence **B** forms an open base of $\tau$.//

By analogy of the criterion of the above Theorem, one defines a local base about a point as follows.

A collection of open sets $\mathbf{B}_p$, each member of which contains a point $p$ in a topological space $(X, \tau)$, is called an open base (or a local base) about the point $p$, if for every open set $G$ containing $p$, there exists a set $V_p \in \mathbf{B}_p$, such that $p \in V_p \subset G$.

It then follows immediately from Theorem 1 that :

**Theorem 2 :** *If **B** is an open base of the topology $\tau$ of a topological space $(X, \tau)$, then the totality of all those members of **B**, which contain a particular point p, forms a local base about the point p.*

A criterion for an aggregate of subsets of a given set $X$ to form an open base of a suitable topology of $X$ is given by :

**Theorem 3 :** *A collection of subsets **B** of a set X forms an open base of a suitable topology of X if, and only if,*

(*i*) *the null set $\Phi \in$ **B**,*

(*ii*) *X is the union of some sets belonging to **B**, and* (*iii*) *the intersection of any two sets belonging to* **B** *is the union of some sets belonging to* **B**.

**Proof :** Let **B** form an open base of a topology $\tau$ of $X$. Since the null set $\Phi$ and the set $X$ are open sets, and $\Phi$ cannot be expressed as the union of non-empty sets, the conditions (*i*) and (*ii*) must hold. Again, since each member of **B** is an open set, the intersection of any two members of **B** is an open set, and is therefore expressible as the union of some sets belonging to **B**. Thus (*iii*) also holds for **B**.

Conversely, let the conditions (*i*), (*ii*) and (*iii*) hold for a given collection of subsets **B**. Let $\tau$ be the family of all those subsets of $X$, which are expressible as unions of some members of **B**. Then $\tau$ forms a topology of $X$, of which $B$ is an open base. In fact, 0 (1) and 0 (2) hold for $\tau$, by virtue of the conditions (*i*) and (*ii*). Also, since every member of $\tau$ is a union of some members of **B**, it follows that the union of any aggregate of members of $\tau$ is expressible as a union of some members of **B**, and is therefore a member of $\tau$. Thus 0 (3)) holds for $\tau$. Finally, the intersection of two members of $\tau$ is the intersection of two unions of members of **B**, which is (by the distributive property for union and intersection of subsets) a union of intersections of pairs of members of **B**; and this is a union of some members of **B**, by (*iii*), *i.e.*, this is a member of $\tau$. Hence 0 (4) is also satisfied for $\tau$.//

A collection of subsets **B** of a set $X$ possessing the properties (*i*), (*ii*) and (*iii*) of Theorem 3, is said to form a *base* on $X$.

The topology $\tau$ obtained, as above, form a base **B** is called the *topology generated by the base* **B**, and is usually denoted by $\tau$ (**B**).

It is sometimes found more convenient to replace the conditions (*ii*) and (*iii*) in Theorem 3 by the following two conditions (*ii*-a) and (*iii*-a), which are equivalent to those two :

(*ii*-a) : *every point of X is contained in at least one member of* **B**;

(*iii*-a) : *if* $p \in V_1$ *and* $p \in V_2$, *where* $V_1, V_2 \in \mathbf{B}$, *then there exists a set* $V_3 \in \mathbf{B}$, *such that* $p \in V_3 \subset V_1 \cap V_2$.

Equivalence of the conditions (*ii*) and (*ii*-a) is quite evident. Also (*iii*-a) follows readily from (*iii*), since $V_1 \cap V_2$ is expressible as a union of some members of **B** of which one at least, say $V_3$, must contain the point $p$, so that $p \in V_3 \subset V_1 \cap V_2$.

On the other hand, arguing as in the converse part of Theorem 1, one obtains (*iii*) from (*iii*-a).

To compare two topologies generated by two bases $\mathbf{B}_1$ and $\mathbf{B}_2$ on a given set $X$, we prove as follows.

**Theorem 4 :** *The topology $\tau_1$, generated by a base $\boldsymbol{B}_1$, is finer than the topology $\tau_2$, generated by a base $\boldsymbol{B}_2$, on the same set X if, and only if, for any $p \in X$ and any $V_2 \in \boldsymbol{B}_2$ such that $p \in V_2$, there always exists a set $V_1 \in \boldsymbol{B}_1$, satisfying $p \in V_1 \subset V_2$.*

**Proof :** Let $G$ be an open set, relative to the topology $\tau_2$. Then, for each $p \in G$, there exists a $V_2 \in \mathbf{B}_2$, such that $p \in V_2 \subset G$ (by Theorem 1). Hence, by the given condition, there exists a $V_1 \in \mathbf{B}_1$, such that $p \in V_1 \subset V_2$, so that $p \in V_1 \subset G$. Hence $G$ is the union of the sets $V_1$, as $p$ runs over $G$; consequently, $G$ is also an open set, relative to the topology $\tau_1$. Thus $\tau_1$ is finer than $\tau_2$. This proves the sufficiency of the given condition.

Conversely, let $\tau_1$ be finer than $\tau_2$. Then, any set $V_2 \in \mathbf{B}_2$ is an open set relative to $\tau_2$, and is therefore also an open set relative to $\tau_1$. As $\mathbf{B}_1$ forms a base of $\tau_1$, the condition of the Theorem follows immediately from Theorem 1.//

The topologies $\tau_1$ and $\tau_2$ are identical, *i.e.*, $\tau_1 = \tau_2$, iff $\tau_2 \subset \tau_1$ and $\tau_1 \subset \tau_2$, *i.e.*, iff $\tau_1$ is finer than $\tau_2$ and $\tau_2$ is also finer than $\tau_1$. Hence, by Theorem 4.

**Theorem 5 :** *Two bases $\boldsymbol{B}_1$ and $\boldsymbol{B}_2$, defined on a set X, generate the same topology on X if, and only if*

(*i*) *given any $V_2 \in \boldsymbol{B}_2$ and any point $q \in V_2$ there exists a $V_1 \in \boldsymbol{B}_1$, such that $q \in V_1 \subset V_2$, and also*

(*ii*) *given any $V_1 \in \boldsymbol{B}_1$ and any point $p \in V_1$ there exists a $V_2 \in \boldsymbol{B}_2$, such that $p \in V_2 \subset V_1$.*

In some cases, when the given set $X$ has already a structure (viz. an order structure or a metric structure) defined on it, it is found convenient to define a topology on the set $X$ in terms of a base. One such case is considered below :

**Topologies on a linearly ordered set :** Let $X$ be a linearly ordered set, where the order relation is denoted by "$\leq$". Let $x < y$ mean $x \leq y$ and $x \neq y$. For $x < y$, $x$ is said to be smaller (or less) than $y$, and $y$ is said to be greater than $x$.

For $a < b$ in $X$, one defines different types of intervals in $X$ as follows:

the *open interval* $(a, b) = \{x \in X : a < x < b\}$;

the *closed interval* $[a, b] = \{x \in X : a \leq x \leq b\}$;

the *left half open interval* $(a, b] = \{x \in X : a < x \leq b\}$;
the *right half open interval* $[a, b) = \{x \in X : a \leq x < b\}$;
the *improper intervals or rays* :

$$(-\infty, b) = \{x \in X : x < b\};\ (-\infty, b] = \{x \in X : x \leq b\};$$
$$(a, +\infty) = \{x \in X : a < x\};\ [a, +\infty) = \{x \in X : a \leq x\};$$
$$(-\infty, +\infty) = X.$$

**Theorem 6 :** *Let X be a linearly ordered set, which has neither any greatest nor any least element, then all open intervals $(a, b)$ together with $\Phi$ constitute a base on X.*

**Proof :** Let **B** be the set of all open intervals of $X$ together with the null set $\Phi$. Then the conditions (*i*), (*ii*-a) and (*iii*-a) for a base are satisfied for **B**. In fact, the condition (*i*) is obvious. Next, let $p \in X$, then by hypothesis, $p$ is neither the least nor the greatest element of $X$. Hence there exist elements $a, b \in X$, such that $a < p$ and $p < b$. Then $p \in (a, b) \in$ **B**. Thus (*ii*-a) also holds for **B**. Finally, let $p \in (a, b)$ and $p \in (c, d)$, then $p \in (e, f) \subset (a, b) \cap (c, d)$, where $e$ is the greater one of $a$ and $c$, and $f$ is the smaller one of $b$ and $d$. This proves (*iii*-a) for **B**.//

The topology, generated by the base **B** in Theorem 6, is called the *order topology* or the *interval topology* on the set $X$.

Topologies can also be generated by bases formed by left half open intervals (or right half open intervals) or improper intervals.

**Theorem 7 :** *Let X be a linearly ordered set, which has no greatest element. Then*

(*i*) *the null set $\Phi$ and all right half open intervals $[a, b)$ form a base on X; and the topology generated by this base is called the lower limit topology (or the right half open interval topology) of X;*

(*ii*) *the null set and all improper intervals $(\infty, b)$ form a base on X; and the topology generated by this base is called the left hand topology of X.*

**Proof :** Left as exercises (the proofs are similar to that for Theorem 6).//

For a linearly ordered set, which does not contain any least element, one can, similarly, define the *upper limit topology* (or the *left half open interval topology*), generated by a base consisting of $\Phi$ and all left half open intervals $(a, b]$, and the *right hand topology*, generated by a base consisting of $\Phi$ and all the improper intervals $(a, +\infty)$.

**Topologies on the set of real numbers :** The set of real numbers, the set of rational numbers and the set of integers, denoted respectively by **R**, **Q** and **Z** are linearly ordered by the arithmetical (weak) inequality relation $\leq$. Also there is neither a greatest nor a least element in **R**, **Q** or **Z**. Hence, for each one of these sets, one obtains :

1. *the older topology, which will be called the usual topology or the natural topology;*
2. *the upper limit topology;*
3. *the lower limit topology;*
4. *the left hand topology, and*
5. *the right hand topology.*

But these topologies may not be distinct in all these cases. For instance, the natural topology, the upper limit topology and the lower limit topology are identical on the set of integers **Z**, since

$$(m,\ n) = (m,\ n-1] = [m+1,\ n).$$

For the sets **R** and **Q**, the upper limit and the lower limit topologies are strictly stronger than the natural topology. In fact, for any $p \in (a, b)$, one has $p \in (a,\ c] \subset (a,\ b)$ and $p \in [d,\ b) \subset (a,\ b)$, where $p \leq c < b$ and $a < d \leq p$. But, for $p \in (a,\ b]$ there does not exist any open interval $(e,\ f)$ satisfying the condition $p \in (e,\ f) \subset (a,\ b]$, when one takes $b = p$. Thus the upper limit topology is strictly stronger than the natural topology for the sets **R** and **Q** ($p, a, b, c, d, e, f$ are real numbers in the case of **R**, and rational numbers in the case of **Q**). Similarly, the lower limit topology is also strictly stronger than the usual topology for the sets **R** and **Q**.

The set of real numbers **R**, with the natural topology, is called the *real number space*. The set **R**, with the lower limit topology, is sometimes called the *Sorgenfrey line*.

**Theorem 8 :** *The null set $\Phi$ and all open intervals $(a,\ b)$, where $a$, $b$ are rational numbers, form a base of the natural topology of the set of real numbers **R**.*

**Proof :** Let $\mathbf{B}_1$ denote the collection of the null set and all open intervals $(a,\ b)$, where $a, b$ are rational numbers. Then $\mathbf{B}_1$ forms a base on the set **R**. In fact, (*i*) $\Phi \in \mathbf{B}_1$; (*ii*-a) for any real numbers $\alpha$, there exist rational numbers $a, b$ satisfying $a < \alpha < b$, and then $\alpha \in (a, b)$, and (*iii*-a) if for a real number $\alpha$, $\alpha \in (a,\ b)$ and $\alpha \in (c,\ d)$, where $a, b, c, d$ are rational numbers, then $\alpha \in (u,\ v) \subset (a,\ b) \cap (c,\ d)$ where $(u, v) \in \mathbf{B}_1$, $u$ being the larger one of $a$ and $c$, and $v$ being the smaller one of $b$ and $d$. Hence, $\mathbf{B}_1$ forms a base on **R**.

The natural topology of **R** is generated by the base **B**, consisting of the null set and all open intervals (α, β), where α, β are real numbers. Let a real number $\mu \in (a, b)$, where $(a, b) \in \mathbf{B}_1$, then $\mu \in (\alpha, \beta) \subset (a, b)$, where $(\alpha, \beta) \in \mathbf{B}$, α, β being real numbers satisfying $a < \alpha < \mu$, and $\mu < \beta < b$. Again, if $\mu \in (\alpha, \beta)$, where $(\alpha, \beta) \in \mathbf{B}$, then there exists $(c, d) \in \mathbf{B}_1$, such that $\mu \in (c, d) \subset (\alpha, \beta)$; in fact $c$, $d$ can be taken to be rational numbers satisfying the conditions $\alpha < c < \mu$ and $\mu < d < \beta$. Thus the two bases **B** and $\mathbf{B}_1$ generate the same topology on the set **R**, by Theorem 5: in other words, $\mathbf{B}_1$ forms a base of the natural topology of **R**.//

## 1.3 Subbase

Let (x, τ) be a topological space. A family of subsets **S** of $X$ is said to form a *subbase of the topology* τ, if the subsets obtained as the intersections of all finite subcollections of **S** constitute a base of τ.

Let **S** be a subbase of the topology τ of a topological space (x, τ). Then the base **B** formed by finite intersections of members of **S** is called the *base generated by the subbase* **S**.

**Theorem 9 :** *A collection of subsets S of a given set X forms a subbase of a suitable topology of X if, and only if,*

(*i*) *either Φ ∈ S, or Φ is the intersection of a finite number of subsets belonging to S;*

(*ii*) *X is the union of the subsets belonging to S.*

**Proof :** Let **S** form a subbase of a topology τ on $X$, and let **B** be the base generated by **S**. As $\Phi \in \mathbf{B}$, either $\Phi \in \mathbf{S}$, or **S** must contain some subsets, finite in number, whose intersection is the null set Φ. Again, since every point of $X$ is contained in at least one member of **B**, every point of $X$ must be contained in at least one member of **S**, *i.e.*, $X$ is the union of the subsets belonging to **S**. Thus the conditions (*i*) and (*ii*) are satisfied for **S**.

Next, let the conditions (*i*) and (*ii*) hold for **S**. Let **B** be the set formed by all finite intersections of members of **S**. Then, it follows immediately from (*i*) that $\Phi \in \mathbf{B}$. Also since $S \subset \mathbf{B}$, it follows from (*ii*) that $X$ is the union of the subsets belonging to **B**. Finally, if a point $p \in V_1$ and $p \in V_2$, where $V_1, V_2 \in \mathbf{B}$, then $V_1 = A_1 \cap A_2 \cap \ldots \cap A_r$ and $V_2 = B_1 \cap B_2 \cap \ldots \cap B_s$, where $A_1, ..., A_r, B_1, ..., B_s \in \mathbf{S}$, so that $p \in V_3 \subset V_1 \cap V_2$ holds and $V_3 = V_1 \cap V_2 = (A_1 \cap A_2 \cap \ldots \cap A_r) \cap (B_1 \cap B_2, \ldots \cap B_s) \in \mathbf{B}$.

Hence **B** forms a base on $X$; and **B** is the base generated by **S**. Thus **S** forms a subbase of a suitable topology τ of $X$; τ is called the topology of $X$, generated by the subbase **S**.//

**Theorem 10 :** *The coarsest topology $\tau$ on a set X containing a given family of subsets **R** of X is the topology generated by the subbase formed by the subsets $\Phi$, X, and all those in **R**.*

**Proof :** Let the subsets $\Phi$ and $X$ together with those in $\mathbf{R}$ form the collection $\mathbf{S}$, then the conditions (*i*) and (*ii*) hold for $\mathbf{S}$ (inclusion of $\Phi$ and $X$ make the conditions (*i*) and (*ii*) obviously true). Thus $\mathbf{S}$ forms a subbase of a topology $\tau$ on $X$; that is, $\tau$ is the topology generated by the subbase $\mathbf{S}$. Let $\tau_1$ be a topology on $X$, such that $\mathbf{R} \subset \tau_1$. Then $\mathbf{S} \subset \tau_1$; and since any topology is closed with respect to finite intersection of its members, it follows that $\mathbf{B} \subset \tau_1$, where $\mathbf{B}$ is the base generated by $\mathbf{S}$. Again, since any topology is closed with respect to union of any aggregate of its members, and $\tau$ is obtained from $\mathbf{B}$ by taking unions of members of all sub-collections of $\mathbf{B}$, it follows that $\tau \subset \tau_1$. Thus $\tau$ is the coarsest topology on $X$ containing the given family of subsets $\mathbf{R}$ of $X$.//

## 1.4 Lattice of topologies

Let $\Lambda$ be the aggregate of all topologies that can be defined on a set $X$.

The *relation of* a topology $\tau_1$ *being weaker* than a topology $\tau_2$ has already been defined (in page 44). This relation can readily be seen to be reflexive (since $\tau \subset \tau$, for every $\tau \in \Lambda$), transitive (since $\tau_1 \subset \tau_2$ and $\tau_2 \subset \tau_3$ imply $\tau_1 = \tau_3$), and anti-symmetric (since $\tau_1 \subset \tau_2$ and $\tau_2 \subset \tau_1$ imply $\tau_1 = \tau_2$). Thus the set $\Lambda$ is *partially ordered* with respect to this relation (also with respect to the inverse relation of one topology *being stronger* than some topology).

Let $\{\tau_\alpha\}$ be a collection of topologies on $X$, then their intersection $\tau = \cap \{\tau_\alpha\}$ is also a topology of $X$. In fact, (*i*) since $\Phi$ and $X$ belong to each $\tau_\alpha$, $\Phi$ and $X$ also belong to $\tau$, (*ii*) if $G, H \in \tau$, then $G$ and $H$ belong to each $\tau_\alpha$, so that their intersection $G \cap H$ belongs to each $\tau_\alpha$, and hence also to $\tau$; and finally, (*iii*) if $\{G_\alpha\}$ is any subcollection of $\tau$, then the sets $G_\alpha$ belong to each $\tau_\alpha$, so that their union $\cup \{G_\alpha\}$ belongs to each $\tau_\alpha$ and hence also to $\tau$. Thus all the conditions 0 (1)-0 (4) hold for $\tau$.

Also, if any topology $\tau'$ is weaker than each $\tau_\alpha$, *i.e.*, $\tau' \subset$ each $\tau_\alpha$, then $\tau' \subset \tau$. Thus $\tau$ is the *g.l.b.* of the family $\{\tau_\alpha\}$. Hence *there exists a g.l.b. of any given sub-collection of* $\Lambda$.

Let $\{\tau_\alpha\}$ be a collection of topologies on $X$, then their union $\cup \{\tau_\alpha\}$ is *not*, in general, a topology on $X$. Let $\tau$ be the topology of $X$, generated by the subbase formed by the subsets $\Phi$, $X$, taken together with all subsets contained in all the given topologies $\tau_\alpha$. Then, by Theorem 10, $\tau$ is the *l.u.b.* of the family $\{\tau_\alpha\}$. Thus there exists an *l.u.b.* of any given subcollection of $\Lambda$.

It then follows, from what have been proved abovc, that :

**Theorem 11 :** *The collection of all topologies on a set X forms a complete lattice, with respect to the relation of "being weaker than" (or the relation of being "stronger than").*

## Art. 2 NEIGHBOURHOODS

Let $(X, \tau)$ be a given topological space. A subset $V$ of $X$ is called a neighbourhood of a point $p \in X$, if there exists an open set $G$, such that $p \in G \subset V$. Any open set $G$, containing the point $p$, is then a *nbd* (abbreviation for neighbourhood) of $p$, since $p \in G \subset G$ holds. The open sets, containing the point $p$, are distinguished as the *open neighbourhoods of the point p.*

If $V$ is an arbitrary neighbourhood of a point $p$, and $q$ is some other point in $V$, then $V$ need *not* be a neighbourhood of the point $q$. But any open set $G$ is a neighbourhood of every point contained in it. In fact.

**Theorem 12 :** *A subset V of a topological space $(X, \tau)$ is a neighbourhood of each point $x \in V$ if, and only if, V is an open set.*

**Proof :** Let $V$ be an open set and $x \in V$. Then it follows directly from the definition that $V$ is a neighbourhood of $x$. Conversely, if $V$ is a set, such that it is a neighbourhood of every point contained in it, then for each point $x \in V$, there exists an open set $G_x$, satisfying the condition $x \in G_x \subset V$; and taking union of all such relations, as $x$ runs over $V$, one obtains $V \subset \cup \{G_x\} \subset V$, so that $V = \cup \{G_x\}$ is an open set.//

**Theorem 13 :** *If p is a point in a topological space $(X, \tau)$, and $\mathbf{N}_p$ denotes the family of all neighbourhoods of p, then*

(*i*) $\mathbf{N}_p$ *is non-empty, and p is contained in every member of* $\mathbf{N}_p$;

(*ii*) *if* $V_1, V_2 \in \mathbf{N}_p$, *then* $V_1 \cap V_2 \in \mathbf{N}_p$;

(*iii*) *if* $V \in \mathbf{N}_p$, *and* $V \subset W$, *then* $W \in \mathbf{N}_p$;

(*iv*) *if* $V \in \mathbf{N}_p$ *then there exists a* $V^* \in \mathbf{N}_p$, *such that* $V \in \mathbf{N}_q$, *for each* $q \in V^*$.

**Proof :** (*i*) It follows from the definition of a *nbd* that $p$ is contained in every *nbd* of $p$. Also there is at least one *nbd* of $p$, viz. the set $X$.

(*ii*) Let $p \in G \subset V_1$ and $p \in H \subset V_2$, where $G$ and $H$ are open sets. Then $p \in G \cap H \subset V_1 \cap V_2$, and as $G \cap H$ is an open set, it follows that $V_1 \cap V_2$ is a *nbd* of $p$, *i.e.*, $V_1 \cap V_2 \in \mathbf{N}_p$.

(*iii*) Let $p \in G \subset V$, where $G$ is an open set, then it follows by $V \subset W$, that $p \in G \subset W$. Hence $W \in \mathbf{N}_p$.

(*iv*) Let $p \in G \subset V$, where $G$ is an open set. As $G$ is an open set, it is a *nbd* of every point contained in it. And since $G \subset V$, it follows by (*iii*) that $V$ is also a *nbd* of every point contained in $G$. Thus $V^* = G$ satisfies the condition required in (*iv*).//

By virtue of the properties (*i*), (*ii*), (*iii*) in Theorem 13, the family $\mathbf{N}_p$ is said to form a filter; and $\mathbf{N}_p$ is called the *neighbourhood filter of the point p*.

**Neighbourhood topology :** Properties (*i*)-(*iv*) in Theorem 13 and Theorem 12 (where open sets are given in terms of neighbourhoods) enable us to characterise a topology in terms neighbourhood filters. In fact :

**Theorem 14 :** *Let X be a given set, and to each point $p \in X$, let there be associated a family of subsets $N_p$ of X, such that :*

$N(1)$ : *$N_p$ is non-empty, and $p \in V$ for every $V \in N_p$;*

$N(2)$ : *if $V_1, V_2 \in N_p$, then $V_1 \cap V_2 \in N_p$;*

$N(3)$ : *if $V \in N_p$, and $V \subset W$, then $W \in N_p$;*

$N(4)$ : *if $V \in N_p$, then there exists a $V^* \in N_p$, such that $V \in N_q$ for every point $q \in V^*$.*

*Then there exists a unique topology $\tau$ of X, such that $N_p$ is the neighbourhood filter of the point p in the topological space $(X, \tau)$.*

**Proof :** Let a subset $G$ of $X$ be called an open set, if either $G = \Phi$, or $G \in \mathbf{N}_x$ for every point $x \in G$ and let $\tau$ denote the aggregate of all such open sets. Then the axioms 0 (1)-0 (4) for a topology are satisfied for $\tau$, so that $\tau$ is a topology of $X$. in fact, (1) that $\Phi$ is an open set follows by the definition of an open set, so that $\Phi \in \tau$. (2) For any point $p \in X$, there exists a $V \in \mathbf{N}_p$, by $N(1)$, and as $V \subset X$, it follows by $N(3)$ that $X \in \mathbf{N}_p$; thus $X$ is an open set, *i.e.*, $X \in \tau$. (3) For any two open sets $G$ and $H$, and for any point $p \in G \cap H$, $G \in \mathbf{N}_p$ (since $p \in G$, and $G$ is an open set), and $H \in \mathbf{N}_p$; hence $G \cap H \in \mathbf{N}_p$ (by $N(2)$), from which it follows that $G \cap H$ is an open set. (4) If $\{G_\alpha\}$ is a family of open sets, and $H$ is their union, then for any point $p \in H$, $p$ belongs to at least one set $G_\alpha$, and since $G_\alpha$ is an open set, $G_\alpha \in \mathbf{N}_p$, from which it follows by $N$ (3) that $H \in \mathbf{N}_p$, since $G_\alpha \subset H$. Hence $H$ is an open set. Thus the conditions 0 (1)-0 (4) hold for $\tau$.

Thus $(X, \tau)$ is a topological space, and to each point $p$ there is associated a neighbourhood filter $\mathbf{N}_p^*$ (say). Let $V \in \mathbf{N}_p^*$, then there exists an open set $G$, such that $p \in G \subset V$. Then $G \in \mathbf{N}_p$ and $G \subset V$ imply $V \in \mathbf{N}_p$, by $N$ (3). Hence, $\mathbf{N}_p^* \subset \mathbf{N}_p$.

Next, let $V \in \mathbf{N}_p$. Let $G$ be the set formed by all those points $x \in X$, such that $V \in \mathbf{N}_x$. Then $G \subset V$, by $N$ (1). Also $p \in G$. Thus $p \in G \subset V$. It will be shown that $G$ is an open set. Let $q$ be any point in $G$; then $V \in \mathbf{N}_q$. By $N$ (4), there exists a set $V^* \in \mathbf{N}_q$, such that $V \in \mathbf{N}_y$, for every point $y \in V^*$. Then all such points $y \in G$ (by the definition of $G$); hence $V^* \subset G$. Now $V^* \in \mathbf{N}_q$ and $V^* \subset G$ imply $G \in \mathbf{N}_q$, by $N$ (3). Thus $G \in \mathbf{N}_a$, for every point $q \in G$; hence $G$ is an open set. Then $p \in G \subset V$ implies that $V \in \mathbf{N}_p^*$. Hence $\mathbf{N}_p \subset \mathbf{N}_p^*$.

Combining the results $\mathbf{N}_p^* \subset \mathbf{N}_p$ and $\mathbf{N}_p \subset \mathbf{N}_p^*$, one obtains $\mathbf{N}_p^* = \mathbf{N}_p$, for every point $p \in X$.//

Conditions $N$ (1)-$N$ (4) are known as *the neighbourhood axioms for a topological space.*

## 2.1 Neighbourhood Basis

A collection $\mathbf{B}(p)$ of *nbds* of a given point $p$ in a topological space $(X, \tau)$ is called a neighbourhood basis of the point $p$, if for every *nbd* $U$ of $p$, there exists a $V \in \mathbf{B}(p)$ such that $p \in V \subset U$. It now follows readily that :

**Theorem 15 :** *If $\mathbf{B}(p)$ is a neighbourhood basis of a point $p$ in a topological space $(X, \tau)$, then*

(*i*) *$\mathbf{B}(p)$ is non-empty, and every member of $\mathbf{B}(p)$ contains the point $p$;*

(*ii*) *if $U$ and $V$ belong to $\mathbf{B}(p)$, then there exists a $W \in \mathbf{B}(p)$, such that $W \subset U \cap V$;*

(*iii*) *if $U \in \mathbf{B}(p)$, then there is a set $W \in \mathbf{B}(p)$, such that, for every $q \in W$, there exists a set $V \in \mathbf{B}(q)$, satisfying $V \subset U$.*

**Proof :** (*i*) Since $\mathbf{N}_p$ is not empty (by $N(1)$), there exists a *nbd* $U$ of $p$. Then there must exist a member $V \in \mathbf{B}(p)$, such that $p \in V \subset U$. Thus $\mathbf{B}(p)$ cannot be empty. Also, since every *nbd* of $p$ contains $p$ (by $N(1)$), it follows that every member of $\mathbf{B}(p)$ must contain $p$.

(*ii*) Since $U, V \in \mathbf{B}(p) \subset \mathbf{N}_p$, $U \cap V \in \mathbf{N}_p$. Then there exists $W \in \mathbf{B}(p)$, such that $p \in W \subset U \cap V$. Thus there exists a set $W \in \mathbf{B}(p)$, such that $W \subset U \cap V$.

(*iii*) As $U \in \mathbf{B}(p) \subset \mathbf{N}_p$, there exists, by $N(4)$, a set $U^* \in \mathbf{N}_p$ such that $U \in \mathbf{N}_y$ for every $y \in U^*$. As $\mathbf{B}(p)$ forms a *nbd* basis, there exists a set $W \in \mathbf{B}(p)$, such that $W \subset U^*$. Now for any $q \in W$, $q \in U^*$ and therefore $U \in \mathbf{N}_q$. Hence, there exists a set $V \in \mathbf{B}(q)$, satisfying $V \subset U$.//

The properties (*i*)-(*iii*) in Theorem 15 characterise a neighbourhood basis. In fact :

**Theorem 16 :** *Let X be a given set, and to each point $p \in X$ let there be associated a family of subsets $\boldsymbol{B}(p)$, satisfying the properties (i)-(iii) of Theorem 15. then there exists a unique topology $\tau$ of X, such that $\boldsymbol{B}(p)$ forms a neighbourhood basis of the point p, for each point $p \in X$.*

**Proof :** Let a collection of subsets $\mathbf{N}_p$ be defined, for each $p \in X$, by $\mathbf{N}_p = \{V : V \supset U, \text{ for some } U \in \mathbf{B}(p)\}$. ...(1)

We shall show that the neighbourhood axiom $N(1)$-$N(4)$ hold for the family of subsets $\mathbf{N}_p$.

1. Since $\mathbf{B}(p) \subset \mathbf{N}_p$ (by Equation (1)), $N(1)$ follows from (*i*).
2. Let $V_1, V_2 \in \mathbf{N}_p$, then $V_1 \supset U_1$ and $V_2 \supset U_2$ for some $U_1, U_2 \in \mathbf{B}(p)$. Then $V_1 \cap V_2 \supset U_1 \cap U_2 \supset W$ by (*ii*), where $W \in \mathbf{B}(p)$. Hence $V_1 \cap V_2 \in \mathbf{N}_p$. Thus $N(2)$ holds for $\mathbf{N}_p$.
3. Let $V \in \mathbf{N}_p$ and $W \supset V$. There exists an $U \in \mathbf{B}(p)$, such that $U \subset V$. Then $U \subset V \subset W$ implies that $W \in \mathbf{N}_p$. Hence $N(3)$ holds for $\mathbf{N}_p$.
4. Let $V \in \mathbf{N}_p$, then there exists a set $U \in \mathbf{B}(p)$, such that $U \subset V$. By (*iii*), there exists a set $V^* \in \mathbf{B}(p)$, such that for every point $q \in V^*$. there is a set $T \in \mathbf{B}(q)$, satisfying $T \subset U$.

As $T \in \mathbf{B}(q) \subset \mathbf{N}_q$ and $T \subset U \subset V$, it follows, by (3), that $V \in \mathbf{N}_q$. Thus $V \in \mathbf{N}_q$ for every $q \in V^*$, where $V^* \in \mathbf{B}(p) \subset \mathbf{N}_p$. This proves $N(4)$.

Thus $\mathbf{N}_p$ forms the *nbd* filter of the point $p$, for each $p \in X$. The *nbd* filters $\mathbf{N}_p$ determine a unique topology $\tau$ on $X$; and it follows from Equation (1) that $\mathbf{B}(p)$ forms a *nbd* basis of the point $p$, for each $p \in X$ in the topological space $(X, \tau)$.//

The topology $\tau$, obtained as above, is said to be *generated by the neighbourhood basis* $\mathbf{B}(p)$. Thus a topology on a set $X$ may be defined either by choosing a *nbd* filter $\mathbf{N}_p$ for every point $p \in X$, or by choosing a *nbd* basis $\mathbf{B}(p)$ for every point $p \in X$.

In view of the property (1), it follows that *in any topological space, the open nbds of a point p form a nbd basis of p*. Also we obtain the following condition for (*topological*) *equivalence* of two *nbd* bases :

**Theorem 17 :** *Two systems of neighbourhood bases* $\boldsymbol{B}_1(p)$ *and* $\boldsymbol{B}_2(p)$ *generate the same topology on a set X if, and only if,*

(*i*) *for each point p and an* $U \in \boldsymbol{B}_1(p)$, *there exists a* $V \in \boldsymbol{B}_2(p)$ *such that* $V \subset U$, *and*

(*ii*) *for each point p and a* $V \in \boldsymbol{B}_2(p)$, *there exists an* $U \in \boldsymbol{B}_1(p)$, *such that* $U \subset V$.

**Proof :** Left as an exercise.//

## 2.2 Accumulation Points and Derived Set

Let $(X, \tau)$ be a topological space, and let $A$ be a subset of $X$. A point $p \in X$ is called an *accumulation point*, or *a limiting point*, of the set $A$ if every *nbd* of $p$ intersects $A$ in at least one point other than $p$ (in case $p$ lies in $A$); *i.e.*, $U \cap (A - \{p\}) \neq \Phi$, for every *nbd* $U$ of the point $p$. Since every *nbd* of $p$ contains an open set containing $p$, and every open set, containing $p$, is a *nbd* of $p$, it follows that $p$ is an accumulation point of $A$ iff every open set containing $p$ intersects $A - \{p\}$.

An accumulation point of $A$ may lie inside $A$ or outside. A point in $A$, which is not an accumulation point of $A$, is called an *isolated point* of the set $A$.

If, in particular, every *nbd* of $p$ intersects $A$ in infinitely many points, then $p$ is called an $\omega$-*accumulation point*, or an $\omega$-*limiting point* of the set $A$. If, further, every *nbd* of $p$ intersects $A$ in a non-countable infinite set of points, then $p$ is called a *point of condensation* of the set $A$; presumably, $A$ must be a non-countable set, in this case.

The set formed by the accumulation points of a set $A$ is called the *derived* set of $A$, and will be denoted by $A'$.

**Example 1.** *In a topological space $(X, \tau)$ the following properties hold for derived sets :*

(*i*) $\phi' = \phi$; (*ii*) $A \subset B$ *implies* $A' \subset B'$;

(*iii*) *if* $p \,\varepsilon A'$, *then* $p \,\varepsilon (A - \{p\})'$; (*iv*) $(A \cup B)' = A' \cup B'$;

(*v*) $A'' \subset A \cup A'$; (*vi*) $(A \cup A')' \subset A \cup A'$;

(*vii*) $(A \cap B)' \subset A' \cap B'$;

*where* $A, B$ *are any subsets of* $X$.

**Example 2.** *Let R denote the set of all real numbers, and let $\tau$ and $\sigma$ denote respectively the usual topology and the upper limit topology of R. Also, let $A = \{1/n;\ n = 1, 2, 3, \ldots\}$, then*

(*i*) *A has only one limiting point, viz., 0, i.e., $A' = \{0\}$, in $(R, \tau)$ but*

(*ii*) *0 is not a limiting point of A in $(R, \sigma)$, since, in this case, $(-1, 0]$ is an open set, containing the point 0, which does not intersect the set A.*

**Example 3.** *If p is a limiting point of a set A in the real number space $(R, \tau)$, then p is an $\omega$-limiting point of A.*

**Solution :** Let $(p - t,\ p + t)$ be any open interval (containing the point $p$). Then it contains at least one point of the set $A$, other than $p$. In fact, this interval must contain infinitely many points of the set $A$. Otherwise, let $q$ be the point of the set $A$, lying closest to $p$. Then the open interval $(p - s\ .\ p + s)$, where $s = |p - q|/2$ would not contain any point of the set $A - \{p\}$. But this is impossible, since $p$ is a limiting point of the set $A$ and $(p - s,\ p + s)$ is an open set containing the point $p$.

**Dense-in-itself and perfect sets :** A subset $A$ of a topological space $(X, \tau)$ is called :

(*a*) a *dense-in-itself set,* if $A \subset A'$;

(*b*) a *perfect set, if* $A = A'$.

Evidently every perfect set is dense-in-itself, but not conversely, as can be seen from the following :

**Example 4.** *Let $(R, \tau)$ be the real number space. Then in $\{R, \tau)$,*

(*i*) *the set of all rational numbers in [0, 1] is dense-in-itself, but not perfect (as the derived set is the entire interval [0, 1]);*

(*ii*) *the closed unit interval [0, 1] is a perfect set.*

**Example 5.** *In any topological space,*

(*i*) *the union of any aggregate of dense-in-itself sets is dense-in-itself;*

(*ii*) ***if D is a dense-in itself set, and $D \subset T \subset D'$, then T is also a dense-in-itself set :***

(*iii*) ***if D is a dense-in-itself set, then $D'$ and $\overline{D}$ are also dense-in-itself;***

(*iv*) ***if D is a dense-in-itself set and G any open set, then $D \cap G$ is either a dense-in-itself set or the null set.***

The union of all dense-in-itself subsets, contained in a set $A$, is called the nucleus of $A$. It follows from Example 5 (*i*) that the nucleus of a set $A$ is a dese-in-itself set. It is the largest dense-in-itself subset of $A$, in the sense that any dense-in-itself subset of $A$ is contained in the nucleus of $A$.

A set, whose nucleus is the null set, is called a *scattered set.* A subset of a scattered set is scattered, and the union of two scattered sets is also a scattered set.

**Discrete and isolated sets :** A subset $A$ of a topological space $(X, \tau)$ is called :

(*i*) a discrete set, iff $A' = \Phi$ (*i.e.*, $A$ has no accumulation point);

(*ii*) an isolated set, iff $A \cap A' = \Phi$ (*i.e.*, every point of $A$ is an isolated point).

Evidently, every discrete set is isolated, and every isolated set is scattered, but not conversely.

**Example 6.** ***In the real number space (R, τ).***

(*a*) ***any finite subset of R and the set of all integers are discrete sets;***

(*b*) ***the set $A = \{1/n : n = 1, 2, \ldots\}$ is isolated but not discrete (as A has an accumulation point 0, which does not lie in A);***

(*c*) ***the set $B = \{0, 1, 1/2, 1/3, \ldots, 1/n, \ldots\}$ is scattered but not isolated.***

## Art 3. CLOSED SETS

Let $(X, \tau)$ be a topological space. A subset $F$ of $X$ is called a *closed set* if its complement $X - F$ is an open set. Thus the complement of an open set is a closed set and, dually, the complement of a closed set is an open set.

The open sets (which constitute the topology $\tau$) are characterised by the properties 0 (1)-0 (4) (page 43). Dualising these, one obtains the following properties for the closed sets;

**Theorem 18 :** *In a topological space $(X, \tau)$ the following properties hold for the closed sets :*

$C(1)$ : *The universal set $X$ is a closed set.*

$C$ (2) : *The null set $\Phi$ is a closed set.*

$C$ (3) : *The intersection of any aggregate of closed sets is a closed set.*

$C$ (4) : *The union of two closed sets is a closed set.*

**Proof :** 1. As $\Phi$ is an open set, by 0 (1), its complement $X$ is a closed set.

2. As $X$ is an open set, by 0 (2), its complement $\Phi$ is a closed set.
3. Let $\{F_\alpha\}$ be an aggregate of closed sets, then $\{X - F_\alpha\}$ is an aggregate of open sets. Now, $X - \cap \{F_\alpha\} = \cup \{X - F\alpha\}$ is an open set, by 0 (3). Hence $\cap \{F_\alpha\}$ is a closed set.
4. Let $F$ and $K$ be two closed sets. Then $X - F$ and $X - K$ are open sets; and $X - (F \cup K) = (X - F) \cap (X - K)$ is an open set, by 0 (4). Hence $F \cup K$ is a closed set.//

The properties $C$ (1)-$C$ (4) completely characterise the family of closed sets in any topological space. In fact :

**Theorem 19 :** *Let $X$ be a given set, and let* **C** *be a family of some subsets of $X$ satisfying the properties :*

$C\,(1) : X \in \mathbf{C}$;

$C\,(2) : \Phi \in \mathbf{C}$;

$C$ (3) : *If $F_\alpha$ is a family of some members of* **C**, *then* $\cap \{F_\alpha\} \in \mathbf{C}$;

$C$ (4) : *If $F, K \in \mathbf{C}$, then $F \cup K \in \mathbf{C}$.*

*Then there exists a unique topology $\tau$ of $X$, such that the closed sets of the topological space $(X, \tau)$, defined in terms of the topology $\tau$, are the same as the members of the given family* **C**.

**Proof :** Let a subset $G$ of $X$ be called an open set, if its complement is a member of **C**; and let $\tau$ denote the collection of all open sets thus determined. Now, dualising the properties $C$ (1)-$C$ (4), which hold for the members of **C**, one obtains the properties 0 (1)-0 (4) for the collection $\tau$ (exactly in the same way as in the proof of Theorem 18). Thus $\tau$ forms a topology on $X$, so that $(X, \tau)$ is a topological space. The closed sets in the topological space $(X, \tau)$, *i.e.*, in terms of the topology $\tau$, being the complements of the open sets, are the same as the members of **C**. The topology $\tau$ is obviously uniquely determined.//

**Example 1.** ***Let X be an infinite set.***

***(a)*** ***Let $C_1$ be the collection consisting of X and all finite subsets of X. Then the conditions C (1)-C (4) hold for the members of $C_1$. The***

***corresponding topology on X is the cofinite topology on X (vide Example 3, page 44).***

(*b*) ***Let $C_2$ be the collection consisting of X and all countable subsets of X (X is supposed to be non-countable). Then the conditions C (1)-C (4) are satisfied for the members of $C_2$, and the corresponding topology is the topology of countable complements on X (vide Example 4, page 44).***

A closed set can be characterised in terms of its drived set as follows:

**Theorem 20 :** *A set F is closed in a topological space* $(X, \tau)$ *if, and only if,* $F' \subset F$.

**Proof :** Let $F$ be a closed set, and let $p \in X - F$. Then $p$ cannot be an accumulation point of $F$, since there exists an open set $X - F$, containing $p$, which does not intersect $F$. Thus no point, lying outside $F$, can be an accumulation point of $F$; hence $F' \subset F$.

Conversely, let $F' \subset F$. Let $p \in X - F$. Then $p$ is not an accumulation point of $F$; hence there exists an open set, $U(p)$ say, containing $p$, such that $F \cap U(p) = \Phi$, so that

$$p \in U(p) \subset X - F \qquad ...(1)$$

Taking union of the relations (1), as $p$ run over $X - F$, one obtains $X - F \subset \cup \{U(p)\} \subset X - F$. Thus $X - F = \cup \{U(p)\}$ is an open set (by 0 (3)). Hence $F$ is a closed set.//

**Corollary :** *A set P is perfect if, and only if, it is closed and dense-in-itself.*

**Proof :** $P = P'$ if and only if $P' \subset P$ and $P \subset P'$.//

**Example 2 (*a*) :** ***The intersection of infinitely many open sets may not be an open set.***

(*b*) ***The union of infinitely many closed sets need not be a closed set.***

**Solution :** Let $(\mathbf{R}, \tau)$ be the real number space. Then (*i*) any open interval $(a, b)$ is an open set, and (*ii*) any closed interval $[a, b]$ is a closed set in $(\mathbf{R}, \tau)$.

(*a*) The sets $G_n = (-1/n, 1/n)$, for $n = 1, 2, \ldots$ are all open. Their intersection $\cap \{G_n\} = \{0\}$ consists of the number 0 only. But the set $\{0\}$ is not an open set, since its complement $\mathbf{R} - \{0\}$ is not a closed set (as, 0, which is an accumulation point of the set $\mathbf{R} - \{0\}$, lies outside it,

(*b*) The sets $F_n = [1/n, 1]$, for $n = 1, 2, \ldots$ are all closed; but their union $\cup \{F_n\} = (0, 1]$ is not a closed set, as 0, which is an accumulation point of the set $(0, 1]$, lies outside the set $(0, 1]$.

**Closed basis :** by dualising, the concept of an open base (or basis), we obtain that of a closed basis (or base). Thus a collection of closed sets **A** in a topological space $(X, \tau)$ is said to form a *closed basis* (or a *closed base*), if every closed set can be expressed as the intersection of some members of **A**.

By dualising Theorems 1 and 3, we obtain :

**Theorem 21 :** *A given collection of closed sets **A** forms a closed basis of the topological space $(X, \tau)$ if, and only if, for any closed set $F$ and any point $p \notin F$ there exists a set $K \in$ **A**, such that $p \notin K$ and $K \supset F$.*

**Theorem 22 :** *A collection of subsets **A** of a set $X$ forms a closed basis of $(X, \tau)$ for a suitable topology $\tau$ on $X$ if, and only if,*

(*i*) $X \in A$; (*ii*) $\cap \{K : K \in A\} = \Phi$, *and*

(*iii*) *if $K_1, K_2 \in$ **A** and $p \notin K_1 \cup K_2$, then there exists a $K_3 \in$ **A**, such that $p \notin K_3 \supset K_1 \cup K_2$.*

Proofs of the Theorems 21 and 22 are obtained by dualising the proofs of the Theorems 1 and 3.//

## 3.1 Closure of a Set

Let $A$ be a subset of a topological space $(X, \tau)$. The intersection of all closed subsets of $X$, containing $A$, is called the closure (or the adherence) of $A$, and is denoted by $\bar{A}$. Since there is at least, one closed subset containing $A$, viz. $X$, the closure of $A$ always exists.

It now follows readily that :

**Theorem 23 :** *For subsets $A$, $D$, $F$ in a topological space $(X, \tau)$,*

(*i*) $\bar{A} \supset A$;

(*ii*) *$\bar{A}$ is a closed set;*

(*iii*) *if $F$ is a closed set, and $F \supset A$, then $F \supset \bar{A}$; i.e., $\bar{A}$ is the smallest closed set containing $A$;*

(*iv*) *$A$ is closed if and only if $\bar{A} = A$;*

(*v*) *$D \supset A$ implies $\bar{D} \supset \bar{A}$.*

**Proof :** (*i*) This follows directly from the definition of $\bar{A}$.

(*ii*) Since the intersection of any collection of closed sets is a closed set, it follows that $\bar{A}$ is necessarily a closed set.

(*iii*) Let $F$ be a closed set, containing $A$. $\bar{A}$ is the intersection of all closed sets containing $A$, and $F$ is one among these; hence $F \supset \bar{A}$.

(*iv*) Let $\overline{A} = A$. Then $A$ is a closed set, since $\overline{A}$ is a closed set (by (*ii*)). Conversely, let $A$ be a closed set. Then, taking $F = A$ in (*iii*), we get $A \supset \overline{A}$. Also $\overline{A} \supset A$ (by (*i*)). Hence $\overline{A} = A$.

(*v*) Let $D \supset A$. Then $\overline{D} \supset D \supset A$. Since $\overline{D}$ is a closed set containing $A$, it follows that $\overline{D} \supset \overline{A}$ by (*iii*).//

If $x \in \overline{A}$, then the point $x$ is said to *adhere* to the subset $A$ or to be a *contact point* of $A$.

**Theorem 24 :** *A point $x$ is a contact point of a subset A, i.e., $x \in \overline{A}$, in a topological space $(X, \tau)$ if, and only if, every nbd of x intersects A.*

**Proof :** Let $x \in \overline{A}$. If, in particular, $x \in A$ then every *nbd* of $x$ obviously intersects $A$. Next, let $x \notin A$, and let $V$ be any *nbd* of $x$. Without any loss of generality, $V$ may be taken to be an open *nbd*. If $V$ does not intersect $A$, then $A \subset \overline{A} - V$. And then $\overline{A} - V = \overline{A} \cap (X - V)$ is a closed set containing $A$, and properly contained in $\overline{A}$ (since $V \neq \Phi$), which is impossible by the definition of $\overline{A}$ (or, by Theorem 23 (*iii*)). Hence $V$ must intersect $A$.

If, on the other hand $x \notin \overline{A}$, then there exists an open *nbd* $X - \overline{A}$ of $x$, which does not intersect $A$.//

**Theorem 25 :** *In a topological space $(X, \tau)$, $\overline{A} = A \cup A'$, for every subset A of X.*

**Proof :** $A \subset \overline{A}$ implies $A' \subset (\overline{A})'$ (since $A \subset B$ implies $A' \subset B'$). But $(\overline{A})' \subset \overline{A}$, by Theorem 20, since $\overline{A}$ is a closed set. Thus $A' \subset \overline{A}$. Also $A \subset \overline{A}$. Hence, $A \cup A' \subset \overline{A}$.

Next, let $x \in \overline{A}$, then every *nbd* of $x$ intersects $A$ (by Theorem 24). Thus either $x \in A$ or $x$ is an accumulation point of $A$, *i.e.*, $x \in A'$. In either case $x \in A \cup A'$. Hence $\overline{A} \subset A \cup A'$.

Now, combining the above two results, we get $\overline{A} = A \cup A'$.//

**Theorem 26 :** *In a topological space $(X, \tau)$,*

$$K(1) : \overline{A} \supset A; \qquad K(2) : \overline{A \cup B} = \overline{A} \cup \overline{B}.$$

$$K(3) : \overline{\overline{A}} = \overline{A}; \qquad K(4) : \overline{\Phi} = \Phi,$$

*where A, B, ... are arbitrary subsets of X.*

**Proof :** $K(1)$ is the same as Theorem 23 (*i*). The result follows directly from the definition of the closure of a set.

$K(2)$ : $A \subset A \cup B$ and $B \subset A \cup B$ imply $\overline{A} \subset \overline{A \cup B}$ and $\overline{B} \subset \overline{A \cup B}$ (by Theorem 23 (*v*)), from which we get $\overline{A} \cup \overline{B} \subset \overline{A \cup B}$. Again $\overline{A} \cup \overline{B}$ is a closed set, containing $A \cup B$; hence $\overline{A \cup B} \subset \overline{A} \cup \overline{B}$ (by Theorem 23 (*iii*)). Thus $\overline{A \cup B} = \overline{A} \cup \overline{B}$.

$K(3)$ follows from Theorem 23 (*iv*), since $\overline{A}$ is a closed set (by Theorem 23 (*ii*)).

$K(4)$ follows from Theorem 23 (*iv*), since $\Phi$ is a closed set.//

**Closure topology :** $A \to \overline{A}$ is an operator on the power set $\mathbf{P}(X)$ of the set $X$. This operator, which assigns to every subset $A \subset X$ a subset $\overline{A} \subset X$, is called a *closure operator* on $X$. We shall now show that the properties $K(1)$-$K(4)$ completely characterise a closure operator.

**Theorem 27 :** *Let X be a given set and let an operator* $A \to \overline{A}$ *be defined on the power set* $\boldsymbol{P}(X)$*, satisfying the properties :*

$K(1) : \overline{A} \supset A;$ $\qquad K(2) : \overline{A \cup B} = \overline{A} \cup \overline{B};$

$K(3) : \overline{\overline{A}} = \overline{A};$ $\qquad K(4) : \overline{\Phi} = \Phi,$

*where A, B, ... are arbitrary subsets of X. Then there exists a unique topology* $\tau$ *of X, such that for every subset A,* $\overline{A}$ *is the closure of A in the topological space* $(X, \tau)$*, i.e., relative to the topology* $\tau$.

**Proof :** Let a subset $F$ of $X$ be called a closed set if $\overline{F} = F$; and let the collection of all closed sets, thus determined, be denoted by $\mathbf{C}$. Then:

(*i*) $\overline{X} \supset X$ (by $K(1)$), which implies $\overline{X} = X$ (as $X$ is the universal set). Hence $X$ is a closed set.

(*ii*) As $\overline{\Phi} = \Phi$ (by $K(4)$), it follows that $\Phi$ is a closed set.

(*iii*) Let $\{F_\alpha\}$ be an aggregate of closed sets, so that $\overline{F_\alpha} = F_\alpha$, for each $F_\alpha$; and let $D$ be their intersection. Then $D \subset$ each $F_\alpha$ implies $\overline{D} \subset \overline{F_\alpha} = F_\alpha$ (for each $F_\alpha$), from which it follows that $\overline{D} \subset D$. But $D \subset \overline{D}$ (by $K(1)$). Hence $\overline{D} = D$, so that $D$ is a closed set. [The result $P \subset Q$ implies $\overline{P} \subset \overline{Q}$ has been used above. This can be deduced from $K(2)$ as follows : $P \subset Q$ implies $Q = P \cup Q$, so that $\overline{Q} = \overline{P \cup Q} = \overline{P} \cup \overline{Q}$ (by $K(2)$); and this implies $\overline{P} \subset \overline{Q}$].

(*iv*) Let $F$ and $M$ be two closed sets, so that $\overline{F} = F$ and $\overline{M} = M$. Then $\overline{F \cup M} = \overline{F} \cup \overline{M}$ (by $K(2)$) $= F \cup M$. Hence $F \cup M$ is a closed set.

Thus the conditions $C(1)$-$C(4)$ of Theorem 19 hold for the family $\mathbf{C}$ of closed sets, as defined above. Now, defining a set $G$ to be open if

$X - G$ is closed, it follows, by Theorem 19, that the collection $\tau$ of all such open sets forms a (unique) topology on $X$, such that the closed sets of the topological space $(X, \tau)$ are the same as the members of **C**.

Next, let the closure of a subset $A$ in the topological space $(X, \tau)$, *i.e.*, determined in terms of the topology $\tau$, be denoted by $A^*$. Then

$$A^* = \text{The intersection of all closed sets in } (X, \tau), \text{ containing the set } A$$
$$= \text{Intersection of all those members of } \mathbf{C}, \text{ which contain the set } A \qquad ...(1)$$

As the property $C(3)$ of Theorem 18 holds for the members of **C** (as shown above), it follows that $A^* \in \mathbf{C}$, and therefore $\overline{A^*} = A^*$. Also $A \subset A^*$ implies $\overline{A} \subset \overline{A^*}$. Hence $\overline{A} \subset A^*$.

On the other hand, $\overline{\overline{A}} = \overline{A}$ (by $K(3)$) implies $\overline{A} \in \mathbf{C}$; and then $A \subset \overline{A}$ (by $K(1)$) shows that $\overline{A}$ is a closed set (*i.e.*, a member of **C**) containing $A$. Hence $A^* \subset \overline{A}$ (by (1)).

Now, combining the results $\overline{A} \subset A^*$ and $A^* \subset \overline{A}$, we get $\overline{A} = A^*$.//

The properties $K(1)$-$K(4)$ are known as Kuratowski's closure axioms, and the topology $\tau$ on $X$, obtained as above, is known as *Kuratowski's closure topology* on $X$.

**Example 3.** ***(i)*** $\overline{A \cap B} \subset \overline{A} \cap \overline{B}$;

***(ii)*** $\overline{A \cap B}$ ***may not be equal to*** $\overline{A} \cap \overline{B}$.

***(iii)*** $\overline{A} - \overline{B} \subset \overline{A - B}$.

**Solution :** (*i*) $A \cap B \subset A$ and $A \cap B \subset B$ imply $\overline{A \cap B} \subset \overline{A}$ and $\overline{A \cap B} \subset \overline{B}$; consequently $\overline{A \cap B} \subset \overline{A} \subset \overline{B}$.

(*ii*) In the real number space $(\mathbf{R}, \tau)$, with the usual topology, let **Q** and **T** denote the subsets formed by the rational numbers and irrational numbers respectively. Then $\overline{\mathbf{Q}} = \mathbf{R}$ and $\overline{\mathbf{T}} = \mathbf{R}$, so that $\overline{\mathbf{Q}} \cap \overline{\mathbf{T}} = \mathbf{R}$. But $\mathbf{Q} \cap \mathbf{T} = \phi$, so that $\overline{\mathbf{Q} \cap \mathbf{T}} = \phi$. Thus $\overline{\mathbf{Q} \cap \mathbf{T}} \neq \overline{\mathbf{Q}} \cap \overline{\mathbf{T}}$.

**Example 4.** ***Let $\tau$ and $\tau'$ be two topologies, defined on a set X, and let the closures of a subset A, relative to the two topologies be denoted by $Cl_\tau(A)$ and $Cl'_\tau(A)$ respectively. Show that :***

***(i)*** $\tau \subset \tau'$ ***if, and only if,*** $Cl_\tau(A) \subset Cl'_\tau(A)$, ***for every subset A of X;***

***(ii)*** $\tau = \tau'$ ***if, and only if,*** $Cl_\tau(A) = Cl'_\tau(A)$, ***for every subset A of X.***

## 3.2 Interior of a Set

The concept of the interior of a set is the dual of that of the closure. Let $A$ be a subset of a topological space $(X, \tau)$. The union of all open subsets of $X$, contained in $A$, is called the *interior* of $A$, and is denoted by Int $(A)$. Since there is at least one open subset contained in $A$, viz. the null set $\Phi$, the interior of a set $A$ always exists.

A point $p$ is called an interior point of a set $A$, if $p \in$ Int $(A)$. It then follows immediately from the definition of the interior of a set that :

**Theorem 28 :** *A point $p$ is an interior point of a set $A$ in a topological space $(X, \tau)$ iff there exists an open set $G$, such that $p \in G \subset A$.*

By dualising Theorem 23, one obtains the following :

**Theorem 29 :** *In a topological space $(X, \tau)$,*

(*i*) *$Int(A) \subset A$;* (*ii*) *$Int(A)$ is an open set;*

(*iii*) *If $G$ is an open set and $G \subset A$, then $G \subset Int(A)$, so that $Int(A)$ is the largest open set contained in $A$;*

(*iv*) *$A$ is an open set if, and only if, $Int(A) = A$;*

(*v*) *$B \supset A$ implies $Int(B) \supset Int(A)$, where $A, B, \ldots$ are arbitrary subsets of $X$.*

**Proof :** (*i*) Follows directly from the definition of Int $(A)$.

(*ii*) Since the union of any collection of open sets is an open set, it follows that Int $(A)$ is an open set.

(*iii*) Let $G$ be any open set contained in $A$. Int $(A)$ is the union of all open sets contained in $A$, and $G$ is one among these; hence $G \subset$ Int $(A)$.

(*iv*) Let Int $(A) = A$. Since Int $(A)$ is an open set (by (*ii*)), it follows that $A$ is an open set. Conversely, let $A$ be an open set. Then, taking $G = A$ in (*iii*), we get $A \subset$ Int $(A)$. But Int $(A) \subset A$ (by (*i*)). Hence Int $(A) = A$.

(*v*) Let $B \supset A$, then Int $(A) \subset A \subset B$. Since Int $(A)$ in an open set contained in $B$, it follows that Int $(A) \subset$ Int $(B)$, by (*iii*).//

**Theorem 30 :** *Let $(X, \tau)$ be a topological space, and $A$ be a subset of $X$, then $Int(A) = X - \overline{X - A}$.*

**Proof :** $X - \overline{X - A}$, being the complement of the closed set $\overline{X - A}$, is an open set. Also $X - A \subset \overline{X - A}$ implies $X - \overline{X - A} \subset A$. Thus $X - \overline{X - A}$ is an open set, contained in $A$.

Next, let $G$ be any open set, contained in $A$. Then $G \subset A$ implies $X - A \subset X - G$, so that $\overline{X - A} \subset \overline{X - G} = X - G$ (since $X - G$ is closed set). Therefore,

$G \subset X - \overline{X - A}$. Hence it follows, by Theorem 29 (*iii*), that $X - \overline{X - A} =$ Int $(A)$.//

**Exterior and Frontier (or Boundary) of a subset :** For a subset $A$ of a topological space $(X, \tau)$, we define

(*i*) the *exterior* of $A$, denoted by Ext $(A)$, by Ext $(A) =$ Int $(X - A)$;

(*ii*) the *frontier* or *border* of $A$, denoted by Fr $(A)$ or Bd $(A)$, by

$$\text{Fr}(A) = \text{Bd}(A) = \overline{A} - \text{Int}(A).$$

It then follows, by Theorems 30 and 29 (*iii*), that

1. Ext $(A)$ is the largest open set, not intersecting $A$.
2. Ext $(A) =$ Int $(X - A) = X - \overline{X - (X - A)} = X - \overline{A}$.
3. Fr $(A) = \overline{A} -$ Int $(A) = \overline{A} \cap (X -$ Int $(A)) = \overline{A} \cap \overline{X - A} =$ Fr $(X - A)$.

A point $p$ is called an *exterior point* or a *boundary point* of a subset $A$, according as $p \in$ Ext $(A)$ or $p \in$ Fr $(A)$.

**Theorem 31 :** *Let $(X, \tau)$ be a topological space, and $A$ be a subset of $X$. Then a point $p \in X$ is :*

(*i*) *an exterior point of $A$, iff there exists an open set $G$, such that $p \in G \subset X - A$;*

(*ii*) *a boundary point of $A$, iff every open set, containing $p$, intersects both the sets $A$ and $X - A$.*

**Proof :** (*i*) Since an exterior point of $A$ is an interior point of $X - A$, (*i*) follows directly from Theorem 28.

(*ii*) Let $p$ be a boundary point of $A$, and let $G$ be any open set containing $p$. As $p \in \overline{A}$, $G$ intersects $A$ (by Theorem 24). Also, as $p \in \overline{X - A}$, $G$ intersects $X - A$. Hence $G$ intersects both $A$ and $X - A$. Conversely, let every open set $G$, containing $p$, intersect both $A$ and $X - A$. Then it follows, by Theorem 24, that $p \in \overline{A}$ and $p \in \overline{X - A}$. Thus

$$p \in \overline{A} \cap \overline{X - A} = \text{Fr}(A),$$

so that $p$ is a boundary point of $A$.//

**Regular open and regular closed sets :** Let $A$ be a subset of $X$ in a topological space $(X, \tau)$, then $A$ is called.

1. a *regular open set* (or *an open domain*), if $A$ is the interior of a closed set (it is evidently an open set);
2. a *regular closed set* (or *a closed domain*), if $A$ is the closure of an open set (it is evidently a closed set).

**Example 5.** ***Give an example of (i) an open set that is not an open domain, and (ii) a closed set that is not a closed domain.***

**Example 6.** ***Show that (i) A is an open domain iff $A = Int\,(\overline{A})$, and (ii) A is a closed domain iff $A = \overline{Int\,(A)}$.***

## 3.3 Dense and Nowhere Dense Sets

Let $(X, \tau)$ be a topological space, and $A$ a subset of $X$. Then $A$ is called

(*a*) *a dense* (or, *everywhere dense*) set, if $\overline{A} = X$;

(*b*) a *nowhere dense set* (or, *a non-dense set*), if $\text{Int}\,(\overline{A}) = \Phi$.

If $A$ is a subset of $X$, and $D$ a subset of $A$, then $D$ is said to be dense *on the subset A, iff* $A \subset \overline{D}$, *i.e.*, iff $D \subset A \subset \overline{D}$.

**Theorem 32 :** *In a topological space* $(X, \tau)$,

(*i*) *any set C, containing a dense set D, is a dense set;*

(*ii*) *if A is a dense set, and B is dense on A, then B is also a dense set.*

**Proof :** (*i*) $D \subset C$ implies $\overline{D} \subset \overline{C}$. But $\overline{D} = X$; hence $X \subset \overline{C}$, so that $\overline{C} = X$. Thus $C$ is a dense set in $(X, \tau)$.

(*ii*) By hypothesis, $\overline{A} = X$ and $A \subset \overline{B}$. $A \subset \overline{B}$ implies $\overline{A} \subset \overline{\overline{B}} = \overline{B}$. Thus $X = \overline{A} \subset \overline{B}$, so that $\overline{B} = X$. Hence $B$ is a dense set in $(X, \tau)$.//

**Theorem 33 :** *In a topological space $(X, \tau)$, a subset N is nowhere dense if, and only if, any one of the following conditions holds:*

(*i*) $X - \overline{N} = X$; (*ii*) $N \subset \overline{X - \overline{N}}$;

(*iii*) *every non-empty open set U contains a non-empty open set A, disjoint with N (i.e., such that $A \cap N = \Phi$);*

(*iv*) $\overline{N}$ *is not dense on any non-empty open set.*

**Proof :** (*i*) $\text{Int}\,(\overline{N}) = \Phi$, iff $X - \overline{X - \overline{N}} = \Phi$, *i.e.*, iff $X \subset \overline{X - \overline{N}}$, *i.e.*, iff $X = \overline{X - \overline{N}}$.

(*ii*) Let $N$ be a nowhere dense set, then $N \subset X = \overline{X - \overline{N}}$. Conversely, let $N \subset \overline{X - \overline{N}}$; then $\overline{N} \subset \overline{\overline{(X - \overline{N})}} = \overline{(X - \overline{N})}$, from which it follows that

$$X = \overline{N} \cup (X - \overline{N}) \subset \overline{(X - \overline{N})} \cup (X - \overline{N}) = \overline{X - \overline{N}},$$

so that $X = \overline{X - \overline{N}}$. Hence $N$ is a nowhere dense set.

(*iii*) Let $N$ be a nowhere dense set. Then Int $(\overline{N}) = \Phi$, *i.e.*, $\overline{N}$ does not contain any non-empty open set. Hence, for any non-empty open set $U$, $U - \overline{N} \neq \Phi$. Thus $A = U - \overline{N}$ is a non-empty open set contained in $U$ and disjoint with $N$.

Conversely, let the given condition hold for the set $N$. That is, for any given non-empty set $U$, there exists a non-empty open set $A$, such that $A \subset U$ and $A \cap N \neq \Phi$. Then $N \subset X - A$, $\overline{N} \subset \overline{X - A} = X - A$, so that $U - \overline{N} \supset U - (X - A) = U \cap A = A \neq \Phi$. Thus $N$ does not contain any non-empty open set, *i.e.*, Int $(\overline{N}) = \phi$. Hence $N$ is a nowhere dense set.

(*iv*) Let $N$ be nowhere dense, and let $U$ be any non-empty open set. If $N$ were dense on $U$, then $U \subset \overline{N}$, *i.e.*, $X - \overline{N} \subset X - U$, from which it follows that $X = \overline{X - \overline{N}} \subset \overline{X - U} = X - U$, which is impossible, since $U$ is non-empty.

Conversely, let $N$ be not dense on any non-empty open set $U$. That is, $U$ is not contained in $\overline{N}$ so that $A = U - \overline{N} \neq \Phi$. Thus, for any non-empty open set $U$, there exists a non-empty open set $A$, such that $A \subset U$ and $A \cap N = \Phi$. Hence $N$ is nowhere dense, by (*iii*).//

**Theorem 34 :** *In a topological space* $(X, \tau)$,

(*i*) *a subset of a nowhere dense set is also nowhere dense;*

(*ii*) *the closure of a nowhere dense set is nowhere dense;*

(*iii*) *the union of two nowhere dense sets is nowhere dense.*

**Proof :** Left as an exercise.//

**Sets of the first and second categories.** A subset $A$ of a topological space is called:

(*a*) a *set of the first category,* if $A$ is expressible as the union of a countable aggregate of nowhere dense sets;

(*b*) a *set of the second category,* if it is not a set of the first category.

**Theorem 35 :** *In a topological space* $(X, \tau)$,

(*i*) *any subset of a set of first category is also a set of the first category;*

(*ii*) *the union of any countable collection of sets of the first category is also a set of the first category.*

**Proof :** (*i*) Let $A$ be a set of the first category in $(X, \tau)$, then $A = \cup \{N_i\}$, for $i = 1, 2, \ldots$, where each $N_i$ is a nowhere dense subset of $X$. Let $B$ be a subset of $A$. Then

$B = B \cap A = B \cap [\cup \{N_i\}] = \cup \{B \cap N_i\}$ is a set of the first category, since each $B \cap N_i$, being a subset of a nowhere dense set $N_i$, is a nowhere dense set, for $i = 1, 2, \ldots$

(*ii*) Let $\{A_i\}$, for $i = 1, 2, \ldots$, be a countable collection of sets of the first category. Then $A_i = \cup \{N_{ij} : j = 1, 2, \ldots\}$, for each $i$, where $N_{ij}$ are nowhere dense sets. Hence

$$\cup \{A_i : i = 1, 2, \ldots\} = \cup [\cup \{N_{ij} : j = 1, 2, \ldots\} : i = 1, 2, \ldots]$$
$$= \cup \{N_{ij} : i = 1, 2, \ldots \text{ and } j = 1, 2, \ldots\},$$

being the union of a countable collection of nowhere dense sets $N_{ij}$, is a set of the first category.//

[A non-empty family **R** of subsets of a given set $X$ is said to form an *ideal*, if it is *hereditary* and *additive*, that is

$A \in \mathbf{R}$ and $B \subset A$ imply $B \in \mathbf{R}$, (*Hereditary property*),

and $A \in \mathbf{R}$ and $B \in \mathbf{R}$ imply $A \cup B \in \mathbf{R}$, (*Additive property*).

If, in particular, the ideal **R** is *countably additive*, that is,

$A_i \in \mathbf{R}$, for $i = 1, 2, \ldots$, imply $\cup \{A_i : i = 1, 2, \ldots\} \in \mathbf{R}$ then **R** is called a $\sigma$-*ideal*.

Now, by virtue of the properties (*i*) and (*iii*) of Theorem 34, and the Theorem 35, we obtain the following:

**Theorem 36 :** *In a topological space*

(*i*) *the family of all nowhere dense subsets forms an ideal;*

(*ii*) *the family of all sets of the first category forms a* $\sigma$-*ideal.*//].

**Example 7.** ***Let $(R, \tau)$ denote the real number space. Since every one-pointic set is, obviously, nowhere dense, and the set $Q$ of the rational numbers is countable, it follows that the rational numbers form a set of the first category in $(R, \tau)$.***

**Solution :** The set **R** itself is not of the first category, by Baire's theorem (which will be proved later). Hence, it follows, by Theorem 35 (*ii*), that the subset formed by the irrational numbers cannot be of the first category, so that it is a set of the second category.

In most of the topological spaces, a set of the first category is a set of measure 0, *i.e.*, it is 'negligibly small' in the sense of measure. Accordingly, a set of the first category is also called a *meagre set*.

The complement of a set of the first category is said to be a *co-meagre set* or a *residual set*.

An important property of the first category (due to Banach) is the following (whose proof is left out):

**Theorem 37 :** *If $P \cap G_\alpha$ is a set of the first category, for every open set $G_\alpha$, belonging to a family of pairwise disjoint open sets $\{G_\alpha\}$, then $P \cap [\cup \{G_\alpha\}]$ is also a set of the first category.//*

## 3.4 $G_\delta$- and $F_\sigma$-Sets

We have seen that the intersection of infinitely many open sets may not be an open set, and (dually) the union of infinitely many closed sets need not be a closed set (vide Example 2, pp. 60). We now introduce the following two concepts :

In a topological space :

(*i*) the intersection of a countable collection of open sets is called a $G_\delta$-set; and

(*ii*) the union of a countable collection of closed sets is called an $F_\sigma$-set.

**Theorem 38 :** *In a topological space $(X, \tau)$,*

(*i*) *the complement of an $F_\sigma$-set is a $G_\delta$-set, and conversely;*

(*ii*) *the union of a countable collection of $F_\sigma$-sets is an $F_\sigma$-set; and, dually, the intersection of a countable collection of $G_\delta$-sets is a $G_\delta$-set;*

(*iii*) *the intersection of two $F_\sigma$-sets is an $F_\sigma$-set; and, dually, the union of two $G_\delta$-sets is a $G_\delta$-set.*

**Proof :** (*i*) Let $A$ be an $F_\sigma$-set, then $A = \cup \{K_i : i = 1, 2, \ldots\}$, each $K_i$ being a closed set. Then $X - A = X - \cup \{K_i\} = \cap \{X - K_i\}$, being the intersection of a countable collection of open sets $X - K_i$ (for $i = 1, 2, \ldots$), is a $G_\delta$-set. The converse follows by dualisation of the first part.

(*ii*) Let $\{A_i : i = 1, 2, \ldots\}$, be a countable collection of $F_\sigma$-sets, and let $B$ be the union of this collection. Now,

$$A_i = \cup \{K_{ij} : j = 1, 2, \ldots\}, \text{ for } i = 1, 2, \ldots$$

where each of the sets $K_{ij}$ is a closed set. Hence

$B = \cup \{A_i : i = 1, 2, \ldots\} = \cup \{K_{ij} : i = 1, 2, \ldots, j = 1, 2, \ldots\}$, being the union of a countable collection of closed sets $K_{ij}$ is an $F_\sigma$-set. The second part follows by dualisation.

(*iii*) Let $A$ and $B$ be two $F_\sigma$-sets. Then

$$A = \cup\{K_i : i = 1, 2, \ldots\}, \quad B = \cup\{M_j : j = 1, 2, \ldots\},$$

where $K_i$ and $M_j$ are closed sets. Then

$$A \cap B = [\cup\{K_i\}] \cap [\cup\{M_j\}] = [\cup\{H_{ij} : i = 1, 2, \ldots \text{ and } j = 1, 2, \ldots\}$$

is the union of a countable collection of sets $H_{ij}$, where each $H_{ij} = K_i \cap M_j$ is a closed set. Hence $A \cap B$ is an $F_\sigma$-set. The second part follows by dualisation of the first part.//

**Theorem 39 :** *In the real number space* ($\boldsymbol{R}$, $\tau$),

(*i*) *every closed set is a* $G_\delta$*-set; and dually, every open set is an* $F_\sigma$*-set;*

(*ii*) *the subset* $\boldsymbol{Q}$*, formed by the rational numbers, is an* $F_\sigma$*-set; and dually, the subset* $\boldsymbol{T}$*, formed by the irrational numbers, is a* $G_\delta$*-set;*

(*iii*) *the subset* $\boldsymbol{Q}$ *is not a* $G_\delta$*-set; and dually, the subset* $\boldsymbol{T}$ *is not an* $F_\sigma$*-set.*

**Proof :** (*i*) Let $K$ be a closed subset of **R**. Let $S_n$ denote the union of all open intervals $(p - 1/n, \ p + 1/n)$, as $p$ runs over $K$, $n$ being any positive integer. Then $S_n$ is an open set, containing $K$, for $n = 1, 2, \ldots$ Let $D = \cap\{S_i : i = 1, 2, \ldots\}$, then $D$ is a $G_\delta$-set. It will be shown that $K = D$.

Since $K \subset S_i$, for $i = 1, 2, \ldots$, it follows that $K \subset D$. Next, let $q \in D$, then $q \in S_i$, for $i = 1, 2, \ldots$ Consequently, there exists an element $p_n \in K$, such that $q \in (p_n - 1/n, \ p_n + 1/n)$, *i.e.*, $p_n \in (q - 1/n, \ q + 1/n)$, for any positive integer $n$. Hence $q \in \overline{K} = K$ (since $K$ is closed). Consequently $D \subset K$. Now, from $K \subset D$ and $D \subset K$, we get $K = D$. The closed set $K$ is thus a $G_\delta$-set.

(*ii*) Every one-pointic subset $\{p\}$ is closed in (**R**, τ), since if $q \neq p$, then $q$ is not a limiting point of $\{p\}$, as the open set $(q - t, \ q + t)$, containing $q$, does not intersect the set $\{p\}$, if we take $t = |p - q|/2$. Hence the countable subset **Q** is $F_\sigma$-set in (**R**, τ). Dually, **T** is a $G_\delta$-set in (**R**, τ).

(*iii*) If possible, let **T** is also an $F_\sigma$-set in (**R**, τ). Then $\mathbf{T} = \cup\{K_i\}$, where $K_i$ is a closed set, for $i = 1, 2, \ldots$ Let $(a, b)$ be any open interval, and let $r$ be any rational number belonging to $(a, b)$. Then $r$ does not belong to any $K_i$, since $r \notin \mathbf{T}$. Hence $r$ cannot be a limiting point of any $K_i$, as each $K_i$ is a closed set. Consequently, there exists an open interval $(c, d) \subset (a, b)$, such that $r \in (c, d)$ and $K_i \cap (c, d) = \Phi$, for $i = 1, 2, \ldots$ Hence each of the sets $K_i$, for $i = 1, 2, \ldots$, is a nowhere dense subset of $\mathbf{R}$. So, the set $\mathbf{T} = \cup \{K_i : i = 1, 2, \ldots\}$ must be a set of the first category, which is contrary to the property that $T$ is a set of the second category (vide Example 7, Page 69). Hence $\mathbf{T}$ cannot be an $F_\sigma$-set. Dually, $\mathbf{Q}$ is not a $G_\delta$-set.//

**Borel sets :** The smallest family of subsets of a topological space $(X, \tau)$ satisfying the following conditions:

(1) every closed set belongs to the family;

(2) if $A$ belongs to the family, then its complement $X - A$ also belongs to the family; and

(3) if $A_n$, for $n = 1, 2, \ldots$, belong to the family, then their union

$$U \{A_n : n = 1, 2, \ldots\}$$

also belongs to the family, is called *the family* $\mathbf{B}$ of Borel sets in $(X, \tau)$.

It follows by the properties (1) and (2) that every open set also belongs to $\mathbf{B}$. And, the family $\mathbf{B}$ may as well be defined to be the smallest family, satisfying the conditions:

(1-a) : every open set belongs to $\mathbf{B}$;

(2-a) : if $A$ belongs to $\mathbf{B}$, then $X - A$ also belongs to $\mathbf{B}$;

(3-a) : if $A_n$, for $n = 1, 2, \ldots$, belong to $\mathbf{B}$, then their intersection $\cap \{A_n \ n = 1, 2, \ldots\}$ also belongs to $\mathbf{B}$.

It follows from the properties (1) and (3) and (1-a) and (3-a) that:

(*i*) every $F_\sigma$-set is a Borel set, and

(*ii*) every $G_\delta$-set is a Borel set.

**Baire property :** A subset $A$ of a topological space, is said to possess the Baire property, if there exists an open set $G$, such that $A - G$ and $G - A$ are both sets of the first category.

In particular, every open set and every closed set have the Baire property. (For a closed set $F$, $F = \text{Int}(F) \cup \text{Fr}(F)$ and $\text{Fr}(F)$ is a nowhere dense set).

One can readily deduce the following:

1. If $\mathbf{B}_1$ denotes the family of subsets of a topological space $(X, \tau)$ possessing the Baire property, then

(i) countable intersection of members of $\mathbf{B}_1$ is a member of $\mathbf{B}_1$;

(ii) countable union of members of $\mathbf{B}_1$ is a member of $\mathbf{B}_1$;

(iii) the complement of a member of $\mathbf{B}_1$ belongs to $\mathbf{B}_1$;

(iv) every Borel set has the Baire property.

2. Each of the following conditions is necessary and sufficient in order that a subset $A$ of a topological space $(X, \tau)$ has the Baire property:

(i) $A$ is expressible as $A = (G - P) \cup R$, where $G$ is open, and $P$ and $R$ are sets of the first category.

(ii) There exists a set $P$ of the first category, such that $A - P$ is both closed and open relative to $X - P$.

(iii) $A$ is the union of a $G_\delta$-set and a set of the first category.

(iv) $A$ is the difference of an $F_\sigma$-set and a set of the first category.

## Art. 4 CONVERGENCE AND LIMIT

**Convergent sequences :** Let $(X, \tau)$ be a topological space. An infinite sequence $\{x_i : i = 1, 2, \ldots\}$, over $X$ (*i.e.*, with elements $x_1, x_2, \ldots$ from $X$), is said to be *convergent*, and a point $p \in X$ is said to be a *limit* of the sequence $\{x_i : i = 1, 2, \ldots\}$, expressed symbolically by $\text{Lt}\ x_n = p$ or $\lim x_n = p$, if corresponding to every open set $U$ containing $p$ (or, equivalently, corresponding to every *nbd* $U$ of $p$), there exists a positive integer $m$, such that $x_n \in U$, for all $n > m$.

Every convergent sequence has at least one limit, and it may also have more than one limit. e.g.

**Example 1.** ***Let X be an infinite set, and let σ be the indiscrete topology on X. Then ϕ and X are the only open sets in this space. Let $\{x_i\}$ be any infinite sequence in X, and let q be any point in X. Then the only open set containing p is X, which contains the entire sequence $\{x_i\}$. Thus***

***(i) every infinite sequence in X is convergent; and***

***(ii) all the points in X are limits of every infinite sequence in X.***

One can however ensure unique limit for every convergent sequence by imposing certain conditions on the topology. In fact, the limit of every convergent sequence is unique in a Hausdroff space (which will be discussed later). In particular, the real number space is a Hausdroff space; hence the limit of a convergent sequence is found to be unique in the real number space.

**Theorem 40 :** *In every topological space $(X, \tau)$, the following properties hold for the convergent sequences:*

$L(1)$ : *For any point $p \in X$, the infinite sequence $\{p, p, \dots\}$ is convergent, having $p$ as a limit.*

$L(2)$ : *Addition of a finite number of terms to a convergent sequence affects neither its convergence, nor the limit or limits to which it converges.*

$L(3)$ : *If the sequence $\{x_1, x_2, \dots\}$ is convergent and converges to a limit $p$, then any infinite sub-sequence $\{x_{n_1}, x_{n_2}\}$, with $n_1 < n_2 < \dots$ is also convergent, and converges to the same limit $p$.*

**Proof :** 1. Any open set $U$, containing $p$, contains all terms of the sequence $\{p, p, \dots\}$. Hence the property $L(1)$ holds.

2. Let $p$ be a limit of the sequence $(x) = \{x_1, x_2, \dots\}$. Then for any open set $U$, containing $p$, there exists a positive integer $m$, such that $x_n \in U$, for all $n > m$. Let now $r$ terms $y_1, y_2, \dots, y_r$ be added to the sequence $(x)$, and let these terms be inserted in between the terms of the sequence $x_1, x_2, \dots$. Let the new sequence thus formed be $\{z_1, z_2, \dots\}$. If the terms $y_1, y_2, \dots, y_r$ occur among the terms $z_1, z_2, \dots, z_s$, then $z_{s+i} = x_{s-r+i}$, for $i = 1, 2, \dots$ Hence $z_n \in U$, for all $n > s + m$. Consequently, the sequence $\{z_1, z_2, \dots\}$ is convergent, and $p$ is also a limit of this sequence. Thus $L(2)$ holds.

3. Let $U$ be any open set containing $p$. Then there exists a positive integer $m$, such that $x_n \in U$, for all $n > m$. As $n_1 < n_2 < \dots$ is a strictly ascending chain of positive integers, there exists an $n_k > m$. Then $x_{n_i} \in U$, for all $n_i > n_k$. Hence the subsequence $\{x_{n_1}, x_{n_2}, \dots\}$ is a convergent sequence, and $p$ is also a limit of this sequence. Thus $L(3)$ also holds.//

It also follows directly from the definition of convergence of a sequence and its limits that:

**Theorem 41 :** *If $G$ and $F$ be an open and a closed set respectively in a topological space $(X, \tau)$, then:*

(*i*) *no infinite sequence lying in $X - G$ has any limit in $G$;*

(*ii*) *every convergent sequence, lying in $F$, has all its limits within $F$.*

**Convergence topology :** It will now be shown that a topology can be defined in terms of convergence and limits, satisfying the conditions $L(1)$-$L(3)$, open sets being determined by the property in Theorem 41 (*i*).

**Theorem 42 :** *Let X be a given set, and let a class of infinite sequences $\Omega$, over X, be distinguished, whose members are called convergent sequences; and let each convergent sequence be associated with elements of X, which are called the limits of the sequence, subject to the conditions L (1)-L (3) of Theorem 40.*

*Let, now, a subset G of X be called an open set, iff no infinite sequence lying in $X - G$ has any limit in G. Then the collection of all open sets, thus obtained, forms a topology $\tau$ on X.*

**Proof :** 1. As $\Phi$ does not contain any point, no convergent sequence can have a point in $\Phi$ as its limit. Hence $\Phi$ is an open set.

2. As there can be no sequence, lying outside $X$, *i.e.*, lying in $X - X = \Phi$, no point in $X$ can ever be a limit of any such sequence. Hence $X$ is also an open set.

3. Let $\{G_v\}$ be a collection of open sets, and let $G$ be their union. If possible, let $G$ be not an open set. Then there exists a convergent sequence $(x) = (x_1, x_2, ...)$, none of whose elements belongs to $G$, while the sequence $(x)$ has a limit $p \in G$. As $G$ is the union of the sets $G_v$, the point $p$ must belong to at least one of these sets, say to $G_k$. Then the sequence $(x)$, lying wholly outside $G_k$, converges to a point $p \in G_k$; this is impossible, since $G_k$ is an open set. Hence $G$ must be an open set.

4. Let $G$ and $H$ be two open sets. If possible, let $G \cap H$ be not an open set. Then there exists an infinite sequence $(x) = \{x_1, x_2, \dots\}$, lying wholly outside $G \cap H$, and converging to a point $p \in G \cap H$. The union $G \cup H$ is an open set (as shown above in (3)). And since $p \in G \cup H$, and $p$ is a limit of the sequence $(x)$, only a finite number of terms of the sequence $(x)$ can lie outside $G \cup H$. Otherwise we obtain an infinite subsequence of $(x)$, lying outside $G \cup H$ and having $p$ as a limit (by $L$ (3)), which is impossible, since $G \cup H$ is an open set. Now, omitting the finite number of terms of $(x)$, which lie outside $G \cup H$, we obtain an infinite sub-sequence, say $(y) = (y_1, y_2, ...)$, lying wholly inside $G \cup H$, and having $p$ as a limit, by $L$ (3). $G \cup H$ is the union of three pair-wise disjoint subsets, $G \cup H = [G \cap (X - H]$ $\cup (G \cap H) \cup [(X - G) \cap H]$. All the elements of the sequence $(y)$ belong to $G \cup H$, but none to $G \cap H$. Hence one at least of the sets $G \cap (X - H)$ and $(X - G) \cup H$ must contain an infinite subsequence $(z)$ of $(y)$; let $(z) \in G \cap (X - H)$. By $L$ (3), the subsequence $(z)$ also converges to the point $p$. Thus $(z)$ is an infinite sequence, lying wholly

outside $H$, but converging to a point $p \in H$. This is impossible, since $H$ is an open set. Hence $G \cap H$ must be an open set.

Thus 0 (1)-0 (4) hold for $\tau$. Hence $\tau$ forms a topology on $X$.//

The topology $\tau$, defined as above (in Theorem 42) is called a *convergence topology* or a *sequential topology* of $X$. And $(X, \tau)$ is often called a *limit space*.

It should however be noted that the family of convergent sequences, determined by the convergence topology $\tau$ for $X$, as above in Theorem 42, need *not* be identical with the given family of convergent sequences, in terms of which the topology $\tau$ was obtained. In fact :

**Theorem 43 :** *If X is a limit space with $\Omega$ as the given collection of convergent sequences, and $\tau$ the resulting convergence topology on X, and if $\Sigma$ is the collection of convergent sequences determined by the topology $\tau$ on X, then $\Omega \subset \Sigma$.*

**Proof :** Let $(x) = \{x_1, x_2, \ldots\} \in \Omega$. Then for any open set $G \in \tau$, containing $p$, where $p$ is a limit of the sequence $(x)$, there exists a positive integer $m$, such that $x_i \in G$ for all $i > m$, since only a finite number of elements of the sequence $(x)$ can lie outside $G$ (as otherwise an infinite subsequence of $(x)$ lying outside $G$ would have the point $p \in G$ as its limit). Hence, it follows that $(x)$ is also a convergent sequence, having $p$ as a limit, in the topological space $(X, \tau)$, so that $(x) \in \Sigma$. Thus one obtains $\Omega \subset \Sigma$.//

On the other hand, one obtains :

**Theorem 44 :** *Let $\Gamma$ be the family of convergent sequences in a topological space $(X, \tau)$, and let $\tau'$ be the convergence topology on X determined by the family $\Gamma$ (which satisfies L (1)-L (3), of Theorem 40), then $\tau \subset \tau'$ (i.e., $\tau'$ is stronger than $\tau$).*

**Proof :** Let $G \in \tau$. Then for any convergent sequence $(x) \in \Gamma$, converging to a point $p \in G$, there exists a positive integer $m$, such that $x_i \in G$ for all $i > m$. That is, at most a finite number of terms of $(x)$ can lie outside $G$. Thus no infinite sequence, lying outside $G$, can converge to the point $p \in G$. Hence $G$ is also an open set in the convergence topology $\tau'$; *i.e.*, $G \in \tau'$. Thus $\tau \subset \tau'$; in other words, $\tau'$ is stronger than $\tau$.//

Theorems 43 and 44 show that if $X$ is a given set, then :

(*i*) for a given topology $\tau$ on $X$, the family of convergent sequences, determined by the topology $\tau$, does not characterise the topology $\tau$; and

(*ii*) for a given family $\Gamma$ of convergent sequences, satisfying $L$ (1)-$L$ (3), the corresponding convergence topology $\tau'$ does not characterise the given family of convergent sequences $\Gamma$.

Thus the axioms $L$ (1)-$L$ (3) for a limit space are not sufficient to characterise a given topology on a set $X$.

For the above reasons, and also for other reasons (which will be pointed out in due course), it was realised that, in topological spaces, the sequences cannot play the same fundamental role as in the Euclidian spaces, and that a satisfactory theory of convergence and limits can be built up by replacing point sequences and set sequences by more general notions of nets and filters respectively. And the classical theory of sequential convergence forms a special case and illustrates the general theory. Theory of nets and filters will be developed in a later chapter.

## Art. 5 FUNCTIONS

[$(X, \tau)$, $(Y, \tau')$, $(Z, \tau'')$ shall denote arbitrary topological spaces throughout this article].

### 5.1 Open and Closed Mappings

Let $f$ be a single-valued mapping of $(X, \tau)$ into $(Y, \tau')$. Then $f$ is said to be

(*i*) an *open mapping,* if $f(G) \in \tau'$, for every $G \in \tau$;

(*ii*) a *closed mapping,* if $f(K)$ is a closed set in $(Y, \tau')$, for every closed set $K$ in $(X, \tau)$.

**Theorem 45 :** *Let $f$ be a single-valued mapping of $(X, \tau)$ into $(Y, \tau')$. Then*

(*i*) *$f$ is an open mapping if, and only if, $f(Int\,(A)) \subset Int\,(f(A))$, for every $A \subset X$;*

(*ii*) *$f$ is a closed mapping if, and only if, $\overline{f(A)} \subset f(\overline{A})$, for every $A \subset X$.*

**Proof :** (*i*) Let $f$ be an open mapping, and let $A \subset X$. Then $f$ (Int $(A)$) is an open set in $(Y, \tau')$, since Int $(A)$ is an open set in $(X, \tau)$. Hence $f(\text{Int}\,(A)) = \text{Int}\,[f(\text{Int}\,(A))] \subset \text{Int}\,(f(A))$, since $f\,(\text{Int}\,(A)) \subset f(A)$ (as Int $(A) \subset A$).

Conversely, let the given condition hold for $f$. Let $G$ be an open set in $(X, \tau)$. Then $f(G) = f(\text{Int}\,(G)) \subset \text{Int}\,(f(G))$ (by hypothesis), which implies that $f(G) = \text{Int}\,(f(G))$, so that $f(G)$ is an open set in $(Y, \tau')$. Hence, $f$ is an open mapping.

(*ii*) Let $f$ be a closed mapping, and let $A \subset X$. Then $f(\overline{A})$ is a closed set in $(Y, \tau')$, since $\overline{A}$ is a closed set in $(X, \tau)$. Hence

$$f(\overline{A}) = \overline{f(\overline{A})} \supset \overline{f(A)}, \text{ since } f(\overline{A}) \supset f(A) \text{ (as } \overline{A} \supset A).$$

Conversely, let the given condition hold for $f$, and let $F$ be a closed set in $(X, \tau)$. Then $f(F) = f(\overline{F}) \supset \overline{f(F)}$. Hence $f(F) = \overline{f(F)}$, so that $f(F)$ is a closed set in $(Y, \tau')$. Consequently, $f$ is a closed mapping.//

Next we shall define the most important type of mappings, viz. the continuous mappings on a topological space.

## 4.2 Continuous Functions

Let $f$ be a single valued function on $(X, \tau)$ into $(Y, \tau')$. Then $f$ is said to be a continuous function, if $f^{-1}(V) \in \tau$, for every $V \in \tau'$.

**Theorem 46 :** *A single-valued function f on $(X, \tau)$ into $(Y, \tau')$ is continuous if, and only if, any one of the following conditions holds :*

(i) *$f^{-1}(K)$ is a closed subset in $(X, \tau)$, for every closed set K in $(Y, \tau')$;*

(ii) *if* **S** *be a sub-base of $(Y, \tau')$, then $f^{-1}(A) \in \tau$, for every $A \in$* **S**;

(iii) *for each $x \in X$ and each neighbourhood W of $f(x)$ in $(Y, \tau')$, there exists a neighbourhood V of x in $(X, \tau)$, such that $f(V) \subset W$;*

(iv) *$f(\overline{A}) \subset \overline{f(A)}$, for every subset A of X;*

(v) *$\overline{f^{-1}(P)} \subset f^{-1}(\overline{P})$, for every subset P of Y;*

(vi) *$f^{-1}(Int(P)) \subset Int(f^{-1}(P))$, for every $P \subset Y$.*

**Proof :** (i) Let $K \subset Y$. Then $f^{-1}(K) = X - f^{-1}(Y - K)$. Let $K$ be a closed subset in $(Y, \tau')$, then $Y - K \in \tau'$. Hence, $f^{-1}(K)$ is a closed set in $(X, \tau)$ if, and only if, $f^{-1}(Y - K)$ is an open set in $(X, \tau)$; that is, if and only if $f$ is continuous.

(ii) Let $f$ be continuous, and let $A \in$ **S**. Then $A \in \tau'$.

Hence $f^{-1}(A) \in \tau$.

Conversely, let the given condition hold for $f$, and let $U$ be an open set in $(Y, \tau')$. Then $U$ is a union of a collection of finite intersections of sets $A$ belonging to **S**. Hence, $f^{-1}(U)$ is a union of a collection of finite intersections of sets $f^{-1}(A)$, where $A \in$ **S**. Since each $f^{-1}(A)$ is open (by hypothesis), and finite intersections and arbitrary unions of open sets are open sets, it follows that $f^{-1}(U)$ is an open set in $(X, \tau)$. Hence, $f$ is continuous.

(iii) Let $f$ be continuous. Let $x \in X$, and $W$ be a *nbd* of $f(x)$ in $(Y, \tau')$. There exists an open set $H \in \tau'$, such that $f(x) \in H \subset W$, and therefore $x \in f^{-1}(H) \subset f^{-1}(W)$. As $f$ is continuous and $H \in \tau'$, it follows that

$f^{-1}(H) \in \tau$, and therefore $V = f^{-1}(W)$ is a *nbd* of $x$ in $(X, \tau)$, such that $f(V) = W \subset W$.

Conversely, let the given condition hold for $f$. Let $U$ be any open set in $(Y, \tau')$, and let $x$ be any point in $f^{-1}(U)$. Then $f(x) \in U$, and $U$ is an open *nbd* of $f(x)$. Hence, by hypothesis, there exists a *nbd* $V$ of $x$, such that $f(V) \subset U$, that is, $V \subset f^{-1}(U)$. Thus $x \in V \subset f^{-1}(U)$. As $V$ is a *nbd* of $x$, there exists an open set $G$ in $(X, \tau)$, such that $x \in G \subset V$, and therefore $x \in G \subset f^{-1}(U)$. Taking union of all such relations, as $x$ runs over $f^{-1}(U)$, one obtains $f^{-1}(U) \subset \cup\{G\} \subset f^{-1}(U)$. Consequently, $f^{-1}(U) = \cup\{G\}$ is an open set in $(X, \tau)$. Hence $f$ is continuous.

(*iv*) Let $f$ be continuous, and let $A$ be any subset of $X$. Let $p \in f(\overline{A})$, then either (1) $p \in f(A)$, or (2) $p \in f(A')$, since $f(\overline{A}) = f(A \cup A') = f(A) \cup f(A')$, where $A'$ is the derived set of $A$. In case (1), $p \in f(A) \subset \overline{f(A)}$. In case (2), $p \in f(A')$. Also, let $p \notin f(A)$; then there exists a point $q \in A' - A$, such that $f(q) = p$. If $U$ is any open set in $(Y, \tau')$ containing $p$, then $f^{-1}(U)$ is an open set containing $q$ in $(X, \tau)$. But as $q$ is an accumulation point of $A$, $f^{-1}(U)$ intersects $A$ in some point $z$, other than $q$. Correspondingly, $f(z) \in U \cap f(A)$, and $f(z) \neq p$, since $p \notin f(A)$. Hence $p$ is an accumulation point of $f(A)$. Thus, in both the cases (1) and (2), $p \in \overline{f(A)}$; consequently,

$$f(\overline{A}) \subset \overline{f(A)}.$$

Conversely, let the given condition hold for $f$, and let $H$ be any closed subset in $(Y, \tau')$. Let $K = H \cap f(X)$, then $f^{-1}(H) = f^{-1}(K)$. Let $p$ be an accumulation point of $f^{-1}(H)$, so that $p \in \overline{f^{-1}(H)} = \overline{f^{-1}(K)}$, from which it follows, by hypothesis, that

$$f(p) \in f(\overline{f^{-1}(K)}) \subset \overline{f(f^{-1}(K))} = \overline{K} \subset \overline{H} = H.$$

Hence, $f(p) \in H \cap f(X) = K$ (since $f(p) \in f(X)$, as $p \in X$). Thus $p \in f^{-1}(K) = f^{-1} H)$. Consequently, $f^{-1}(H)$ is a closed subset in $(X, \tau)$. Hence, $f$ is continuous, by (*i*).

(*v*) Let $f$ be continuous, and let $P$ be any subset of $Y$. Let $f^{-1}(P) = A$. Then, by (*iv*),

$$f(\overline{A}) \subset \overline{f(A)} = \overline{f(f^{-1}(P))} = \overline{P \cap f(X)} \subset \overline{P},$$

so that $\overline{A} \subset f^{-1}(\overline{P})$; thus $\overline{f^{-1}(P)} \subset f^{-1}(\overline{P})$.

Conversely, let the given condition hold for $f$, and let $B$ be a closed subset in $(Y, \tau')$. Then, by hypothesis, $\overline{f^{-1}(B)} \subset f^{-1}(\overline{B}) = f^{-1}(B)$, from which it follows that $\overline{f^{-1}(B)} = f^{-1}(B)$ is a closed subset in $(X, \tau)$. Hence, $f$ is continuous, by $(i)$.

$(vi)$ Let $f$ be continuous, and $P \subset Y$. Then, using $(v)$, we get

$$f^{-1}(\text{Int}(P)) = f^{-1}(Y - \overline{Y - P}) = X - f^{-1}(\overline{Y - P}) \subset X - \overline{f^{-1}(Y - P)}$$
$$= X - \overline{X - f^{-1}(P)} = \text{Int}(f^{-1}(P)).$$

Conversely, let the condition $(vi)$ hold for $f$. Let $P$ be an open set in $(Y, \tau')$. Then $P = \text{Int}(P)$, and it follows by $(vi)$ that

$$f^{-1}(P) = f^{-1}(\text{Int}(P)) \subset \text{Int}(f^{-1}(P)),$$

and so $f^{-1}(P) = \text{Int}(f^{-1}(P))$. Hence $f^{-1}(P)$ is an open set in $(X, \tau)$; and $f$ is therefore continuous.//

**Theorem 47 :** *If $f$ is a continuous function on $(X, \tau)$ into $(Y, \tau')$ and $g$ is a continuous function on $(Y, \tau')$ into $(Z, \tau'')$, then $g.f.$ is a continuous function on $(X, \tau)$ into $(Z, \tau'')$.*

**Proof :** Let $M \in \tau''$, then $g^{-1}(M) = U \in \tau'$. Also $f^{-1}(U) = G \in \tau$. Thus $(g.f)^{-1}(M) = f^{-1}(g^{-1}(M)) = f^{-1}(U) = G \in \tau$. Hence, $g.f$ is a continuous function on $(X, \tau)$ into $(Z, \tau'')$.//

**Continuity as a local property :** A single-valued function $f$ on a topological space $(X, \tau)$, with values in a topological space $(Y, \tau')$, is said to be continuous at a point $x$ in $X$, if corresponding to any *nbd* $W$ of $f(x)$ in $(Y, \tau')$, there exists a *nbd* $V$ of $x$ in $(X, \tau)$, such that $f(V) \subset W$.

It now follows directly from Theorem 46 $(iii)$ that :

**Theorem 48 :** *A single-valued function $f$ on $(X, \tau)$ into $(Y, \tau')$ is continuous (on $X$) if, and only if, it is continuous at every point $x$ in $X$.*

A necessary condition for the continuity of a function in terms of convergence of sequences is given by :

**Theorem 49 :** *If $f$ is a continuous function on $(X, \tau)$ into $(Y, \tau')$, and $\{x_n\}$ is an infinite sequence of points converging to a point $p$ in $(X, \tau)$, then the sequence $\{f(x_n)\}$ converges to the point $f(p)$ in $(Y, \tau')$; that is,*

$$Lt\, x_n = p \text{ in } (X, \tau) \text{ implies } Lt\, f(x_n) = f(p) \text{ in } (Y, \tau').$$

**Proof :** Let $U$ be an open set containing $f(p)$ in $(Y, \tau')$. Then $G = f^{-1}(U)$ is an open set containing $p$ in $(X, \tau)$. As the sequence $\{x_n\}$ converges to $p$, there exists a positive integer $m$, such that $x_k \in G$

for all $k > m$. Corresponding, $f(x_k) \in f(G) = U$ for all $k > m$. Hence the sequence $\{f(x_n)\}$ converges to $f(p)$ in $(Y, \tau')$.//

The converse of Theorem 49 is, however, *not true*, in general as shown in the following example :

**Example 1.** ***Example of a single-valued functions*** $f : (X, \tau) \to (Y, \tau')$, ***such that*** $Lt\, x_n = p$ ***in*** $(X, \tau)$ ***implies*** $Lt\, f(x_n) = f(p)$ ***in*** $(Y, \tau')$, ***but*** $f$ ***is not continuous on*** $X$.

**Solution :** Let $X$ and $Y$ be the closed intervals $[0, 3]$ and $[1, 2]$ respectively. Let $\tau$ consist of all those subsets of $X$ which are either empty or have countable complements in $X$. And let $\tau'$ consist of all subsets $Y \cap G$, where $G \varepsilon \tau$. Then $(X, \tau)$ and $(Y, \tau')$ are topological spaces (verification left as an exercise).

Let a function $f : (X, \tau) \to (Y, \tau')$ be defined by

$$f(x) = x \text{ when } x \varepsilon Y, \text{ and } f(x) = 1 \text{ when } x \varepsilon X - Y.$$

A sequence $\{x_n\}$ converges to a point $p$ in $X$ if and only if there exists a positive integer $m$, such that $x_i = p$ for all $i > m$. And, then the sequence $\{f(x_n)\}$ necessarily converges to $f(p)$, since $f(x_i) = 1$ for all $x_i > m$, and $f(p) = 1$. Thus $\{f(x_n)\}$ converges to $f(p)$, whenever $\{x_n\}$ converges to $p$.

But $f$ is not continuous on $X$. In fact, if $U = (1, 2)$, then

$U = Y \cap [X - (1) \cup (2)] \in \tau'$, but $f^{-1}(U) = U$ does not belong to $\tau$.

If $f$ is a single-valued mapping of set $X$ into a set $Y$, then the inverse $f^{-1}$ is not defined as a mapping of $Y$ into $X$, unless $f$ is surjective; and even then $f^{-1}$ is not single-valued unless $f$ is also injective. Also, for a subset $A$ of $X$,

1. $f(A)$ and $f(X - A)$ may not be disjoint, unless $f$ is injective, and
2. $f(A) \cup f(X - A) = Y$ may not be true, unless $f$ is surjective.

Consequently, an open mapping is, in general, different from a closed mapping, and the inverse of a continuous mapping need not be an open or a closed mapping. Open, closed and continuous mappings are, in general, distinct that can be seen from the following example.

**Example 2.** ***Let*** $X = \{a, b, c\}$, ***and*** $\tau = \{\phi, \{a\}, \{a, b\}, \{a, c\}, X\}$,

***and*** $Y = \{x, y, z\}$ ***and*** $\tau' = \{\phi, \{x\}, \{y, z\}, Y\}$.

***It can be readily verified that*** $(X, \tau)$ ***and*** $(Y, \tau')$ ***are topological spaces. Then***

**(i)** ***$f : (X, \tau) \rightarrow (Y, \tau')$, defined by $f(a) = f(b) = f(c) = y$, is continuous but neither open nor closed;***

**(ii)** ***$g : (Y, \tau') \rightarrow (X, \tau)$, defined by $g(x) = g(y) = g(z) = a$, is continuous and open, but not closed;***

**(iii)** ***$h : (Y, \tau') \rightarrow (X, \tau)$, defined by $h(x) = h(y) = h(z) = c$, is continuous and closed, but not open.***

For continuous functions which are either injective or surjective, we have the following properties :

**Theorem 50 :** *If $f : (X, \tau) \rightarrow (Y, \tau')$ is a continuous, injective function, and*

(i) *if $p$ is an accumulation point of a subset $A$ in $(X, \tau)$, then $f(p)$ is an accumulation point of $f(A)$ in $(Y, \tau')$;*

(ii) *if a subset $D$ is dense-in-itself in $(X, \tau)$, then $f(D)$ is dense-in-itself in $(Y, \tau')$.*

**Proof :** (i) Let $p \in A'$, where $A'$ is the derived set of $A$, then $p \in \overline{A - (p)}$, Hence $f(p) \in f(\overline{A - (p)}) \subset \overline{f(A - (p))}$ (since $f$ is continuous) $= \overline{f(A) - f(p)}$ (since $f$ is injective), from which it follows that $f(p)$ is an accumulation point of the set $f(A)$ in $(Y, \tau')$.

(ii) Let $D$ be dense-in-itself in $(X, \tau)$. Then every point of $D$ is an accumulation point of $D$ in $(X, \tau)$. Hence, by (i), every point of $f(D)$ is an accumulation point of $f(D)$ in $(Y, \tau')$; in other words, the set $f(D)$ is dense-in-itself in $(Y, \tau')$.//

**Theorem 51 :** *If $f : (X, \tau) \rightarrow (Y, \tau')$ is a continuous, surjective function, and if $A$ is dense in $(X, \tau)$, then $f(A)$ is dense in $(Y, \tau')$.*

**Proof :** By hypothesis, $\overline{A} = X$. Hence, $f(\overline{A}) = f(X) = Y$ (since $f$ is surjective). But $f(\overline{A}) \subset \overline{f(A)}$, since $f$ is continuous. Thus $Y \subset \overline{f(A)}$, from which it follows that $\overline{f(A)} = Y$. Hence $f(A)$ is dense in $(Y, \tau')$.//

For a bijective mapping $f$ of a set $X$ onto a set $Y$, $f^{-1}$ is also a bijective mapping of $Y$ onto $X$, and $f(X - A) = Y - f(A)$ for every subset $A$ of $X$. Hence we obtain readily the following :

**Theorem 52 :** *Let $(X, \tau)$ and $(Y, \tau')$ be two topological spaces, and $f$ be a bijective mapping of $X$ onto $Y$. Then*

(i) *$f$ is an open mapping of $(X, \tau)$ onto $(Y, \tau')$ if, and only if, $f$ is also a closed mapping;*

(ii) *$f$ is an open mapping (and so also a closed mapping) if, and only if, $f^{-1}$ is a continuous mapping of $(Y, \tau')$ onto $(X, \tau)$;*

(*iii*) *$f$ is a continuous mapping of $(X, \tau)$ onto $(Y, \tau')$ if, and only if, $f^{-1}$ is an open mapping (and also a closed mapping) of $(Y, \tau')$ onto $(X, \tau)$.*

But a bijective, continuous mapping need not be an open (or a closed) mapping, and conversely. e.g.

**Example 3.** ***Let $\tau$ and $\tau'$ be two topologies defined on the same set X. The identity mapping, $i(x) = x$, for every $x \in X$, is a bijective mapping of X onto itself. The mapping i of $(X, \tau)$ onto $(X, \tau')$ is then***

***(i) an open mapping (and also a closed mapping) if and only if $\tau \subset \tau'$; and this is not a continuous mapping, unless $\tau = \tau'$;***

***(ii) a continuous mapping if and only if $\tau' \subset \tau$; and this is not an open mapping (and also not a closed mapping), unless $\tau = \tau'$.***

## 4.3 Homeomorphisms

A bijective mapping of $(X, \tau)$ onto $(Y, \tau')$ which is both continuous and open is called a *homeomorphism* (or a *topological mapping*) of $(X, \tau)$ onto $(Y, \tau')$.

It follows immediately from the definition of a homeomorphism and Theorems 45, 46, 50 and 52 that :

**Theorem 53 :** *A bijective mapping $f : X \to Y$ is a homeomorphism of $(X, \tau)$ onto $(Y, \tau')$ if, and only if, any one of the following (equivalent) sets of conditions holds :*

(*i*) *$f$ and $f^{-1}$ are both continuous;*

(*ii*) *$f$ and $f^{-1}$ are both open or both closed;*

(*iii*) *$f$ is both continuous and closed;*

(*iv*) *$f(\overline{A}) = \overline{f(A)}$, for every subset A of X;*

(*v*) *$f^{-1}(\overline{C}) = \overline{f^{-1}(C)}$, for every subset C of Y;*

(*vi*) *$f(Int(A)) = Int(f(A))$, for every subset A of X;*

(*vii*) *$p$ is an accumulation point of a subset A in $(X, \tau)$ if, and only if, $f(p)$ is an accumulation point of $f(A)$ in $(X, \tau)$.*

It now follows from Theorem 53 (*i*) (or (*ii*)) that :

**Theorem 54 :** *A mapping f of $(X, \tau)$ onto $(Y, \tau')$ is a homeomorphism if, and only if, $f^{-1}$ is a homeomorphism of $(Y, \tau')$ onto $(X, \tau)$.*

Further, we have :

**Theorem 55 :** *If f is a homeomorphism of $(X, \tau)$ onto $(Y, \tau')$ and g is a homeomorphism of $(Y, \tau')$ onto $(Z, \tau'')$, then g.f is a homeomorphism of $(X, \tau)$ onto $(Z, \tau'')$.*

**Proof :** Since $f$ and $g$ are bijective mappings, their product $g.f$ is also a bijective mapping of $X$ onto $Z$. Again, since $f$ and $g$ are continuous mappings, their product $g.f$ is a continuous mapping of $(X, \tau)$ onto $(Z, \tau'')$, by Theorem 47. Similarly, since $g^{-1}$ and $f^{-1}$ are continuous mappings, their product $f^{-1} . g^{-1} = (g.f)^{-1}$ is also a continuous mapping of $(Z, \tau'')$ onto $(X, \tau)$. Hence $g.f$ is a homeomorphism of $(X, \tau)$ onto $(Z, \tau'')$.//

**Example 4.** ***A bijective mapping, $f : X \to Y$, is a homeomorphism of $(X, \tau)$ onto $(Y, \tau')$ if, and only if, $\tau'$ is the finest topology on $Y$ for which $f$ is continuous.***

**Example 5.** ***Give an example of a set $X$, and two different topologies $\tau_1$ and $\tau_2$ on $X$, such that $(X, \tau_1)$ and $(X, \tau_2)$ are homeomorphic.***

Let $f$ be a homeomorphism of $(X, \tau)$ onto $(Y, \tau')$. Then $f$ establishes a (1, 1)-correspondence between the elements of $X$ and $Y$, since $f$ is bijective. As $f$ is both open and continuous, there exists a (1, 1)-correspondence between the open sets in $\tau$ and the open sets in $\tau'$, since for every $G \in \tau$, there exists a unique $U \in \tau'$, such that $U = f(G)$, and for every $U \in \tau'$, there exists a unique $G \in \tau$, such that $f(G) = U$.

Thus a homeomorphism establishes a (1, 1)-correspondence between the elements of $X$ and $Y$, also a (1, 1)-correspondence between the open sets in $(X, \tau)$ and the open sets in $(Y, \tau')$. Consequently, a homeomorphism may be considered as an abstract equality among topological spaces.

The relation for a topological space to be homeomorphic to a topological space is an equivalence relation. In fact, the relation is reflexive, since identity mapping is a homeomorphism. The relation is symmetric, since the inverse of a homeomorphism is a homeomorphism (by Theorem 54); and the relation is transitive, since the product of two homeomorphisms is a homeomorphism (by Theorem 55). Hence, the aggregate of all topological spaces is partitioned into disjoint classes of homeomorphic spaces.

Properties common to all members of the same class, *i.e.*, common to all spaces homeomorphic to each other, are said to be *topological properties*. Thus a property of a topological space, which remains invariant under all homeomorphisms of the space, is called a *topological property*. For instance, the property of a given subset of a topological space to be dense-in-itself, or scattered, or isolated is a topological property.

Accordingly, the subject of topology may be fitted broadly in the definition of a geometry in the sense of Felix Klein, by saying that "the subject of topology is the study of topological properties".

## 4.4 Initial Topology or Weak Topology

Let $X$ be a given (non-empty) set, and let $(Y, \tau_1)$ be a topological space. Also let $f$ be a single-valued function on $X$ into $Y$. The problem is to choose a topology $\tau$ of $X$, such that $f$ becomes a continuous function on $(X, \tau)$ into $(Y, \tau_1)$. This can of course be done in many ways. The simplest way would be to take $\tau$ as the discrete topology of $X$. If $f$ is continuous for some topology $\tau$ of $X$, then $f$ is continuous for all topologies of $X$ which are finer than $\tau$. So, let us try to find the coarsest topology $\tau$ of $X$, for which $f$ is continuous.

The minimum requirement for $f$ to be continuous is that $f^{-1}(U)$ should belong to $\tau$ for every $U \in \tau_1$. So, the coarsest topology, for $f$ to be continuous, is the coarsest topology which contains all sets $f^{-1}(U)$, where $U \in \tau_1$. In other words, $\tau$ is the topology generated by the subbase formed by the sets $f^{-1}(U)$ for all $U \in \tau_1$. (The conditions for the sets $f^{-1}(U)$ to form a subbase can be easily verified). It can readily be seen that the aggregate **S**, formed by the sets $f^{-1}(U)$, where $U \in \tau_1$, constitutes a topology of $X$. In fact, $f^{-1}(\Phi) = \Phi$, $f^{-1}(Y) = X$, $f^{-1}(U) \cap f^{-1}(V) = f^{-1}(U \cap V)$, and $\cup \{f^{-1}(V)\} = f^{-1}\{\cup U\}$, show that the conditions 0 (1)-0 (4) hold for **S**. Hence, $\tau = \mathbf{S}$; thus :

**Theorem 56 :** *The coarsest topology $\tau$ of $X$, for which a given mapping $f$ of $(X, \tau)$ into $(Y, \tau_1)$ is continuous is given by*

$$\{\tau = \{f^{-1}(U) : U \in \tau_1\}.$$

Next, let us consider a more general case.

**Theorem 57 :** *Let $X$ be a non-empty set, and let $(Y_\nu, \tau_\nu)$ be a given family of topological spaces, for $\nu \in I$, where $I$ is an indexing set. Let $f_\nu$ be a single-valued mapping of $X$ into $Y_\nu$, for each $\nu \in I$. Then*

(*i*) *the coarsest topology $\tau$ of $X$, for which $f_\nu$ is a continuous mapping of $(X, \tau)$ into $(Y_\nu, \tau_\nu)$, for each $\nu \in I$, is the topology generated by the subbase $\mathbf{S} = \{f_\nu^{-1}(U_\nu) : U_\nu \in \tau_\nu, \ \nu \in I\}$;*

(*ii*) *a mapping $g$ of a topological space $(Z, \tau')$ into $(X, \tau)$ is continuous at a point $z \in Z$ if, and only if, the mapping $f_\nu \cdot g$ of $(Z, \tau')$ into $(Y_\nu, \tau_\nu)$ is continuous at the point $z$, for each $\nu \in I$.*

**Proof :** (*i*) The coarsest topology $\tau$ of $X$, for which $f_\nu$ is a continuous mapping of $(X, \tau)$ into $(Y_\nu, \tau_\nu)$, for every $\nu \in I$, is the coarsest topology of $X$, such that $\mathbf{S} \subset \tau$, where the family $\mathbf{S} = \{f_v^{-1}(U_v) : U_v \in \tau_v;\ v \in I\}$. Then $\mathbf{S}$ forms a subbase, and $\tau$ is the topology generated by the subbase $\mathbf{S}$ (by Theorem 10). (It may be noted that $\mathbf{S}$ does not form a topology of $X$ in the general case).

(*ii*) If $g$ is continuous at a point $z \in Z$, then the product $f_v \cdot g$ is continuous at the point $z$, $f_v \cdot g$ being considered as a mapping of $(Z, \tau')$ into $(Y_v, \tau_v)$, for each $v \in I$. Conversely, let the mapping $f_v \cdot g$ of $(Z, \tau')$ into $(Y_v, \tau_v)$ be continuous at a point $z \in Z$, for each $v \in I$. Let $V$ be a *nbd* of $g(z)$ in $(X, \tau)$. Since $\mathbf{S}$ forms a subbase of $\tau$, the finite intersections of members of $\mathbf{S}$ form a base $\mathbf{B}$ of $\tau$. Hence

$$g(z) \in \cap \{f_v^{-1}(U_v) : v \in J\} \subset V,$$

$J$ being a finite subset of $I$, so that

$$z \in g^{-1}[\cap \{f_v^{-1}(V_v) : v \in J\}] \subset g^{-1}(V).$$

Consequently, $g^{-1}(V)$ is a *nbd* of the point $z$ in $(Z, \tau')$, since, by hypothesis, each of the sets $g^{-1}(f_v^{-1}(U_v))$, for $v \in J$, is a *nbd* of $z$ in $(Z, \tau')$. Hence $g$ is continuous at the point $z$ in $(Z, \tau')$.//

The topology $\tau$, as determined in (*i*) above, is called the *initial topology* or the *weak topology* of $X$ determined by the family of functions

$$f_v : X \to Y_v, \text{ for } v \in I.$$

The above results have many fruitful and significant applications.

It has been shown in page 52 that any given collection of topologies $\tau_v$ ($v \in I$, where $I$ is an indexing set) on a set $X$ has a least upper bound in the set of all topologies of $X$. The principle of initial topologies may be applied to determine this topology $\tau$, which is the *l.u.b.* of the given family of topologies. Let $Y_v = X$, for each $v \in I$, and let $f_v$ be the identity mapping of $X$ onto $Y_v$, *i.e.*, $f_v(x) = x$, $x \in X$. Then the required topology $\tau$ is the coarsest topology of $X$, for which $f_v$ is a continuous mapping of $(X, \tau)$ into $(Y_v, \tau_v)$, for every $v \in I$.

The principle of initial topologies will find some important applications in the next article.

## Art. 6 CONSTRUCTION OF TOPOLOGIES

In this article some methods for constructing new topological spaces from some given topological spaces will be described.

## 6.1 Subspace Topology (or Relative Topology)

Let $A$ be a given subset of a topological space $(X, \tau)$, and let

$$\tau_A = \{A \cap U : U \in \tau\}.$$

Then $\tau_A$ forms a topology on $A$. In fact,

(*i*) $\Phi = A \cap \Phi \in \tau_A$, (*ii*) $A \cap X = A \in \tau_A$,

(*iii*) Let $G_\nu \in \tau_A$ for all $\nu \in I$ (where $I$ is an indexing set). Then $G_\nu = A \cap U_\nu$ where $U_\nu \in \tau$, for each $\nu$; and

$\cup\{G_\nu : \nu \in I\} = \cup\{A \cap U_\nu\} = A \cap [\cup\{U_\nu\}] \in \tau_A$ (since $\cup\{U_\nu\} \in \tau$, as each $U_\nu \in \tau$).

(*iv*) Let $G, H \in \tau_A$; then $G = A \cap U$, $H = A \cap V$, where $U, V \in \tau$. Now $G \cap H = (A \cap U) \cap (A \cap V) = A \cap (U \cap V) \in \tau_A$ (since $U \cap V \in \tau$, as $U, V \in \tau$).

$\tau_A$ is called *the topology on A relative to the topology* $\tau$ *on* $X$, or, simply, *the relative topology on A, or the subspace topology on A* in $(X, \tau)$; and $(A, \tau_A)$ is called a *subspace* of the topological space $(X, \tau)$.

It can readily be seen that the subspace topology $\tau_A$ is the initial topology of $A$, induced by the canonical injection $i$ of $A$ into $X$; in fact,

$$\{i^{-1}(U) : U \in \tau)\} = \{U \cap A : U \in \tau\} = \tau_A.$$

**Theorem 58 :** *Let $(A, \tau_A)$ be a subspace of a topological space $(X, \tau)$, then*

(*i*) *If $\boldsymbol{B}$ is a base of $\tau$, then $\boldsymbol{B}_A = \{A \cap U : U \in \boldsymbol{B}\}$ is a base of $\tau_A$; and if $\boldsymbol{S}$ is a subbase of $\tau$, then $\boldsymbol{S}_A = \{A \cap W : W \in \boldsymbol{S}\}$ is a subbase of $\tau_A$.*

(*ii*) *A set $N \subset A$ is a nbd of a point $x \in A$ in $(A, \tau_A)$ if, and only if, $N = A \cap M$, where $M$ is a nbd of $x$ in $(X, \tau)$.*

(*iii*) *A subset $F$ is closed in $(A, \tau_A)$ if, and only if, $F = A \cap K$, where $K$ is a closed subset in $(X, \tau)$.*

(*iv*) *A point $x$ is an accumulation point of a subset $B \subset A$ in $(A, \tau_A)$ if, and only if, $x$ is an accumulation point of $B$ in $(X, \tau)$.*

(*v*) *The closure of a subset $C \subset A$ in $(A, \tau_A)$ is given by $A \cap \overline{C}$, where $\overline{C}$ is the closure of $C$ in $(X, \tau)$.*

(*vi*) *The interior of a subset $B \subset A$ in $(A, \tau_A)$ contains the interior of $B$ in $(X, \tau)$.*

**Proof :** (*i*) As $\Phi \in \mathbf{B}$, $\Phi = A \cap \Phi \in \mathbf{B}_A$. Let $x \in A$, then there exists an $U \in \mathbf{B}$, such that $x \in U$; and then $x \in A \cap U$, where $A \cap U \in \mathbf{B}_A$. Next, let $x \in V_1 \cap V_2$, where $x \in A$ and $V_1, V_2 \in \mathbf{B}_A$. Then $V_1 = A \cap U_1$ and $V_2 = A \cap U_2$, where $U_1, U_2 \in \mathbf{B}$. As $\mathbf{B}$ forms a base of $\tau$, and $x \in U_1 \cap U_2$, there exists an $U_3 \in \mathbf{B}$, such that $x \in U_3 \subset U_1 \cap U_2$. Correspondingly, $x \in V_3 \subset V_1 \cap V_2$, where $V_3 = A \cap U_3 \in \mathbf{B}_A$. Thus all the three conditions for a base are satisfied for $\mathbf{B}_A$; so, $\mathbf{B}_A$ forms a base of $\tau_A$.

Similarly, the conditions for a subbase can be readily verified for $\mathbf{S}_A$; accordingly, $\mathbf{S}_A$ forms a subbase for $\tau_A$.

(*ii*) Let $N$ be a *nbd* of $x$ in $(A, \tau_A)$. Then there exists a $V \in \tau_A$, such that $x \in V \subset N$. But $V = A \cap U$ for some $U \in \tau$. Let $M = U \cup N$, then $x \in U \subset M$, from which it follows that $M$ is a *nbd* of $x$ in $(X, \tau)$; also $N = A \cap M$.

Conversely, let $M$ be a *nbd* of $x$ in $(X, \tau)$, and let $N = A \cap M$. There exists an $U \in \tau$, such that $x \in U \subset M$. Also $x \in A$; hence $x \in V \subset A \cap M = N$, where $V = A \cap U \in \tau_A$. Consequently, $N$ is a *nbd* of $x$ in $(A, \tau_A)$.

(*iii*) Let $F$ be a closed set in $(A, \tau_A)$; then $A - F \in \tau_A$. Hence $A - F = A \cap U$ for some $U \in \tau$. Let $X - U = K$; then $K$ is a closed set in $(X, \tau)$. Also

$$A \cap K = A \cap (X - U) = A - A \cap U = A - (A - F) = F.$$

Conversely, let $F = A \cap K$, where $K$ is a closed set in $(X, \tau)$. Then $X - K \in \tau$, and therefore $(X - K) \cap A \in \tau_A$. Also

$$A \cap (X - K) = A - A \cap K = A - F;$$

hence $A - F \in \tau_A$. Consequently, $F$ is a closed set in $(A, \tau_A)$.

(*iv*) Let $x$ be an accumulation point of a subset $B \subset A$ in $(A, \tau_A)$. Let $U$ be any open set in $(X, \tau)$ containing $x$. Then $A \cap U$ is an open set in $(A, \tau_A)$ containing the point $x$. Hence, by hypothesis, $A \cap U$ intersects the set $B$ in some point $y$ other than $x$; and therefore $U$ also intersects $B$ in the point $y\,(\neq x)$. Consequently $x$ is an accumulation point of a subset $B$ in $(X, \tau)$.

Conversely, let $x$ be an accumulation point of a subset $B \subset A$ in $(X, \tau)$. Let $G$ be any open set in $(A, \tau_A)$ containing the point $x$. Then $G = A \cap U$, for some $U \in \tau$. $U$ is an open set in $(X, \tau)$ containing the point $x$. Hence, by hypothesis, $U$ intersects $B$ in a point $y$ other than $x$. Now, $y \in U$ and $y \in B \subset A$ imply that $y \in A \cap U = G$. Thus $G$ intersects $B$ in a point $y\ (\neq x)$. Hence $x$ is also an accumulation point of the set $B$ in $(A, \tau_A)$.

(*v*) $A \cap \bar{B} = B \cup$ (all those accumulation points of $B$ in $(X, \tau)$ which lie inside $A$), as $B \subset A$,

$= B \cup$ (all the accumulation points of $B$ in $(A, \tau_A)$), by (*iv*)

$=$ The closure of $B$ in $(A, \tau_A)$.

(*vi*) The interior of $B$ in $(A, \tau_A) = \cup \{G \subset B : G \in \tau_A\}$

$= \cup \{A \cap U \subset B : U \in \tau\} \supset [\cup \{U : U \subset B \text{ and } U \in \tau\}] \cap A$

$=$ [interior of $B$ in $(X, \tau)$] $\cap A =$ interior of $B$ in $(X, \tau)$,

since the interior of $B$ in $(X, \tau) \subset B \subset A$.//

**Example 1.** ***Let A be a subset of a topological space $(X, \tau)$; then:***

(*i*) ***If A is a closed subset in $(X, \tau)$, then a subset F of A is closed in $(X, \tau)$ if and only if F is closed in $(A, \tau_A)$. The closure of a subset B of A in $(A, \tau_A)$ is the same as the closure of B in $(X, \tau)$.***

(*ii*) ***If A is an open set in $(X, \tau)$, then a subset G of A is open in $(A, \tau_A)$ if and only if G is open in $(X, \tau)$. The interior of a subset B of A in $(A, \tau_A)$ is the same as the interior of B in $(X, \tau)$.***

[**Hints :** (*i*) Let $A$ be closed in $(X, \tau)$. A subset $F$ of $A$ is closed in $(A, \tau_A)$ if and only if $F = A \cap K$, where $K$ is closed in $(X, \tau)$. As $A$ is closed in $(X, \tau)$, $A \cap K$ is also closed in $(X, \tau)$; that is, $F$ is closed in $(X, \tau)$. Conversely, let $F$ be a closed set in $(X, \tau)$. Then $A \cap F = F$ (since $F \subset A$) is also a closed set in $(X, \tau)$.

The closure of $B$ in $(A, \tau_A)$ is $A \cap \bar{B}$, where $\bar{B}$ is the closure of $B$ in $(X, \tau)$. Again, $B \subset A$ implies $\bar{B} \subset \bar{A} = A$ (since $A$ is closed in $(X, \tau)$); consequently, $A \cap \bar{B} = \bar{B}$. Thus the closure of $B$ in $(A, \tau_A)$ is $\bar{B} =$ the closure of $B$ in $(X, \tau)$].

**Theorem 59 :** *If $f$ is a continuous function on $(X, \tau)$ with values in $(Y, \tau')$, and $A$ is a subset of $X$, then $f|_A$ (the restriction of $f$ to $A$) is a continuous function on $(A, \tau_A)$ into $(Y, \tau')$.*

**Proof :** Let $h = f|_A$. Let $x \in A$, and let $U$ be a *nbd* of $h(x)$ in $(Y, \tau')$. As $f$ is continuous at the point $x$, and $h(x) = f(x)$, there exists a *nbd* $V$

of $x$ in $(X, \tau)$, such that $f(V) \subset U$. Then $A \cap V$ is a *nbd* of the point $x$ in $(A, \tau_A)$, by Theorem 58 (*ii*); and $h(A \cap V) \subset f(V) \subset U$. Hence $h$ is continuous at the point $x$ in $(A, \tau_A)$. So, $h = f|_A$ is a continuous function on $(A, \tau_A)$ into $(Y, \tau')$.//

**Theorem 60 :** *Let $(X, \tau)$ be a topological space, and $B \subset A \subset X$. Then the subspace $(B, (\tau_A)_B)$ of $(A, \tau_A)$ is also a subspace of $(X, \tau)$; that is, $(\tau_A)_B = \tau_B$.*

**Proof :** Let $G \in (\tau_A)_B$, then $G = B \cap M$, where $M \in \tau_A$; and $M = A \cap U$, where $U \in \tau$. Thus $G = B \cap M = B \cap (A \cap U) = (B \cap A) \cap U = B \cap U \in \tau_B$. Next, let $H \in \tau_B$, then $H = B \cap V$, where $V \in \tau$. But $B = B \cap A$; hence $H = B \cap V = (B \cap A) \cap V = B \cap (A \cap V) = B \cap W$, where $W = A \cap V \in \tau_A$; consequently, $H \in (\tau_A)_B$. Thus $(\tau_A)_B = \tau_B$.//

A property of a topological space is said to be *hereditary* (or, *cogradient*), if every subspace also possesses this property.

A topological space $(Y, \tau')$ is said to be *embeddable* in a topological space $(X, \tau)$, if there exists a subset $A$ of $X$ such that $(Y, \tau')$ is homeomorphic to the subspace $(A, \tau_A)$ of $(X, \tau)$.

## 6.2 Sum of Topological Spaces

Let $\{(X_\nu, \tau_\nu) : \nu \in I\}$ be a family of topological spaces ($I$ being an indexing set). Let $X$ denote the union of all the sets $X_\nu$, *i.e.*, $X = \cup \{X_\nu : \nu \in I\}$. And, let $\tau$ denote a collection of subsets of $X$, defined by

$$\tau = \{U \subset X : U \cap X_\nu \in \tau_\nu, \text{ for all } \nu \in I\}.$$

Then $\tau$ forms a topology on $X$. In fact,

(*i*) $\Phi \in \tau$, since $\Phi \cap X = \Phi \in \tau_\nu$, for all $\nu \in I$.

(*ii*) $X \in \tau$, since $X \cap X_\nu = X_\nu \in \tau_\nu$, for all $\nu \in I$.

(*iii*) Let $\{U_\mu\}$ be a family of sets belonging to $\tau$, then $\cup \{U_\mu\} \in \tau$, since $[\cup \{U_\mu\}] \cap X_\nu = \cup \{U_\mu \cap X_\nu\} \in \tau$, as $U_\mu \cap X_\nu \in \tau_\nu$, for each $\mu$; and this is true for all $\nu \in I$.

(*iv*) Finally, let $U, V \in \tau$, then $U \cap V \in \tau$, since

$$(U \cap V) \cap X_\nu = (U \cap X_\nu) \cap (V \cap X_\nu) \in \tau_\nu$$

as $U \cap X_\nu \in \tau_\nu$ and $V \cap X_\nu \in \tau_\nu$; and this holds for all $\nu \in I$.

The topological space $(X, \tau)$ is called the *topological sum* or, simply, the *sum* of the family of the topological spaces $(X_\nu, \tau_\nu)$. And, we denote this by $(X, \tau) = \Sigma \{(X_\nu, \tau_\nu) : \nu \in I\}$.

**Theorem 61 :** *Let $\{(X_\nu, \tau_\nu) : \nu \in I\}$ be a family of topological spaces, and let $(X, \tau)$ be their topological sum. Let $i_\nu$ denote the inclusion mapping of $X_\nu$ into $X$, for each $\nu \in I$. Then :*

(*i*) *$i_\nu$ is a continuous mapping of $(X_\nu, \tau_\nu)$ into $(X, \tau)$;*

(*ii*) *$i_\nu$ is also both open and closed if, the sets $X_\nu$ of the given family are pair-wise disjoint.*

**Proof :** (*i*) Let $U \in \tau$, then $i_\nu^{-1}(U) = U \cap X_\nu \in \tau_\nu$. Hence $i_\nu$ is a continuous mapping of $(X_\nu, \tau_\nu)$ into $(X, \tau)$.

(*ii*) Let the sets of the family $\{X_\nu : \nu \in I\}$ be pair-wise disjoint. Let $G \in \tau_\nu$, then $i_\nu(G) = G$. Also $G \cap X_\nu = G \in \tau_\nu$ and $G \cap X_\mu = \Phi \in \tau_\mu$, for all $\mu \neq \nu$; hence $G \in \tau$, *i.e.*, $i_\nu(G) \in \tau$. Consequently $i_\nu$ is an open mapping of $(X_\nu, \tau_\nu)$ into $(X, \tau)$.

Next, let $F$ be any closed set in $(X_\nu, \tau_\nu)$, and let $V = X - F$. Then $V \cap X_\nu = (X - F) \cap X_\nu = X_\nu - F \in \tau_\nu$, and $V \cap X_\mu = (X - F) \cap X_\mu = X_\mu \in \tau_\mu$. Hence $V \in \tau$, so that $F$ is closed in $(X, \tau)$. Thus $i_\nu(F) = F$ is closed in $(X, \tau)$. So, $i_\nu$ a closed mapping on $(X_\nu, \tau_\nu)$ into $(X, \tau)$.//

## 6.3 Topological Products

Let $(X_\nu, \tau_\nu)$ be a family of topological spaces, for $\nu \in I$, where $I$ is an indexing set. Let $X$ denote the Cartesian product of the sets $X_\nu$, for all $\nu \in I$ (see page 13); that is, $X = \pi\{X_\nu . \nu \in I\}$. Any point $x \in X$ is a function, having $I$ as its domain, such that $x_\nu = x(\nu) \in X_\nu$, for each $\nu \in I$. Then $x = \{x_\nu\}$, where $x_\nu \in X_\nu$ for each $\nu \in I$; $x_\nu$ is called the $\nu$th coordinate of the point $x$.

Let $p_\nu$ be the single-valued mapping of $X$ into $X_\nu$, defined by $p_\nu(x) = x_\nu$, for every $\nu \in I$. Then $p_\nu$ is called the *projection mapping* of the product set $X$ into the $\nu$th co-ordinate set $X_\nu$, for each $\nu \in I$.

**Finite Products :** Let the indexing set $I$ be a finite set; say $I = \{1, 2, ..., n\}$. Let **B** be a family of subsets of the product set $X = X_1 \times X_2 \times \ldots \times X_n$, defined by

$$\mathbf{B} = \{U_1 \times U_2 \times \ldots \times U_n : U_i \in \tau_i, \text{ for } i = 1, 2, ..., n\}.$$

Then **B** forms a base for a topology on $X$. In fact,

(*i*) $\Phi \in \mathbf{B}$, since $\Phi \in \tau_i$, for $i = 1, 2, ..., n$ (assuming

$$\Phi \times \Phi \times \ldots \times \Phi = \Phi).$$

(*ii*) $X = X_1 \times X_2 \times \ldots \times X_n \in \mathbf{B}$, since $X_i \in \tau_i$, for $i = 1, 2, \ldots, n$; hence every point of $X$ is contained in at least one member of **B** (viz. $X$).

(*iii*) Let $x \in U \cap V$, where $U$, $V \in \mathbf{B}$. Now, $U = U_1 \times \ldots \times U_n$ and $V = V_1 \times V_2 \times \ldots \times V_n$ where $U_i$, $V_i \in \tau_i$, for $i = 1, 2, \ldots, n$ so that

$$U \cap V = (U_1 \times U_2 \times \ldots \times U_n) \cap (V_1 \times V_2 \times \ldots \times V_n)$$
$$= (U_1 \cap V_1) \times (U_2 \cap V_2) \times \ldots \times (U_n \cap V_n) \in \mathbf{B},$$

since $U_i \cap V_i \in \tau_i$, for $i = 1, 2, \ldots, n$. Thus $x \in W \subset U \cap V$, where $W = U \cap V \in \mathbf{B}$. Hence the conditions (*i*), (*ii*-a) and (*iii*-a) for a base (vide pages 45-46) are satisfied for **B**.

The topology $\tau(\mathbf{B})$, generated by the base **B**, is called the *product topology* on $X$, and $(X, \tau(\mathbf{B}))$ is called the *topological product* of the family of topological spaces $\{(X_i, \tau_i) : i = 1, 2, \ldots, n\}$.

**Theorem 62 :** *Let $(X, \tau)$ be the topological product of a family of topological spaces $\{(X_i, \tau_i) : i = 1, 2, \ldots, n\}$, and let $p_j$ denote the projection mapping : $(X, \tau) \to (X_j, \tau_j)$, for $j = 1, 2, \ldots, n$. Then, for each $j$,*

(*i*) *$p_j$ is an open mapping of $(X, \tau)$ on $(X_j, \tau_j)$,*

(*ii*) *$p_j$ is a continuous mapping of $(X, \tau)$ on $(X_j, \tau_j)$,*

(*iii*) *the product topology $\tau$ is the coarsest topology on the product set $X$, for which each projection mapping $p_j$ is continuous.*

**Proof :** (*i*) Let $U \in \tau$, and let $x_j \in p_j(U) \subset X_j$. Let $x$ be a point in $U$, such that $x_j$ is the $j$th co-ordinate of $x$. Since the family of sets

$$\{V_1 \times V_2 \times \ldots \times V_n : V_i \in \tau_i,\ i = 1, 2, \ldots, n\}$$

forms a base for the product topology $\tau$, it follows that

$$x \in V_1 \times V_2 \times \ldots \times V_n \subset U,$$

for suitable $V_i \in \tau_i$ (for $i = 1, 2, \ldots, n$). Correspondingly, $x_j \in V_j \subset p_j(U)$, and then taking union of the corresponding relations, one obtains $p_j(U) \subset \cup \{V_j\} \subset p_j(U)$, so that $p_j(U) = \cup \{V_j\} \in \tau_j$. Hence $p_j$ is an open mapping of $(X, \tau)$ on $(X_j, \tau_j)$.

(*ii*) Let $U_j \in \tau_j$. Then

$$p_j^{-1}(U_j) = X_1 \times \ldots \times X_{j-1} \times U_j \times X_{j+1} \times \ldots \times X_n \in \mathbf{B} \subset \tau,$$

where **B** is the base which generates the product topology $\tau$ on X. Hence $p_j$ is a continuous mapping of $(X, \tau)$ on $(X_j, \tau_j)$, for $j = 1, 2, \ldots, n$.

(*iii*) It has been proved above in (*i*) that for the product topology $\tau$, each projection mapping $p_j : (X, \tau) \to (X_j, \tau_j)$ is continuous.

Let $\sigma$ be also a topology on the product set $X$, such that each projection mapping $p_j$ is a continuous mapping of $(X, \sigma)$ on $(X_j, \tau_j)$. Let **B** denote the base which generates the product topology $\tau$, and let $W \in \mathbf{B}$. Then $W = V_1 \times V_2 \times \ldots \times V_n$, where $V_i \in \tau_i$, for $i = 1, 3, \ldots, n$. Since $p_j$ is a continuous mapping of $(X, \sigma)$ on $(X_j, \tau_j)$ it follows that $p_j^{-1}(V_j) \in \sigma$, for $j = 1, 2, \ldots, n$. Hence, $p_1^{-1}(V_1) \cap \ldots \cap p_n^{-1}(V_n) \in \sigma$. But

$$p_1^{-1}(V_1) \cap \ldots \cap p_n^{-1}(V_n) = V_1 \times V_2 \times \ldots \times V_n = W,$$

since a point $x \in X$ belongs to the set $p_1^{-1}(V_1) \cap \ldots \cap p_n^{-1}(V_n)$ if, and only if, $x_i \in V_i$, for $i = 1, 2, \ldots, n$; and this is exactly the criterion for the point $x$ to belong to the set $V_1 \times V_2 \times \ldots \times V_n$. Thus $W \in \sigma$, so that $\mathbf{B} \subset \sigma$. Consequently, $\tau = \tau(\mathbf{B}) \subset \sigma$; hence the product topology $\tau$ is coarser than the topology $\sigma$.//

**General case :** Let $\{(X_\nu, \tau_\nu)\} : \nu \in I\}$, where $I$ is an indexing set, be a family of topological spaces, and let $X$ denote the Cartesian product of the sets $X_\nu$, for all $\nu \in I$. Let **S** be a family of subsets of $X$, defined by $\mathbf{S} = \{p_\nu^{-1}(U_\nu) : U_\nu \in \tau_\nu,\ \nu \in I\}$, where $p_\nu$ denotes the projection mapping of $X$ onto $X_\nu$, for each $\nu \in I$. Then **S** forms subbase of a topology on $X$. In fact, taking $U_\nu = \Phi$ we have $p_\nu^{-1}\,\Phi = \Phi$ (since $p_\nu$ is a surjective mapping); thus $\Phi \in \mathbf{S}$. Also, for $U_\nu = X_\nu$, we have $p_\nu^{-1}(X_\nu) = X$ (as $p_\nu$ is a surjective mapping); thus $X \in \mathbf{S}$.

The topology on $X$, generated by the subbase **S**, is called the *product topology* (or the *Tychonoff topology*) on $X$. And, if $\tau$ denotes this product topology on $X$, then $(X, \tau)$ is called the *product space* or the *topological product* of the given family of topological spaces.

Let **B** denote the base generated by the subbase **S**. Then a subset $V$ of $X$ belongs to the base **B** if, and only if, $V$ is the intersection of finitely many members of the subbase **S**; that is, $V = \cap\{p_\nu^{-1}(U_\nu) : U_\nu \in \tau_\nu,\ \nu \in J\}$, $J$ being some finite subset of $I$. And, since

$$p_\nu^{-1}(U) = \pi\,\{V_\mu : V_\nu = U_\nu \text{ and } V_\mu = X_\mu \text{ for } \mu \neq \nu\},$$

it follows that every member of the base **B** is a set of the form $\pi\,\{U_\nu : \nu \in I\}$, where, for each $\nu \in I$, $U_\nu \in \tau_\nu$, and for all but a finitely many $\nu' s$, $U_\nu = X_\nu$.

In particular, when the indexing set is finite, say $I = \{1, 2, ..., n\}$, then the base **B** consists of all products $U_1 \times U_2 \times ... \times U_n$, where $u_i \in \tau_i$, for $i = 1, 2, ..., n$. This shows that the product topology, defined earlier, for a finite number of topological spaces, is the same as in the general case.

The properties (*i*), (*ii*) and (*iii*), proved for a finite product, in Theorem 62, hold also in the general case. In fact, (*i*) it can be proved in exactly the same way as in the case of finite products (*i.e.*, as in the proof of Theorem 62 (*i*)) that *each projection mapping is an open mapping.* (*ii*) and (*iii*) : Each $p_\nu$ is continuous if and only if $p_\nu^{-1}(U_r) \in \tau$; and this certainly holds, since, for each $\nu$, $p_\nu^{-1}(U_\nu) \in \mathbf{S} \subset \tau$, for each $U_\nu \in \tau_\nu$. And, since the product topology $\tau$ is generated by the subbase **S**, $\tau$ is the coarsest topology containing the subbase; *i.e.*, *$\tau$ is the coarsest topology on X, for which all projection mappings are continuous.*

Thus the product topology $\tau$ can as well be defined as the initial topology on the product set $X$, induced by the family of projection mappings, $p_\nu : X \to X_\nu$, for all $\nu \in I$. Hence, it follows from Theorem 57 (*ii*), that:

**Theorem 63 :** *Let $(X, \tau)$ be the topological product of a family of topological spaces $\{(X_\nu, \tau_\nu), \nu \in I\}$, where I is an indexing set. Then a mapping f of a topological space $(Y, \tau')$ into the product space $(X, \tau)$ is continuous if, and only if, $p_\nu . f$ is a continuous mapping of $(Y, \tau')$ into the co-ordinate space $(X, \tau_\nu)$, for each $\nu \in I$.*

Let $f_\nu$ be a single valued mapping of a topological space $(Y, \tau')$ into the co-ordinate space $(X_p, \tau_\nu)$, for each $\nu \in I$. Then the mapping $e$ of $(Y, \tau')$ into the product space $(X, \tau)$, defined by $e(y) = \{x_\nu\}$, for $y \in Y$, where $x_\nu = f_\nu(y)$, is called the evaluation mapping of $(Y, \tau')$ into the product space $(X, \tau)$ with respect to the family of mappings $\{f_\nu\}$ of $(Y, \tau')$ into the co-ordinate spaces $(X_\nu, \tau_\nu)$.

**Theorem 64 :** *If e is the evaluation mapping of a topological space $(Y, \tau')$ into the product space $(X, \tau) = \pi(X_\nu, \tau_\nu)$, with respect to a family of mappings $\{f_\nu\}$ of $(Y, \tau')$ into the co-ordinate spaces $(X_\nu, \tau_\nu)$, then e is continuous if, and only if, each $f_\nu$ is continuous.*

**Proof :** It follows from the definition of the evaluation mapping $e$ that $p_\nu . e(y) = f_\nu(y)$, for each $\nu$. Hence, by Theorem 63, $e$ is continuous if, and only if, each $f_\nu$ is continuous.//

**Theorem 65 :** *An infinite sequence* $\{S_n\}$ *converges to a point* $\{x_\nu\}$ *in the product space* $(X, \tau) = \pi(X_\nu, \tau_\nu)$ *if, and only if, the sequence* $p_\nu\{S_n\}$ *converges to* $x_\nu$ *in the co-ordinate space* $(X_\nu, \tau_\nu)$, *for each* $\nu \in I$, *where* $I$ *is the indexing set.*

**Proof :** Let the sequence $\{S_n\}$ converge to a point $x = \{x_\nu\}$ in the product space $(X, \tau)$. Since each projection mapping $p_\nu$ is continuous, it follows, by Theorem 49, that the sequence $p_\nu\{S_n\}$ converges to

$$p_\nu(x) = x_\nu \text{ in } (X_\nu, \tau_\nu), \text{ for each } \nu \in I.$$

Conversely, let the sequence $p_\nu\{S_n\}$ converge to $x_\nu$, for each $\nu \in I$, where $\{S_n\}$ is a given infinite sequence in the product space $(X, \tau)$. Let $V$ be an open set, containing the point $\{x_\nu\}$ in $(X, \tau)$, and such that $V$ belongs to the subbase **S** which generates the product topology $\tau$. Then $V = p_\mu^{-1}(U_\mu)$, for some $\mu \in I$ and some $U_\mu \in \tau_\mu$. As $\{x_\nu\} \in V$, it follows that $x_\mu = p_\mu\{x_\nu\} \in U_\mu$. Since $U_\mu$ is an open set containing the point $x_\mu$ in $(X_\mu, \tau_\mu)$, and the infinite sequence $p_\mu\{S_n\}$ converges to $x_\mu$, there exists a positive integer $k$, such that $p_\mu\{S_n\} \in U_\mu$, for all $n > k$. Correspondingly, $S_n \in p_\mu^{-1}(U_\mu) = V$, for all $n > k$. Hence the infinite sequence $\{S_n\}$ converges to $\{x_\nu\}$ in the product space $(X, \tau)$.//

By virtue of the property proved in Theorem 65 (and also for similar properties for nets and filters), the convergence in the product space is called *coordinate-wise convergence* or *pointwise convergence*. These expressions are used more frequently in the particular case when all the coordinate spaces are identical. Thus if $X_\nu = Y$, for all $\nu \in I$ (where $I$ is an indexing set), then the product set $X$ consists of all functions $x$ mapping $I$ into $Y$, and is usually denoted by $Y^I$. The corresponding product topology $\tau$ is called the *weak topology of the function space* $Y^I$ or the *topology of point-wise convergence* in $Y^I$ (sometimes it is also known as the simple topology of $Y^I$).

Let $(\mathbf{R}, \sigma)$ denote the real number space; then the topological product of $(\mathbf{R}, \sigma)$ with itself $n$ times, *i.e.*, the product space $(\mathbf{R}, \sigma) \times (\mathbf{R}, \sigma) \times \ldots \times (\mathbf{R}, \sigma)$ ($n$ copies), is called the *n-dimensional Euclidean space* or the *Euclidean n-space*, and is denoted by $\mathbf{R}^n$. In particular, $\mathbf{R}^1$ or $\mathbf{R}$ is called the *Euclidean line*, and $\mathbf{R}^2$ is called the *Euclidean plane*.

**Example 2.** ***Show that the family of open discs forms an open base of the topology of the Euclidean plane.***

[**Hints :** The open rectangles (with sides parallel to the coordinate axes) form an open base of the topology of the Euclidean plane. The family of open discs forms a base, which is topologically equivalent to the base formed by the family of open rectangles (apply Theorem 5)].

Let $I$ denote a closed interval $[a, b]$ in the real number space, with its relative topology. Then the product $I^n$ of $n$ copies ($n$ is finite) of $I$ is called the *n-dimensional cube*. It is a subspace of $\mathbf{R}^n$.

Let $\alpha$ be the cardinal number of an arbitrary set $A$. Then the topological product $I^\alpha = \pi \{I_v : v \in A\}$, where $I_v = I$ for all $v \in A$, is called the *cube* $I^\alpha$.

The topological product of the countably many closed intervals $[0, 1/n]$, for $n = 1, 2, \ldots$ is called the *Hilbert cube*.

If, in particular, $I$ denotes the closed unit interval $[0, 1]$, then $I^n$ is called the *unit n-cube* in the Euclidean space $\mathbf{R}^n$.

The subset of $\mathbf{R}^n$, consisting of all points $(x_1, x_2, \ldots, x_n)$ for which $x_1^2 + x_2^2 + \ldots + x_n^2 \leq 1$, is called the *n-dimensional* (*unit*) ball or *solid sphere.*

The subset of $\mathbf{R}^{n+1}$, consisting of all points $(x_1, x_2, \ldots, x_{n+1})$ for which $x_1^2 + x_2^2 + \ldots + x_{n+1}^2 = 1$, is called the *n-dimensional* (*unit*) *sphere* (or *open sphere*), and is denoted by $S^n$. In particular, for $n = 1$, $S^1$ or $S$ is called a *circle* on the Euclidean plane $\mathbf{R}^2$.

$S' \times S'$ is called a *torus*, and $S^1 \times I$ is called a *cylinder* in the Euclidean 3-space $\mathbf{R}^3$.

**Example 3.** ***Let*** $\{(X_v, \tau_v) : v \in I\}$ ***be a given family of topological spaces, and*** $(X, \tau)$ ***their topological product. Then each*** $(X_v, \tau_v)$ ***can be embedded in the product space*** $(X, \tau)$.

[**Hints :** Let $\mu \in I$; it will be proved that $(X_\mu, \tau_\mu)$ is embeddable in the product space $(X, \tau)$. Fix a point $z = \{z_v\}$ in $(X, \tau)$. Let $Y$ be the subset of $X$, consisting of all those points $x = \{x_\alpha\} \in X$, such that $x_\alpha = z_\alpha$ for all $\alpha \neq \mu$, and $x_\mu$ runs over $X_\mu$. Let $f$ be the restriction of the projection mapping $p_\mu$ to the subset $Y$. Then $f$ is a homeomorphism of $(Y, \tau_Y)$ onto $(X_\mu, \tau_\mu)$].

**Example 4.** ***Let*** $(X, \tau)$ ***be the topological product of a family of topological spaces*** $(X_v, \tau_v)$, ***for*** $v \in I$ ***(where I is an indexing set), and let*** $Y_v \subset X_v$, ***for each*** $v \in I$. ***Then***

(*i*) $\overline{\pi Y_\nu} = \pi \overline{Y_\nu}$, *for all* $\nu \in I$;

(*ii*) $\pi Y_\nu$ *is closed in* $(X, \tau)$ *if, and only if,* $Y_\nu$ *is closed in* $(X_\nu, \tau_\nu)$, *for each* $\nu \in I$;

(*iii*) $\pi Y_\nu$ *is open in* $(X, \tau)$ *if, and only if,* $Y_\nu$ *is open in* $(X_\nu, \tau_\nu)$, *for each* $\nu \in I$, *and if, moreover,* $Y_\nu = X_\nu$, *for all but finitely many indices* $\nu \in I$;

(*iv*) *if* $(Y, \tau')$ *be the product of the family of topological spaces* $(Y_\nu, \tau(Y_\nu))$ *for all* $\nu \in I$, *(where* $\tau(Y_\nu)$ *denotes the subspace topology on* $Y_\nu$ *in* $(X, \tau)$*), then* $\tau' = \tau_Y$;

(*v*) $\pi Y_\nu$ *is dense in* $(X, \tau)$ *if, and only if,* $Y_\nu$ *is dense in* $(X_\nu, \tau_\nu)$, *for all* $\nu \in I$;

(*vi*) $Int\,(\pi \{Y_\nu : \nu = 1, 2, ..., n\}) = \pi \{Int\,(Y_\nu) : \nu = 1, 2, ..., n\}$.

**Example 5.** *Let* $(Y, \tau)$ *be the topological product of a family of topological spaces* $\{(Y_\alpha, \tau_\alpha) : \alpha \in I\}$, *where* $I$ *is an indexing set. Let* $(X, \sigma)$ *be any topological space, and let a mapping,* $f_\alpha : X \to Y_\alpha$, *be defined for each* $\alpha \in I$, *and let* $f : X \to Y$ *be defined by* $f(x) = \{f_\alpha(x)\}$. *Then* $f$ *in continuous if, and only if, each* $f_\alpha$ *is continuous.*

## 6.4 Quotient Spaces

Let $(X, \tau)$ be a given topological space, and let $\sigma$ be an equivalence relation on $X$. Then $X$ is partitioned into disjoint classes, and the set formed by these classes is called the quotient set $X/\sigma$. Let $D_\alpha$ denote the class which contains the element $a \in X$; then $D_\alpha = D_b$ iff $a$ and $b$ belong to the same class, *i.e.*, iff $(a, b) \in \sigma$ (*i.e.*, iff $a\sigma b$ holds).

Let $h$ be a mapping of the set $X$ onto the quotient set $X/\sigma$, defined by $h(x) = D_x$, for the every $x \in X$. Then $h$ is obviously a surjective mapping. Let $\tau_\sigma$ denote a family of subsets of $X/\sigma$, defined by

$$\tau_\sigma = \{M \subset X/\sigma : h^{-1}(M) \in \tau\}.$$

Then $\tau_\sigma$ forms a topology on $X/\sigma$. In fact,

(*i*) as $h$ is surjective, $h^{-1}(\Phi) = \Phi \in \tau$, and so $\Phi \in \tau_\sigma$, and

(*ii*) $h^{-1}(X/\sigma) = X \in \tau$ implies $X/\sigma \in \tau_\sigma$.

(*iii*) Let $\{M_\alpha\}$ be a family of members of $\tau_\sigma$ so that $h^{-1}(M_\alpha) \in \tau$, for each $\alpha$. Then $h^{-1}[\cup \{M_\alpha\}] = \cup \{h^{-1}(M_\alpha)\} \in \tau$, so that $\cup \{M_\alpha\} \in \tau_\sigma$.

(*iv*) Let $M, N \in \tau_\sigma$ then $h^{-1}(M \cap N) = h^{-1}(M) \cap h^{-1}(N) \in \tau$, since $h^{-1}(M)$, $h^{-1}(N) \in \tau$; hence $M \cap N \in \tau_\sigma$.

The topology $\tau_\sigma$ on the quotient set $X/\sigma$, as defined above, is called the *quotient topology*, and the topological space $(X/\sigma, \tau_\sigma)$ is called the *quotient space* or the *decomposition space*, determined by the equivalence relation $\sigma$ on $X$.

The mapping $h$, which is evidently continuous, is called the *canonical (or natural) mapping* of $X$ onto the quotient set $X/\sigma$.

**Quotient maps :** Let $(X, \tau)$ be a topological space, $Y$ a set, and $f: X \to Y$ a surjective mapping. Then the family of subsets $\tau_f$ of $Y$, defined by $\tau_f = \{V \subset Y : f^{-1}(V) \in \tau\}$, forms a topology on $Y$ (verify); it is called the *quotient topology on Y, induced by f.*

**Theorem 66 :** *The quotient topology $\tau_f$ is the finest topology on Y for which f is continuous.*

**Proof :** The mapping $f: (X, \tau) \to (Y, \tau_f)$ is evidently continuous. Let $f: (X, \tau) \to (Y, \tau')$ be also continuous, for some topology $\tau'$ on $Y$. Let $G \in \tau'$, then $f^{-1}(G) \in \tau$; and therefore $G \in \tau_f$. Thus $\tau' \subset \tau_f$; and this proves the theorem.//

Next, let $(X, \tau)$, $(Y, \tau')$ be two topological spaces. Then a surjective mapping $f: (X, \tau) \to (Y, \tau')$ is called a quotient map, if $\tau' = \tau_f$, the quotient topology on $Y$, induced by $f$.

**Theorem 67 :** *A surjective mapping $f: (X, \tau) \to (Y, \tau')$ is a quotient map if,*

(*i*) *f is both open and continuous, or*

(*ii*) *f is both closed and continuous.*

**Proof :** (*i*) Let $f$ be both open and continuous. Then, for every $G \in \tau'$, $f^{-1}(G) \in \tau$, since $f$ is continuous; and then $G \in \tau_f$. Hence $\tau' \subset \tau_f$. Next, let $V \in \tau_f$, then $f^{-1}(V) \in \tau$; hence $V = f[f^{-1}(V)] \in \tau'$, since $f$ is open. Thus $\tau_f \subset \tau'$. Consequently, $\tau' = \tau_f$.

(*ii*) Let $f$ be both closed and continuous. Then $\tau' \subset \tau_f$, as in case (*i*), since $f$ is continuous. Conversely, let $U \in \tau_f$ then $f^{-1}(U) \in \tau$, so that $X - f^{-1}(U)$ is closed in $(X, \tau)$. Hence $f[X - f^{-1}(U)] = Y - U$ is closed in $(Y, \tau')$, since $f$ is closed, so that $U \in \tau'$. Thus $\tau_f \subset \tau'$. Hence $\tau' = \tau_f$.//

The conditions (*i*) and (*ii*) of Theorem 67 are sufficient for $f$ to be a quotient map, but not necessary.

**Theorem 68 :** *Let* $(X, \tau)$, $(Y, \tau')$, $(Z, \tau'')$ *be three topological spaces,* $f: (X, \tau) \to (Y, \tau')$ *be a quotient map, and* $g: (Y, \tau') \to (Z, \tau'')$ *be an arbitrary mapping. Then g is continuous if, and only if, g.f. is continuous.*

**Proof :** If $g$ is continuous, then $g.f.$ is certainly continuous, since the quotient map $f$ is always continuous. Conversely, let $g.f.$ be continuous. Let $H \in \tau''$, then $f^{-1}[g^{-1}(H)] = (g.f.)^{-1}(H) \in \tau$, from which it follows that $g^{-1}(H) \in \tau'$ (since $\tau'$ is the quotient topology $\tau_f$). Hence $g$ is continuous.//

Next, we shall establish the equivalence of the concepts of quotient topology, determined; (1) by an equivalence relation, and (2) by a surjective mapping, in the following theorem.

**Theorem 69 :** *Let* $(X, \tau)$ *be a topological space, Y a set,* $f: X \to Y$ *a surjective mapping and* $\tau_f$ *the quotient topology on Y, induced by f, then* $(Y, \tau_f)$ *is homeomorphic to the quotient space* $(X/\mu, \tau_\mu)$, *where* $\mu$ *is the equivalence relation on X, such that* $(x_1, x_2) \in \mu$ *iff* $f(x_1) = f(x_2)$.

**Proof :** $\mu$ is evidently an equivalence relation on $X$. Correspondingly, one obtains the quotient space $(X/\mu, \tau_\mu)$.

As $f$ is surjective, the mapping $g$ of the quotient set $X/\mu$ onto $Y$, defined by $g(D_x) = f(x)$ is bijective (where $D_x$ denotes the class in $X/\mu$ which contains the element $x \in X$).

Let $h$ be the canonical mapping of $(X, \tau)$ onto $(X/\mu, \tau_\mu)$. Then, by Theorem 68, $g$ is continuous, since $g.h. = f$ is continuous, and $g^{-1}$ is continuous, since $g^{-1}.f = h$ is continuous. Hence $g$ is a homeomorphism of $(X/\mu, \tau_\mu)$ onto $(Y, \tau')$.//

**Identification topology :** We shall now introduce the concepts of identification mapping and identification topology.

For a given topological space $(X, \tau)$, the quotient spaces $(X/\mu, \tau_\mu)$, are determined by different equivalence relations $\mu$ on $X$. An equivalence relation $\mu$ on $X$ may as well be determined by a given decomposition (*i.e.*, partition) of $X$ into disjoint classes. Intuitively, it means that the space $(X/\mu, \tau_\mu)$ is obtained from $(X, \tau)$ by squeezing the various equivalence classes to points, or by identifying the various points of the same equivalence class. For this reason, the quotient topology $\tau_\mu$ is called the *identification topology*, and the natural mapping $h$ of $X$ onto $X/\mu$ is called the *identification mapping*, corresponding to the equivalence relation $\mu$.

**Example 6.** ***The real projective space $P^n$ : Let $(X, \tau)$ be the subspace of the Euclidean space $R^{n+1}$ where $X = R^{n+1} - \{0\}$ and $\{0\} = (0, 0, ..., 0)$. Let $\mu$ be a binary relation on X, such that $(a_1, a_2, ..., a_{n+1})\ \mu\ (b_1, b_2, ..., b_{n+1})$ holds iff $b_i = ta_i$ $(i = 0, 1, ..., n + 1)$, for some non-zero real number t. Then $\mu$ is an equivalence relation on X (verify). The quotient space $(X/\mu, \tau_\mu)$ is called the real projective space $P^n$. The points of $P^n$ are the straight lines in $R^{n+1}$ passing through the origin O.***

**Final Topologies :** The principle of defining a topology, as in the case of identification topology, may be generalised as follows:

**Theorem 70 :** *Let Y be a set, and $\{(X_\alpha, \tau_\alpha) : \alpha \in I\}$ be a family of topological spaces. For each $\alpha$, let $f_\alpha$ be a mapping of $X_\alpha$ into Y, and let $\sigma = \{V \subset Y : f_\alpha^{-1}(V) \in \tau_\alpha$, for each $\alpha \in I\}$. Then*

(*i*) *$\sigma$ is the finest topology on Y for which the mappings $f_\alpha$, $\alpha \in I$, are continuous;*

(*ii*) *a mapping g of $(Y, \sigma)$ into a topological space $(Z, \tau')$ is continuous if, and only if, the mapping $g.f_\alpha : (X_\alpha, \tau_\alpha) \to (Z, \tau')$, is continuous, for each $\alpha \in I$.*

**Proof :** (*i*) Since $f_\alpha^{-1}(\Phi) = \Phi \in \tau_\alpha$ and $f_\alpha^{-1}(Y) = X_\alpha \in \tau_\alpha$, for each $\alpha \in I$, it follows that $\Phi$, $Y \in \sigma$. Also, if $G, H \in \sigma$, so that $f_\alpha^{-1}(G)$, $f_\alpha^{-1}(H) \in \tau_\alpha$, and therefore $f_\alpha^{-1}(G \cap H) = f_\alpha^{-1}(G) \cap f_\alpha^{-1}(H) \in \tau_\alpha$, for each $\alpha \in I$, it follows that $G \cap H \in \sigma$. Next, let $\{G_\nu\}$ be a family of members of $\sigma$, so that $f_\alpha^{-1}(G_\nu) \in \tau_\alpha$, for every $\nu$. Then

$$f_\alpha^{-1}[\cup \{G_\nu\}] = \cup \{f_\alpha^{-1}(G_\nu)\} \in \tau_\alpha,$$

for every $\alpha \in I$; hence $\cup \{G_\nu\} \in \sigma$. Consequently, $\sigma$ forms a topology on Y.

It is obvious that, for each $\alpha \in I$, $f_\alpha : (X_\alpha, \tau_\alpha) \to (Y, \sigma)$ is continuous. Also, if for a topology $\sigma'$ on Y, $f_\alpha : (X_\alpha, \tau_\alpha) \to (Y, \sigma')$ is continuous, for each $\alpha \in I$, then for any $W \in \sigma'$, $f_\alpha^{-1}(W) \in \tau_\alpha$, and therefore $W \in \sigma$. Thus $\sigma' \subset \sigma$; that is, $\sigma$ is the finest topology on Y, for which $f_\alpha$ is continuous for each $\alpha \in I$.

(*ii*) Next, let $g : (Y, \sigma) \to (Z, \tau')$ be continuous, then the product $g.f_\alpha$ of two continuous mappings is certainly a continuous mapping of

$(X_\alpha, \tau_\alpha)$ into $(Z, \tau')$, for each $\alpha \in I$. Conversely, let each $g.f_\alpha$ be continuous, and let $V \in \tau'$. Then $(g.f_\alpha)^{-1}(V) = f_\alpha^{-1}(g^{-1}(V)) \in \tau_\alpha$, for each $\alpha \in I$; hence $g^{-1}(V) \in \sigma$. Thus $g$ must be continuous.//

The topology $\sigma$ on $Y$, as defined in Theorem 70, is called the *final topology* or the *strong topology* on $Y$, induced by the family of functions, $f_\alpha : X_\alpha \to Y$.

The following are important examples of a final topology :

1. A *quotient topology* is a final topology (vide Theorem 66).
2. The topology, defined in the case of a *topological sum* (of topological spaces), is a final topology. In fact, if $(X, \tau) = \Sigma\{(X_\nu, \tau_\nu) : \nu \in I\}$, where $I$ is an indexing set, then $\tau$ is the final topology on $X = \cup\{X_\nu : \nu \in I\}$, induced by the family of canonical injections,
$$i_\nu : X_\nu \to X, \text{ for } \nu \in I.$$
3. The greatest lower bound (or *g.l.b.*) of a given collection of topologies $\{\tau_\alpha : \alpha \in I\}$ on a set $X$ is a topology $\tau$ on $X$, given by $\tau = \cap\{\tau_\alpha : \alpha \in I\}$, $I$ being an indexing set (vide page 51-52). $\tau$ is the final topology on $X$, induced by the family of identity mappings $f_\alpha : (Y_\alpha, \tau_\alpha) \to (X, \tau)$, where $Y_\alpha = X$, for each $\alpha \in I$.

## MISCELLANEOUS EXERCISES I

**1.** Let **R** denote the real number space. Prove the following in **R** :

(*a*) Let **Q** denote the set of all rational numbers, and **T** that of all irrational numbers; then $\text{Int}(\mathbf{Q}) \cup \text{Int}(\mathbf{T}) \neq \text{Int}(\mathbf{Q} \cup \mathbf{T})$.

(*b*) Let $A = \left[0, \frac{1}{2}\right]$ and $B = \left[\frac{1}{2}, 1\right]$, then

(*i*) $\text{Int}(A) \cup \text{Int}(B) \neq \text{Int}(A \cup B)$.

(*ii*) $A$ and $B$ are regular closed sets, but $A \cap B = \left\{\frac{1}{2}\right\}$ is not so.

(*c*) Let $C = \left(0, \frac{1}{2}\right)$ and $D = \left(\frac{1}{2}, 1\right)$, then

(*i*) $\overline{C \cap D} \neq \overline{C} \cap \overline{D}$.

(*ii*) $C$ and $D$ are regular open sets, but $C \cup D$ is not so.

2. Let $X$ denote the subspace $[0, 1]$ of the real number space. Show that the set of all numbers $m/2^n$, where $m$ is a non-negative integer and $n$ a positive integer, is dense in $X$.

3. Let $X$ denote the open unit interval $(0, 1)$, and let $\tau$ consist of $\Phi$, $X$ and all sets of the form $U_n = \left(0, 1 - \dfrac{1}{n}\right)$, for $n = 1, 2, \ldots$ Show that $(X, \tau)$ is a topological space.

4. Let $X$ denote the closed interval $[-1, 1]$.
   (*a*) Let $\tau$ consist of
   (*i*) all those subsets of $X$ which do not contain 0, and
   (*ii*) all subsets of $X$ which contain the open interval $(-1, 1)$. Show that $\tau$ is a topology (called the *either-or topology*) on $X$. Find the closed sets in $(X, \tau)$.
   (*b*) Let **A** consist of
   (*i*) sets of the form $[-1, r)$, $0 < r < 1$ and
   (*ii*) sets of the form $(r, 1]$, $-1 < r < 0$. Find the coarsest topology on $X$, containing the family of subsets **A**.

5. A topological space $(X, \tau)$ is called a door space, if every subset of it is either open or closed. Verify that $(X, \tau)$ is a door space, where $X = \{a, b, c\}$ and $\tau = \{\Phi, \{a\}, \{b\}, \{a, b\}, X\}$. Find the derived sets of all subsets of $X$.

6. Let $X = \{u, v, w\}$ and $\tau = \{\Phi, \{u\}, \{v, w\}, X\}$. Show that $(X, \tau)$ is a topological space.
   (*a*) Prove the following properties in $(X, \tau)$:
   (*i*) Every subset in $\tau$ is both open and closed.
   (*ii*) If $A = \{v\}$ and $B = \{w\}$, then $\overline{A \cap B} \neq \bar{A} \cap \bar{B}$.
   (*iii*) Let $x_1 = v$, $x_2 = w$, and $x_i = u$ for $i > 2$. Then the sequence $\{x_n\}$ converges to $u$, but does not converge to either $v$ or $w$. On the other hand, $v$ and $w$ are accumulation points of $\{x_n\}$, whereas $u$ is not so.
   (*b*) Find in $(X, \tau)$,
   (*i*) the derived sets of all subsets of $X$, and
   (*ii*) $\text{Int}(\{v\})$.

**7.** Let **N** denote the set of all natural numbers.

(*a*) Let $\tau$ denote the cofinite topology on **N**, and let

$$G_n = \{1, n, n+1, n+2, \ldots\},$$

for $n = 1, 2, \ldots$ Then each $G_n \in \tau$, but

$$\cap \{G_n : n = 1, 2, \ldots\} = \{1\} \notin \tau.$$

[Thus the intersection of infinitely many open sets may not be open].

(*b*) Let $\tau$ consist of **N**, and all those subsets of **N** which contain neither 1 nor 2, or contain all but finitely many elements of **N**. Show that (**N**, $\tau$) is a topological space. Let

$$A_m = \{3, 4, \ldots, (m+2)\}, \text{ for } m = 1, 2, \ldots;$$

then show that $\overline{\cup \{A_m\}} \neq \cup \{\overline{A}_m\}$, for $m = 1, 2, \ldots$

(*c*) Let $\tau$ consist of $\Phi$, **N**, and the subsets $H_m$, for $m = 1, 2, \ldots$ where $H_m = \{1, 2, \ldots, m\}$. Show that (**N**, $\tau$) is a topological space.

(*d*) Let $\tau$ consist of $\Phi$ and the subsets $G_m$, for $m = 1, 2, \ldots$, where $G_m = \{m, m+1, m+2, \ldots\}$. Then show that :

(*i*) (**N**, $\tau$) is a topological space.

(*ii*) If **B** consists of all those subsets of **N**, which contain some $G_m$, then **B** forms a base of a topology on **N**. Describe the closure operation in this space.

(*e*) Let **B** consist of $\Phi$ and the subsets $A_m = \{2m-1, 2m\}$, for $m = 1, 2, \ldots$ Show that **B** forms a base of a topology on **N**. (The resulting topology is called the odd-even topology on **N**).

(*f*) Let **S** consist of the subsets $\mathbf{B}_k = \{k, k+1\}$, for $k = 1, 2, \ldots$ Show that **S** forms a subbase of a topology on **N**; describe the resulting topology.

(*g*) Let **B** consist of $\Phi$ and the subsets

$$V_m(n) = \{n + rm \in \mathbf{N};\ r \in \mathbf{Z}\},$$

where $m, n \in \mathbf{N}$ and $(m, n) = 1$. Show that **B** forms a base of a topology $\tau$ on **Z**, and that :

(*i*) for each prime $p \in \mathbf{N}$, the set $\{mp : m \in \mathbf{N}\}$ is closed in (**N**, $\tau$);

(*ii*) if $P$ is the set of all primes in **N**, then $\text{Int}(P) = \Phi$.

**8.** Let **Z** denote the set of all integers and let $\tau$ consist of $\Phi$, and all those subsets $G$ of **Z**, such that $x \pm 2, x \pm 4, ..., x \pm 2n, ...$ belong to $G$, whenever $x \in G$. Show that $(\mathbf{Z}, \tau)$ is a topological space.

**9.** Let **R** denote the set of all real numbers.

(*a*) Let **S** consist of all sets of form $(-\infty, a)$, and all sets of the form $(b, \infty)$, for arbitrary real numbers $a$, $b$. Show that **S** forms a subbase of a topology on **R**.

(*b*) Let **B** consist of $\Phi$ and all right open intervals $[a, r)$, where $a \in \mathbf{R}$ and $r$ is rational $(a < r)$. Show that

(*i*) **B** forms a base of a topology on **R**;

(*ii*) Every member of **B** is both open and closed in the resulting topological space $(\mathbf{R}, \tau)$.

(*c*) Let **B** consist of $\Phi$ and all left-closed improper intervals $[a, \infty)$, $a \in \mathbf{R}$. Show that **B** forms a base of a topology on **R**. (The resulting topology $\tau$ is called the right topology on **R**). Show that in $(\mathbf{R}, \tau)$.

(*i*) the intersection of any collection of open sets is open;

(*ii*) the closure of $\{a\}$ is the interval $(-\infty, a)$, for any $a \in \mathbf{R}$.

(*d*) Let **B** consist of $\Phi$ and all closed intervals $[a, b]$, $a, b \in \mathbf{R}$ and $a < b$. Show that **B** forms a base of a topology on **R**, which is different from the usual topology on **R**.

(*e*) Let $\tau$ consist of

(*i*) all those subsets of **R** which do not contain 0, and

(*ii*) the four subsets $\mathbf{R} - \{1, 2\}$, $\mathbf{R} - \{1\}$, $\mathbf{R} - \{2\}$ and **R**. Show that $(\mathbf{R}, \tau)$ forms a topological space.

(*f*) Let $\sigma$ denote the topology of countable complements on **R**, and let **T** be the subset formed by the irrational numbers. Prove that:

(*i*) Every rational number is a limiting point of the set **T**.

(*ii*) No rational number is the limit of a sequence of irrational numbers.

(*iii*) The identity mapping $i$ of $(\mathbf{R}, \sigma)$ onto the real number space $(\mathbf{R}, \tau)$ maps convergent sequences to convergent sequences, although $i$ is not continuous.

**10.** Let $X$ be any non-empty set.

(*a*) Let $A \subset X$, and let $\tau$ consist of $\Phi$ and all those subsets of $X$ which contain $A$. Show that $(X, \tau)$ is a topological space. Describe the closure and the interior operations in $(X, \tau)$.

[If, in particular, $X = \{0, 1\}$, and $A = \{0\}$, then $(X, \tau)$ is called a *Sierpinski space*].

(*b*) Let $p$ be an arbitrary, but fixed point, of $X$, and let $\tau$ consist of $X$ and all those subsets of $X$ which do not contain $p$. Show that $(X, \tau)$ is a topological space.

(*c*) A closure operator $A \to \overline{A}$ is defined for the subsets $A$ of $X$ by $\overline{\Phi} = \Phi$, and $\overline{A} = X$ when $A \neq \Phi$.

Verify the closure axioms, and show that the resulting topology is the discrete topology on $X$.

(*d*) Define a closure operator for subsets of $X$, for which the resulting topology is the indiscrete topology on $X$.

(*e*) (*i*) Let $B$ be a fixed subset of $X$, and let a closure operator be defined for subsets of $X$ by $\overline{\Phi} = \Phi$ and $\overline{A} = A \cup B$ when $A \neq \Phi$.

Verify the closure axioms, and describe the open sets in the resulting topology on $X$.

(*ii*) If, in particular, $B = \{p\}$, then show that the only one-pointic closed set is $\{p\}$, and that the other one-pointic sets are open but not closed.

(*f*) Show that Kuratowski's four closure axioms may be replaced by the following single one :

$$A \cup \overline{A} \cup \overline{\overline{B}} = \overline{A \cup B} - \overline{\Phi}, \text{ for arbitrary subsets } A, B \text{ of } X.$$

(*g*) Let **P** be a partition of $X$ into disjoint subsets. Let **B** consist of $\Phi$ and the disjoint subsets of the partition **P**. Show that **B** forms a base of a topology on $X$. (The resulting topology is known as a *partition topology on X*).

(*h*) The family of all closed sets in a topological space $(X, \tau)$ does not, in general, form any topology on $X$; but it does so if $X$ is a finite set.

**11.** Let $X$ be an arbitrary infinite set.

(*a*) Let $p$ be an arbitrary, but fixed, point of $X$. Let $\tau$ consist of; (1) all those subsets of $X$ which do not contain $p$, and (2) all those subsets of $X$ which are complements of finite subsets. Show that $(X, \tau)$ is a topological space. (It is known as a *Fort space*). Prove the properties in $(X, \tau)$ as follows.

(*i*) $\{p\}$ is a closed set, but not open.

(*ii*) Every one-pointic set, different from $\{p\}$ is both open and closed.

(*iii*) No two closed infinite sets are disjoint.

(*iv*) The family consisting of the sets $\{x\}$ for every $x \neq p$, the sets $X - A$, where $A$ is a finite subset, and $\Phi$ forms an open base of the topology $\tau$.

(*v*) The family consisting of the sets $\{x\}$ for $x \neq p$, and the sets $X - \{x\}$ form a subbase of the topology $\tau$.

(*vi*) For any subset $A$ of $X$,

$\overline{A} = A$ if $A$ is finite, and $\overline{A} = A \cup \{p\}$ if $A$ is an infinite set; $\text{Int}\,(A) = A$ if $A$ is finite, and $\text{Int}\,(A) = A - \{p\}$, if $A$ is infinite.

(*b*) Let $X$ be an uncountable set, and $p$ be an arbitrary, but fixed, point in $X$. Let $\tau$ consist of

(*i*) all those subsets of $X$ which do not contain $p$, and

(*ii*) all those subsets of $X$ which are complements of countable subsets. Show that $(X, \tau)$ is a topological space. (It is called a *Fortissimo space*).

(*c*) Let $X$ be the union of an infinite set $K$ and two distinct one-pointic sets $\{p\}$ and $\{q\}$, disjoint with $K$. Let $\tau$ consist of

(*i*) all subsets of $K$, and

(*ii*) all those subsets $G$ of $X$, such that $G$ contains $p$ or $q$ and all but a finite number points of $K$. Show that $(X, \tau)$ is a topological space. (It is known as a *modified Fort space*).

(*d*) Let a closure operator be defined for the subsets $A$ of $X$

by $\quad \overline{A} = A$ when $A$ is a finite subset of $X$,

and $\quad \overline{A} = X$ when $A$ is an infinite subset of $X$.

Verify the closure axioms, and show that the resulting topology is the cofinite topology on $X$.

**12.** (*a*) Let $\mathbf{B}\,(x)$ denote the local base at the point $x$ in a topological space $(X, \tau)$. Prove that

*LB* (1) : $\mathbf{B}\,(x) \neq \Phi$, and $x \in U$ for every $U \in \mathbf{B}\,(x)$.

*LB* (2) : If $U \in \mathbf{B}\,(x)$, then there exists a $V \in \mathbf{B}\,(x)$, such that $V \subset U$.

*LB* (3) : If $U \in \mathbf{B}\,(x)$ and $V \in \mathbf{B}\,(x)$, then there exists a $W \in \mathbf{B}\,(x)$, such that $W \subset U \cap V$.

Conversely, let $X$ be a non-empty set, and let us associate with each point $x \in X$ a family $\mathbf{B}(x)$ of subsets of $X$, satisfying the properties $LB$ (1)-$LB$ (3). Let $\tau$ denote the family of sets which are unions of subfamilies of the family $\cup \{\mathbf{B}(x) : x \in X\}$. Then show that $\tau$ is a topology on $X$, and that $\mathbf{B}(x)$ forms a local base at the point $x$ in the topological space $(X, \tau)$.

(*b*) Let $X = R^I$, be the set of all real-valued functions on the closed unit interval $I$.

(*i*) For each $f \in X$, each finite subset $A$ of $I$, and each positive real number $\delta$, let $U(f, A, \delta) = \{g \in X : |g(x) - f(x)| < \delta$, for each $x \in A\}$. Show that the sets $U(f, A, \delta)$ form a local base at $f$.

(*ii*) For each $f \in X$, and each $\varepsilon > 0$, let

$$V(f, \varepsilon) = \{g \in X : |g(x) - f(x)| < \varepsilon, \text{ for each } x \in I\}.$$

Show that the sets $V(f, \varepsilon)$ form a local base at $f$.

**13.** (*a*) Show that the following properties are satisfied by the interior operator in a topological space $(X, \tau)$:

$I(1)$: $\quad \text{Int}(A) \subset A$;

$I(2)$: $\quad \text{Int}(A \cap B) = \text{Int}(A) \cap \text{Int}(B)$;

$I(3)$: $\quad \text{Int}(X) = X$;

$I(4)$: $\quad \text{Int}[\text{Int}(A)] = \text{Int}(A)$;

where $A, B$ are arbitrary subsets of $X$.

Conversely, let $X$ be a non-empty set, and let $A \to \text{Int}(A)$ be an operator defined for every subset $A$ of $X$, satisfying the properties $I(1)$-$I(4)$. Let $\tau = \{A \subset X : \text{Int}(A) = A\}$. Show that $\tau$ forms a topology on $X$, and that the interior of a subset $A$ in the topological space $(X, \tau)$ is the set $\text{Int}(A)$.

(*b*) Let $X$ be an arbitrary set with more than one element. An operator $A \to \text{Int}(A)$, is defined for every subset $A$ of $X$ by

$$\text{Int}(X) = X, \text{ and } \text{Int}(A) = \Phi \text{ when } A \neq X.$$

Verify that the properties $I(1)$-$I(4)$ are satisfied; and show that the induced topology is the indiscrete topology on $X$.

**14.** Let $X$ be a non-empty set, and let $A \to A'$ be an operator, defined for every subset $A$ of $X$, satisfying the following axioms (due to F. Riesz):

$D(1)$: $\quad (A \cup B)' = A' \cup B'$;

$D(2)$: $\quad A'' \subset A'$;

$D(3)$: $\quad (\{x\})' = \Phi$, for every $x \in X$,

$A, B$ being arbitrary subsets of $X$.

For every subset $A$ of $X$, we now define a closure operator, $A \to \overline{A}$, by $\overline{A} = A \cup A'$. Verify that this closure operator satisfies Kuratowski's closure axioms, and that if $\tau$ denotes the resulting closure topology on $X$, then the derived set of a subset $A$ in the topological space $(X, \tau)$ is the same as the set $A'$.

**15.** Let $(X, \tau)$ be a topological space, and $A, B, D, F, G, H, \ldots$ are subsets of $X$. Then prove that :

(*a*) If $G$ is an open set, and $G \cap A = \Phi$, then $G \cap \overline{A} = \Phi$.

(*b*) If $G$ is an open set, and $D$ a dense subset, then

$$\overline{G \cap D} = \overline{G}, \text{ (and in this case, } \overline{G \cap D} = \overline{G} \cap \overline{D}).$$

(*c*) If $G$ is an open set, then $\text{Fr}(G) = \overline{G} - G$.

(*d*) $G$ is an open set if, and only if, $G \cap \text{Fr}(G) = \Phi$.

(*e*) $G$ is an open set if, and only if, $\text{Fr}(G)$ is closed and nowhere dense.

(*f*) If $G$ and $H$ are open sets, then $\text{Int}(\overline{G \cap H}) = \text{Int}(\overline{G} \cap \overline{H})$.

(*g*) If $G$ and $H$ are open sets, and $G \cap H = \Phi$, then

$$\overline{G} \cap H = \Phi, \text{ and } G \cap \overline{H} = \Phi.$$

(*h*) $A$ is both open and closed if, and only if, $\text{Fr}(A) = \Phi$.

(*i*) If $G$ is an open set, then $G \subset \text{Int}(\overline{G})$, and

if $F$ is a closed set, then $\overline{\text{Int}(F)} \subset F$.

(*j*) If $A$ has no isolated points, then the same is true for $\overline{A}$.

(*k*) If $A$ is nowhere dense, then $\text{Int}(A) = \Phi$; is the converse true?

(*l*) $\text{Int}(A \cup B) \subset \text{Int}(A) \cup \text{Int}(B)$.

(*m*) $\text{Int}(A) = A - \text{Fr}(A)$; $\overline{A} = A \cup \text{Fr}(A)$.

(*n*) $\text{Int}[A \cap \text{Fr}(A)] = \Phi$. Is $\text{Int}[\text{Fr}(A)] = \Phi$ ?

(*o*) $\text{Int}(A) \cup \text{Fr}(A) \cup \text{Ext}(A) = X$, and $\text{Int}(A) \cap \text{Fr}(A) \cap \text{Ext}(A) = \Phi$.

(*p*) $\text{Fr}(A \cup B) \subset \text{Fr}(A) \cup \text{Fr}(B)$; $\text{Fr}(A \cap B) \subset \text{Fr}(A) \cup \text{Fr}(B)$.

(*q*) $\text{Fr}[\text{Fr}(A)] \subset \text{Fr}(A)$; $\text{Fr}(\overline{A}) \subset \text{Fr}(A)$; $\text{Fr}[\text{Int}(A)] \subset \text{Fr}(A)$.

(*r*) $(A \cap B) \cap \text{Fr}(A \cap B) = (A \cap B) \cap [\text{Fr}(A) \cup \text{Fr}(B)]$.

**16.** (*a*) Let $(X, \tau)$ be a topological space, and $A_\nu \subset X$, $\nu \in I$, $I$ being an indexing set. Show that

$$\overline{\cap A_\nu} \subset \cap \overline{A_\nu}, \text{ and } \overline{\cup A_\nu} \supset \cup \overline{A_\nu}.$$

(*b*) Give an example of a topological space, in which the closure operator is completely additive, that is, where $\overline{\cup A_\nu} = \cup \overline{A}_\nu$, $\nu \in I$, $I$ being an (arbitrary) indexing set.

**17.** Let $(X, \tau)$, $(Y, \tau')$ and $(Z, \tau'')$ be three topological spaces, and $f : X \to Y$ and $g : Y \to Z$ be two mappings. Show that :

(*a*) If $f$ and $g$ are both open (respectively both closed), then so also is $g.f.$

(*b*) If $g.f.$ is open (respectively closed) and $f$ is continuous and surjective, then $g$ is open (respectively closed).

(*c*) It $g.f.$ is open (respectively closed) and $g$ is continuous and injective, then $f$ is open (respectively closed).

**18.** Let $(X, \tau)$ and $(Y, \tau')$ be two topological spaces, and $f : X \to Y$. Prove the following :

(*a*) If $f$ is a closed mapping, then given any subset $B \subset Y$ and any open set $G$ containing $f^{-1}(B)$, there exists an open set $V$ containing $B$, such that $f^{-1}(V) \subset G$.

(*b*) If $f$ is an open mapping, then given any subset $B \subset Y$ and any closed set $F$ containing $f^{-1}(B)$, then there exists a closed set $K$ containing $B$, such that $f^{-1}(K) \subset F$.

(*c*) $f$ is an open mapping if, and only if, any one of the following conditions holds :

(*i*) $f^{-1}[\text{Fr}(B)] \subset \text{Fr}[f^{-1}(B)]$ for each $B \subset Y$.

(*ii*) For each $x \in X$ and a *nbd* $V$ of $x$ in $(X, \tau)$, there exists a *nbd* $W$ of $f(x)$ in $(Y, \tau')$, such that $W \subset f(V)$.

(*d*) $f$ is a continuous mapping if, and only if, any one of the following conditions holds :

(*i*) $f(A') \subset \overline{f(A)}$ for each $A \subset X$.

(*ii*) $\text{Fr}[f^{-1}(B)] \subset f^{-1}[\text{Fr}(B)]$ for each $B \subset Y$.

(*e*) If $f$ is continuous, then it remains continuous, if $X$ is given a finer topology and/or $Y$ is given a coarser topology.

(*f*) If $f$ is continuous, then $f : (X, \tau) \to (W, \sigma)$ is also continuous for all $W$, satisfying $Y \subset W$ and $\tau' = \sigma_Y$.

**19.** Prove the following :

(*a*) If $Y \subset X$, and $(X, \tau)$ be a topological space, then the inclusion mapping of $(Y, \tau_Y)$ into $(X, \tau)$ is continuous.

(*b*) Let $X = A \cup B$, and $f: (X, \tau) \to (Y, \tau')$, where $(X, \tau)$, $(Y, \tau')$ are two topological spaces. If $f|_A$ and $f|_B$ are continuous at a point $x \in A \cap B$, then $f$ is also continuous at $x$. (See Pasting Lemma in Chapter 9).

(*c*) Let $X = \{0, 1\}$, and

(*i*) $\tau_1$ be the topology on $X$, for which it is a Sierpinski space, and

(*ii*) $\tau_2$ be the discrete topology on $X$, and let $f$ be the identity mapping on $X$. Then $f: (X, \tau_1) \to (X, \tau_2)$ is not continuous, but $f^{-1}: (X, \tau_2) \to (X, \tau_1)$ is continuous.

**20.** (*a*) Let $f: (X, \tau) \to (Y, \tau')$. Show that $\tau$ is the discrete topology on $X$ if, and only if, $f$ is continuous for all topologies $\tau'$ on $Y$.

(*b*) Let $g: (Y, \tau') \to (X, \tau)$. Show that $\tau$ is the indiscrete topology on $X$ if, and only if, $f$ is continuous for all topologies $\tau'$ on $Y$.

(*c*) Let $(X, \tau)$ be a topological space, and $A$ be a non-empty, proper subset of $X$. A real-valued function $f$ on $X$, defined by

$$f(x) = 1 \text{ if } x \in A, \text{ and } f(x) = 0 \text{ if } x \in X - A,$$

is called the *characteristic function of A* in $(X, \tau)$. Show that:

(*i*) the characteristic function is continuous on $X$ if, and only if, $A$ is both open and closed in $(X, \tau)$;

(*ii*) the characteristic function is discontinuous on the boundary of $A$, *i.e.*, on Fr $(A)$.

(*d*) Let **2** denote the discrete space $\{0, 1\}$, and $I$ be the subspace $[0, 1]$ of the real number space $(\mathbf{R}, \tau)$. If $f$ denotes the characteristic function of the subset $\left[0, \frac{1}{2}\right]$ in $I$, then show that $f: I \to \mathbf{2}$ is surjective, open and closed, but not continuous.

(*e*) Let $(X, \tau)$ and $(Y, \tau')$ be two topological spaces, and $(Z, \tau'')$ be a subspace of $(Y, \tau')$. Show that, for a mapping $f: X \to Y$, $f: (X, \tau) \to (Z, \tau'')$ is continuous if, and only if, $f: (X, \tau) \to (Y, \tau')$ is continuous.

**21.** Let $X$ be an infinite set, $\tau$ the cofinite topology on $X$, and $f$ an injective mapping of $X$ into itself. Show that the mapping

$$f: (X, \tau) \to (Y, \tau_Y),$$

where $Y = f(X)$, is open and continuous.

**22.** Let $(\mathbf{R}, \tau)$ denote the real number space.

($a$) Let $I$ denote the subspace $[0, 1]$ of $(\mathbf{R}, \tau)$. Then the mapping $r : \mathbf{R} \to I$, defined by $r(x) = 0$ if $x \leq 0$,

$$r(x) = x \text{ if } 0 \leq x, \leq 1, \text{ and } r(x) = 1 \text{ if } x \geq 1,$$

is called a *retraction*. Show that the retraction mapping is closed but not open.

($b$) Let $A$ denote the subspace $(-1, 1)$ of $(\mathbf{R}, \tau)$. Show that the mapping $f : \mathbf{R} \to A$, defined by $f(x) = x/[1 + |x|]$, is a homeomorphism.

($c$) ($i$) All open intervals, including the unbounded intervals, with their respective relative topologies in the real number space $(\mathbf{R}, \tau)$, are homeomorphic to each other.

($ii$) All bounded closed intervals, having more than one point, with their respective relative topologies in the real number space $(\mathbf{R}, \tau)$, are homeomorphic to each other.

($iii$) Let $A = (0, 1)$ and $B = [0, 1]$, both having their respective relative topologies in $(\mathbf{R}, \tau)$. Then $A$ is homeomorphic to the subspace $(1/2, 2/3)$ of $B$, and $B$ is homeomorphic to the subspace $[1/4, 3/4]$ of $A$, but $A$ and $B$ are not homeomorphic.

(Thus the analogue of *Schroeder-Bernstein principle does not hold for homeomorphisms*).

[**Hints :** ($i$) (1) $f(x) = (x - a)/(b - a)$ is a homeomorphism of $(a, b)$ onto $(0, 1)$, (2) $g(x) = a/x$ is a homeomorphism of $(a, \infty)$ onto $(0, 1)$, (3) $h(x) = -x$ is a homeomorphism of $(-\infty, a)$ onto $(a, \infty)$, and (4) $k(x) = \tan^{-1} x$ is a homeomorphism of $(-\infty, \infty)$ onto $(-\pi/2, \pi/2)$.

($ii$) $f(x) = (x - a)/(b - a)$ is a homeomorphism of $[a, b]$ onto $[0, 1]$].

**23.** Let $\mathbf{R}$ denote the set of all real numbers. Let a mapping, $f : \mathbf{R} \to \mathbf{R}$, be defined by $f_a(x) = ax$, for some fixed real number $a \neq 0$. Then show that $f_a$ is a homeomorphism, for each one of the following topologies on $\mathbf{R}$ : (1) the indiscrete topology, (2) the discrete topology, (3) the cofinite topology, (4) the usual topology, (5) the topology of Example 11 ($a$), and (6) the topology of Example 10 ($e$-ii).

**24.** *Semicontinuity :* Let $(X, \tau)$ be a topological space, and $(\mathbf{R}, \sigma)$ the real number space (with usual topology).

A function $f : (X, \tau) \to (\mathbf{R}, \sigma)$ is said to be upper-semicontinuous if, for each $a \in R$, $f^{-1}(-\infty, a)$ is an open set in $(X, \tau)$.

A function $f: (X, \tau) \to (\mathbf{R}, \sigma)$ is said to be *lower-semicontinuous* if, for each $a \in \mathbf{R}$, $f^{-1}(a, \infty)$ is an open set in $(X, \tau)$.

Prove the following results :

(*a*) $f$ is continuous if, and only if, it is both upper-semicontinuous and lower semicontinuous.

(*b*) Let $f$ be the characteristic function of a subset $A$ in $(X, \tau)$. Then

(*i*) $f$ is upper-semicontinuous if, and only if, $A$ is closed, and

(*ii*) $f$ is lower semicontinuous if and only if, $A$ is open.

(*c*) (*i*) $f$ is upper-semicontinuous if, and only if, it is continuous with respect to the left-hand topology on $\mathbf{R}$, and

(*ii*) $f$ is lower-semicontinuous if, and only if, it is continuous with respect to the right-hand topology on $\mathbf{R}$.

(*d*) In particular, let $(X, \tau)$ be also the real number space $(\mathbf{R}, \sigma)$. Let $f: (\mathbf{R}, \sigma) \to (\mathbf{R}, \sigma)$ be defined by

$f(x) = 0$, if $x$ is irrational, and

$f(x) = 1/q$, if $x$ is a rational number $p/q$, with $(p, q) = 1$, $q > 0$.

Show that $f$ is upper-semicontinuous, but not lower-semicontinuous.

**25.** Let $X$ be a linearly ordered set, such that there exists an *l.u.b.*, for every subset bounded above, and let $\tau$ be the order (or interval) topology on $X$. Show that a subset $G$ is open in $(X, \tau)$ if, and only if, it is expressible as a union of a family of *disjoint* open intervals.

**26.** Let $(X, \tau)$, $(Y, \tau')$, $(Z, \tau'')$ and $(X_i, \tau_i)$, $(Y_i, \tau_i')$ for $i = 1, 2$, denote topological spaces.

(*a*) If $(X_i, \tau_i)$ is homeomorphic to $(Y_i, \tau_i')$ for $i = 1$ and 2, prove that $X_1 \times X_2$ is homeomorphic to $Y_1 \times Y_2$, with respect to their respective product topologies.

(*b*) If $f_i : (X_i, \tau_i) \to (Y_i, \tau_i')$ is continuous, for $i = 1, 2$, then the mapping, $h : X_1 \times X_2 \times Y_1 \times Y_2$, defined by $H(x_1, x_2) = (f_1(x_1), f_2(x_2))$, is also continuous, with respect to the product topologies for $X_1 \times X_2$ and $Y_1 \times Y_2$.

(*c*) Let $(Z, \tau'')$ be the topological product of $(X, \tau)$ and $(Y, \tau')$. If $A \subset X$ and $B \subset Y$, then show that

(*i*) $\text{Int}_X(A) \times \text{Int}_Y(B) = \text{Int}_Z(A \times B)$;

(*ii*) $(\overline{A})_X \times (\overline{B})_Y = \overline{(A \times B)}_Z$;

(*iii*) $\mathrm{Fr}_Z(A \times B) = [\mathrm{Fr}_X(A) \times (\overline{B})_Y] \cup [(\overline{A})_X \times \mathrm{Fr}_Y(B)]$.

[**Note :** The result (*ii*) can be extended to infinite products, but (*i*) and (*iii*) hold for finite products only].

(*d*) Prove that, with their respective product topologies

(*i*) $(X \times Y) \times Z$ is homeomorphic to $X \times (Y \times Z)$;

(*ii*) $X \times Y$ is homeomorphic to $Y \times X$;

(*iii*) $X \times Y$ is dense-in-itself if, and only if, at least one of the spaces $(X, \tau)$ and $(Y, \tau')$ is dense-in-itself.

[**Note :** The results (*i*) and (*ii*) can be extended to infinite products. Also,

1. it is possible that $X \times Y$ is homeomorphic to $X \times Z$ but $(Y, \tau')$ is not homeomorphic to $(Z, \tau'')$ and
2. it is possible that $(X, \tau)$ and $(Y, \tau')$ are not homeomorphic, but $X \times X$ and $Y \times Y$ (with their respective product topologies) are homeomorhic].

(*e*) Let the mapping, $f : (X, \tau) \times (Y, \tau') \to (Z, \tau'')$, be continuous.

(*i*) For each $x \in X$, let $g_x : (Y, \tau') \to (Z, \tau'')$ be defined by $g_x(y) = f(x, y)$. Show that $g_x$ is continuous for each $x \in X$.

(*ii*) For each $y \in Y$, let $g_y : (X, \tau) \to (Z, \tau'')$ be defined by $g_y(x) = f(x, y)$. Show that $g_y$ is continuous for each $y \in X$.

The above results may be stated in the form : If $f$ is continuous in both the variables $x$ and $y$, then it is also continuous in each of the variables $x$ and $y$. Is the converse of this true?

**27.** Let $X$ be the set of all pairs of real numbers $(x, y)$ (*i.e.*, the set of all points on the Euclidean plane). Show that

(*i*) The family of all subsets of the form

$$\{(x, y) : a \le x < b \,.\, c \le y < d\},\ a, b, c, d$$

being real numbers, forms a base of a topology $\tau$ on $X$.

(*ii*) $(X, \sigma)$ is the topological product $(\mathbf{R}, \tau_1) \times (\mathbf{R}, \tau_1)$, where $\tau_1$ is the lower limit topology on the set of real numbers $\mathbf{R}$ (that is, $(\mathbf{R}, \tau_1)$ is the Sorgenfrey line). [The topology $\sigma$ is known as *Sorgenfrey's half-open square topology*].

**28.** *Cantor discontinuum.* Cantor discontinuum or Cantor's ternary set $K$ is defined to be the set of all those real numbers $a$, which are expressible in the form, $a = a_1/3 + a_2/3^2 + \ldots + a_n/3^n + \ldots$, where $a_1, a_2, \ldots a_n, \ldots$ assume the values 0 and 2 only (vide page 22). Let $\tau$ denote the subspace topology on $K$ in the real number space.

Let $A$ denote the set $\{0, 2\}$, and $\sigma$ be the discrete topology on $A$. Let the topological product $(A, \sigma) \times (A, \sigma) \times \ldots$ (denumerable copies) be denoted by $(C, \tau')$.

(*a*) Show that the mapping $f: (C, \tau') \to (K, \tau)$, defined by

$$f(\{a_i\}) = \sum_{i=1}^{\infty} a_i/3^i,$$

is a homeomorphism.

(*b*) Prove the following :

(*i*) If $I$ denotes the closed unit interval $[0, 1]$, with the subspace topology in the real number space, then

$$g(\{a_i\}) = \sum_{i=1}^{\infty} a_i/2^{i+1}$$

is a continuous surjection of $(C, \tau')$ onto $I$.

(*ii*) For any given positive integer $n$, there exists a continuous surjection of $(C, \tau')$ onto $I^n$.

(*iii*) There exists a continuous surjection of $I$ onto $I^n$, for every positive integer $n$. (This proves the existence of a *space-filling curve*; such a curve is known as a *Peano curve*).

(*c*) Show that $K$ is a nowhere dense, perfect set in $I$, and that $K$ has a non-empty derived set.

**29.** Let $X = [0, 2\pi]$, with its subspace topology in the real number space, and $Y = \{(x, y) \in \mathbf{R}^2 : x^2 + y^2 = 1\}$ with the subspace topology in the Euclidean plane $\mathbf{R}^2$. Let a mapping $f: X \to Y$ be defined by

$$f(x) = (\cos x, \ \sin x).$$

Show that $f$ is a continuous, closed surjection of $X$ onto $Y$ (but not an open mapping), so that $f$ is a quotient map and $Y$ is a quotient space of $X$.

**30.** Let $X$ denote the product space $[0, 2\pi] \times [0, 2\pi]$, where $[0, 2\pi]$ has the subspace topology in the real number space. We obtain different quotient spaces, which have interesting properties in Algebraic Topology, from $X$ as follows:

(*a*) **The cylinder** $S^1 \times [0, 2\pi]$ is obtained from $X$,

(*i*) by the quotient map $f(x, y) = ((\cos x, \sin x), y)$, or

(*ii*) by identifying each point $(0, y)$ with the point $(2\pi, y)$.

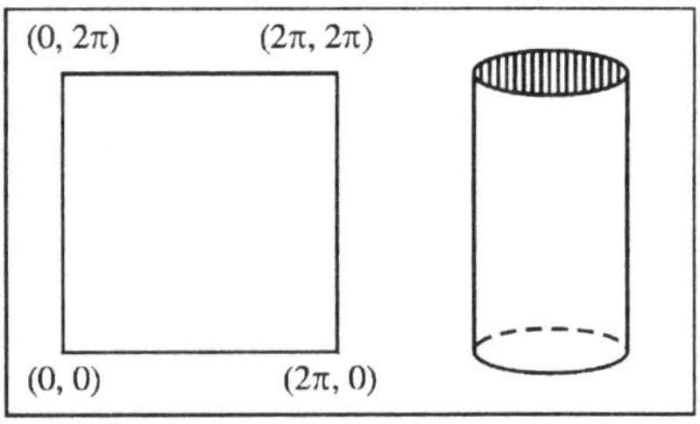


(*b*) The **torus** $S^1 \times S^1$ is obtained from $X$,

(*i*) by the quotient map $f(x, y) = ((\cos x, \sin x), (\cos y, \sin y))$, or

(*ii*) by identifying each point $(0, y)$ with the point $(2\pi, y)$, and also each point $(x, 0)$ with the point $(x, 2\pi)$.

(*c*) The **Mobius strip** is obtained from $X$, by identifying each point $(x, 0)$ with the point $(2\pi - x, 2\pi)$.

(*d*) The **Klein's bottle** is obtained from $X$ by identifying each point $(0, y)$ with the point $(2\pi, y)$, and also each point $(x, 0)$ with the point $(2\pi - x, 2\pi)$.

**31.** **Cones :** Let $X$ be any topological space, and $I$ the closed unit interval $[0, 1]$. Then the quotient space $(X \times I)/\rho$, where $\rho$ is the equivalence relation on $X \times I$, defined by $((x, 1), (x', 1)) \in \rho$, for all $x, x' \in X$, is called the cone TX over $X$. Thus it is obtained by identifying all the points $(x, 1)$ in $X \times I$ with a single point.

Show that if $f: X \to Y$ is continuous then : $g: TX \to TY$, defined by $g(x, t) = (f(x), t)$, is also continuous.

**32.** **Suspensions :** Let $X$ be any topological space, and $H$ the closed interval $[-1, 1]$. Then the quotient space $(X \times H)/\rho$, where $\rho$ is the equivalence relation on $X \times H$, defined by $((x, 1), (x', 1)) \in \rho$, and $((x, -1), (x', -1)) \in \rho$, for all $x, x' \in X$, is called *the suspension SX over X*. Thus it is obtained by identifying all the points $(x, 1)$ in $X \times H$ to a single point, and all the points $(x, -1)$ to another point. Show that:

(*i*) $SX$ is homeomorphic to the quotient space $TX/X$.

(*ii*) If $f: X \to Y$ is continuous, then $g: SX \to SY$, defined by

$$g(x, t) = (f(x), t),$$

is also continuous.

**33.** **Attaching of spaces :** Let $X$ and $Y$ be two disjoint topological spaces, and $X + Y$ be their topological sum. Let $A$ be a closed subset of $X$, and $f: A \to Y$ be continuous. Then the quotient space $(X + Y)/\rho$ is said to be "*X attached to Y by f*", and is denoted by $X \cup_f Y$, where $\rho$ is the equivalence relation on $X + Y$, such that $(a, f(a)) \in \rho$, for all $a \in A$; $f$ is called the *attaching map.* Thus $X \cup_f Y$ is obtained from $X + Y$, by identifying each $a \in A$ with its image $f(a) \in Y$.

In particular, let $Y = \{p\}$, with its discrete topology. Then

(*i*) For a closed subset $A \subset X$, and $f(A) = \{p\}$, $X \cup_f \{p\}$ is homeomorphic to the quotient space $X/A$.

(*ii*) (*a*) The cone $TX$ is obtained by attaching $X \times I$ to a point $\{p\}$ by $f(X \times I) = \{p\}$.

(*b*) The suspension $SX$ is obtained by attaching $X \times [-1, 1]$ to $\{p\} \cup \{q\}$ by $f(X \times 1) = \{p\}$ and $f(X \times (-1)) = \{q\}$.

*Chapter 3*

# SEPARATION AXIOMS

## Art. 1 SEPARATION BY OPEN SETS

**Weak and strong separation :** Let $(X, \tau)$ be a topological space. Two non-empty subsets $A$ and $B$ are said to be

(*i*) *Weakly separated*, or (simply) *separated* in $(X, \tau)$ if there exist two open sets $U$ and $V$, such that

$$A \subset U,\ B \subset V,\ A \cap V = \Phi, \text{ and } B \cap U = \Phi;$$

(*ii*) Strongly separated in $(X, \tau)$, if there exist two open sets $G$ and $H$, such that $A \subset G$, $B \subset H$, and $G \cap H = \Phi$.

It now follows readily from the definitions that :

**Theorem 1 :** *If two subsets are strongly separated, then they are also weakly separated.*

**Theorem 2 :** *Two non-empty subsets A and B are weakly separated in a topological space* $(X, \tau)$ *if, and only if, any one of the following conditions holds :*

(*i*) $A \cap B = \Phi$, $A \cap B' = \Phi$ and $A' \cap B = \Phi$;

(*ii*) $(\overline{A} \cap B) \cup (A \cap \overline{B}) = \Phi$. *(Hausdorff-Lennes condition)*

**Proof :** Let $A$, $B$ be weakly separated by the open sets $U$, $V$. Then $A \subset U$ and $B \cap U = \Phi$ imply $A \cap B = \Phi$. Also $U$ is an open set, containing $A$, and $U$ does not intersect $B$, hence no point in $A$ can be an accumulation point of $B$, that is, $A \cap B' = \Phi$. Similarly, $A' \cap B = \Phi$. Thus (*i*) holds.

Next, let (*i*) hold. $\overline{A} \cap B = (A \cup A') \cap B = (A \cap B) \cup (A' \cap B) = \Phi$, and similarly, $A \cap \overline{B} = \Phi$. Hence $(\overline{A} \cap B) \cup (A \cap \overline{B}) = \Phi$. Thus (*i*) implies (*ii*).

Finally, let (*ii*) hold. Then $\overline{A} \cap B = \Phi$ and $A \cap \overline{B} = \Phi$, so that

$$B \subset X - \overline{A} \text{ and } A \subset X - \overline{B},$$

Now taking $U = X - \overline{B}$ and $V = X - \overline{A}$, we get $A \subset U$, $B \subset V$, $A \cap V = \Phi$ and $B \cap U = \Phi$; hence $A$ and $B$ are weakly separated in $(X, \tau)$.//

**Corollary :** *If C and D are non-empty subsets of two weakly separated subsets A and B respectively, then C and D are also weakly separated in* $(X, \tau)$.

*And, similar result holds for strong separation also.*

**Proof :** $C \subset A \subset U$, $D \subset B \subset V$, $C \cap V \subset A \cap V = \Phi$, and $D \cap U \subset B \cap U = \Phi$. This proves the first part of the corollary. The second part follows from the fact that $C \subset A \subset G$, $D \subset B \subset H$, and $G \cap H = \Phi$.//

## Art. 2 SEPARATION AXIOMS AND $T_t$-SPACES

We shall now impose certain conditions on the topology to obtain some particular types of topological spaces. These conditions, called *Trennungs-axiome* or *separation axioms* by Alexandroff and Hopf, are as follows :

**Axiom $T_0$** (due to *Kolmogoroff*) : *For every pair of distinct points in* $(X, \tau)$, *there exists on open set, which contains one only of the two given points.*

Or, equivalently, *for every pair of distinct points, there exists a nbd of one of these points which does not contain the other point.*

**Axiom $T_1$** (due to *Frechet*) : *For every pair of distinct points p, q in X, there exist two open sets U, V, such that*

$$p \in U,\ q \in V,\ p \notin V, \text{ and } q \notin U.$$

In other words, *every pair of distinct points is weakly separated in* $(X, \tau)$.

Or, equivalently, *for every pair of distinct point p, q in X, there exist a nbd of p which does not contain q, and a nbd of q which does not contain p.*

**Axiom $T_2$** (due to *Hausdorff*) : *Any two distinct points are strongly separated in* $(X, \tau)$.

Or, equivalently, *distinct points have disjoint nbds.*

**Axiom $T_3$** (due to Vietoris) : *Any closed set F and any point p, not in F, are always strongly separated in* $(X, \tau)$.

**Axiom $T_4$** (due to *Tietze*) : *Any two disjoint closed sets are strongly separated in* $(X, \tau)$.

**Axiom $T_5$ :** *If two sets are weakly separated, then they are also strongly separated in* $(X, \tau)$.

A topological space $(X, \tau)$ is said to be a

(*i*) **$T_0$-space** if, the axiom $T_0$ holds in it;

(*ii*) **$T_1$-space** if, the axiom $T_1$ holds in it (it is also known as a **Frechet space**);

(*iii*) **Hausdorff space** if, the axiom $T_2$ holds in it;

(*iv*) **regular space** if, the axiom $T_3$ holds in it;

(*v*) **normal space** if, the axiom $T_4$ holds in it;

(*vi*) **completely normal space** if, the axiom $T_5$ holds in it.

Following Alexandroff and Hopf, a $T_1$-space $(X, \tau)$ is called a $Ti$-space if the axiom $T_i$ holds in it for $i = 2, 3, 4$ and 5. Thus

(*vii*) a $T_1$ Hausdorff space is a $T_2$-*space*;

(*viii*) a $T_1$ regular space is a $T_3$-*space*;

(*ix*) a $T_1$ normal space is a $T_4$-*space*;

(*x*) a $T_1$ completely normal space is a $T_5$-*space*.

If $(X, \tau)$ is a $T_i$-space, then $\tau$ is, correspondingly, called a $T_i$ topology. Likewise, $\tau$ is called a Hausdorff, regular, normal, or a completely normal topology, according as $(X, \tau)$ is a Hausdorff, regular, normal, or a completely normal space respectively.

## 2.1 Mutual Dependence

It now follows readily from the above definitions that :

**Theorem 3 :** (*i*) *Every $T_1$-space is a $T_0$-space.*

(*ii*) *Every Hausdorff space is a $T_1$-space, and is, therefore, a $T_2$-space.*

(*iii*) *Every completely normal space is a normal space.*

(*iv*) *Every $T_5$-space is a $T_4$-space.*

**Proof :** (*i*) The result follows immediately from the axioms $T_0$ and $T_1$.

(*ii*) The result is an immediate consequence of Theorem 1.

(*iii*) Let $(X, \tau)$ be a completely normal space, and $P$ and $Q$ be a pair of disjoint closed sets in it. Then $P$ and $Q$ are weakly separated by the two open sets $X - Q$ and $X - P$ in $(X, \tau)$. Since $(X, \tau)$ is a completely normal space, the two weakly separated sets $P$ and $Q$ are also strongly separated in $(X, \tau)$. Hence $(X, \tau)$ is a normal space.

(*iv*) The result follows directly from (*iii*).//

But a regular space may not be Hausdorff, it may not even be a $T_0$-space; and a normal space may not be regular and may not even be a $T_0$-space, as can be seen from the following two examples.

**Example 1.** ***Example of a regular space which is not a $T_0$-space, and is therefore neither a $T_1$-space nor a Hausdorff space.***

**Solution :** Let $X = \{a, b, c\}$, and $\tau = \{\phi, \{a\}, \{b, c\}, X\}$. Then $\tau$ constitutes a topology on $X$ (verify). There are just 4 closed sets in $(X, \tau)$, viz. $\phi$, $\{a\}$, $\{b, c\}$ and $X$. If $F$ denotes a closed set and $p$ a point lying outside $F$, then there exist just 3 such pairs $[F, p]$ in $(X, \tau)$; and each of these 3 pairs is separated strongly in $(X, \tau)$ as follows. Both the pairs $[\{a\}, b]$ and $[\{a\}, c]$ are separated strongly by the two disjoint open sets $\{a\}$ and $\{b, c\}$; and the pair $[\{b, c\}, a]$ is separated strongly by the two disjoint open sets $\{b, c\}$ and $\{a\}$. Hence $(X, \tau)$ is a *regular space.*

But the axiom $T_0$ is not satisfied for the pair of distinct points $b$ and $c$ in $(X, \tau)$. In fact, any open set that contains $b$ also contains $c$, and any open set that contains $c$ also contains $b$. Consequently, $(X, \tau)$ is not a $T_0$-*space*. Hence, by Theorem 3 (*i*) and (*ii*), $(X, \tau)$ can *neither* be *a $T_1$-space nor a Hausdorff space.*

**Example 2.** ***Example of a normal space which is not a regular space, and not even a $T_0$-space, and therefore neither a $T_1$-space nor a Hausdorff space.***

**Solution :** Let $X = \{a, b, c, d, e, f\}$, and $\tau = \{\phi, \{e\}, \{f\}, \{e, f\}, \{a, b, e\}, \{c, d, f\}, \{a, b, e, f\}, \{c, d, e, f\}, X\}$. Then $(X, \tau)$ is a topological space (verify). There are just 4 pairs of disjoint closed sets in $(X, \tau)$; viz. $[\{a, b\}, \{c, d\}]$, $[\{a, b\}, \{c, d, f\}]$, $[\{a, b, e\}, \{c, d\}]$, and $[\{a, b, e\}, \{c, d, f\}]$, and each of these pairs is separated strongly by the pair of disjoint open sets $\{a, b, e\}$ and $\{c, d, f\}$. Hence $(X, \tau)$ *is a normal space.*

But the closed set $\{a, b\}$ and the point $e$, lying outside it, are not strongly separated; hence $(X, \tau)$ is *not a regular space.*

Also, the axiom $T_0$ does not hold for the pair of distinct points $a$ and $b$. Hence $(X, \tau)$ is *not a $T_0$-space.* Consequently, it can *neither* be a *$T_1$-space nor a Hausdorff space.*

It is however, true that every $T_3$-space is a $T_2$-space, and every $T_4$-space is a $T_3$-space. This will be proved below in theorem 4. The following Lemma is required for the purpose.

**Lemma :** *A topological space* $(X, \tau)$ *is a $T_1$-space if, and only if, every one-pointic set is closed in* $(X, \tau)$.

**Proof :** Let $(X, \tau)$ be a $T_1$-space, and $\{x\}$ a one-pointic subset of $X$. Let $y$ be any point in $X$ other than $x$. Then by axiom $T_1$, there exists an open set $V$, such that $y \in V$ and $x \notin V$. Hence, $y$ cannot be an accumulation point of the set $\{x\}$; consequently, $\{x\}$ is a closed set. Thus every one-pointic set is closed $(X, \tau)$.

Conversely, let $(X, \tau)$ be a topological space, in which every one-pointic set is closed. Let $x$ and $y$ be two distinct points in $X$. Then $x$ and $y$ are weakly separated by the two open sets $X - \{y\}$ and $X - \{x\}$ in $(X, \tau)$. Hence $(X, \tau)$ is a $T_1$-space.//

**Theorem 4 :** (*i*) *Any $T_3$-space is a $T_2$-space, and*

(*ii*) *Any $T_4$-space is a $T_3$-space.*

**Proof :** (*i*) Let $(X, \tau)$ be a $T_3$-space, and let $x, y$ be any two distinct points in $(X, \tau)$. As $(X, \tau)$ is a $T_1$-space, the one pointic set $\{x\}$ is closed in $(X, \tau)$. As $y$ is a point lying outside the closed set $\{x\}$, $\{x\}$ and $y$ can be strongly separated in $(X, \tau)$, since the axiom $T_3$ holds in $(X, \tau)$. Thus the points $x$ and $y$ can be strongly separated in $(X, \tau)$; and $(X, \tau)$ is therefore a $T_2$-space.

(*ii*) Let $(X, \tau)$ be a $T_4$-space, and let $F$ be a closed set and $p$ a point lying outside $F$ in $(X, \tau)$. As $(X, \tau)$ is a $T_1$-space, the one pointic set $\{p\}$ is closed, and $F$ and $\{p\}$ are two disjoint closed sets in $(X, \tau)$. And, as the axiom $T_4$ holds in $(X, \tau)$, $F$ and $\{p\}$ can be strongly separated in $(X, \tau)$. Thus the closed set $F$ and the point $p$, lying outside $F$, can be strongly separated in $(X, \tau)$. Hence, $(X, \tau)$ is a $T_3$-space.//

Now, by combining Theorems 4 (*i*) and (*ii*) and 3 (*i*), (*ii*) and (*iv*), we obtain the following dependence among the $T_i$-spaces :

**Theorem 5 :** *Every $T_i$-space is necessarily a $T_{i-1}$-space, for $i = 1$, 2, 3, 4 and 5.*

On the other hand, it will be shown by suitable counter-examples that:

(*i*) a $T_{i-1}$-space is not, in general, a $T_i$-space, for $i = 1, 2, 3, 4$ and 5;

(*ii*) a topological space need not be a $T_0$-space;

(*iii*) a Hausdorff space need not be regular, or normal;

(*iv*) a regular space need not be normal, or completely normal, and

(*v*) a normal space need not be completely normal.

The above ones, and several other results on mutual dependence of the $T_i$ axioms, will be illustrated by suitable counter-examples. A few such counter-examples are given below, and others will be given later on.

## 2.2 Examples of Some $T_i$-Spaces

(*A*) Let $\tau_d$ be the discrete topology on a set $X$, with 2 or more elements. Then every subset of $X$ is both open and closed in $(X, \tau_d)$. Any two disjoint subsets $A$ and $B$ are separated strongly by the two disjoint open sets $A$ and $B$. Thus *every discrete space is a $T_5$-space.*

(*B*) Let $\tau_i$ be the indiscrete topology on a set $X$, with 2 or more elements. Then $\Phi$ and $X$ are the only open sets in $(X, \tau_i)$.

For any two distinct points $p, q \in X$, the only open set $X$, which contains $p$, also contains $q$, and conversely. Hence none of the axioms $T_0$, $T_1$ and $T_2$ holds in $(X, \tau_i)$.

There are only two closed sets, viz. $\Phi$ and $X$, in $(X, \tau_i)$. Hence there does not exist

(*i*) any pair $[F, p]$, where $F$ is a closed set and $p$ a point lying outside $F$,

(*ii*) any pair of disjoint (non-empty) closed sets, or

(*iii*) any pair of weakly separated sets in $(X, \tau_i)$. Hence the axioms $T_3$, $T_4$ and $T_5$ are, vacuously, satisfied in $(X, \tau_i)$.

Thus, *any indiscrete space is (vacuously) regular, normal and completely normal, but it is not* $T_0$, $T_1$, $T_2$, $T_3$, $T_4$ or $T_5$.

(*C*) (*a*) **Order (or interval) topology on a linearly ordered set.** Let $(X, \tau)$ be a linearly ordered set, which does not possess any greatest or least element, and in which every subset bounded above has a least upper bound. Let $\tau$ be the order (or interval) topology on $X$.

Let $p$ and $q$ be any two distinct points in $X$, where $p < q$ (say). As $p$ is not the least element in $X$, and $q$ is not the greatest element in $X$, there exist two elements $a, b \in X$, such that $a < p$ and $q < b$. Then $p$ and $q$ are separated (weakly) by the open intervals $(a, q)$ and $(p, b)$. Thus $(X, \tau)$ is a $T_1$-space.

We shall now show that the axiom $T_5$ also holds in $(X, \tau)$. Let $A$, $B$ be a pair of (weakly) separated sets in $(X, \tau)$, so that $A \cap \overline{B} = \overline{A} \cap B = \Phi$, *i.e.*, $A \subset X - \overline{B}$ and $B \subset X - \overline{A}$. The open set $G = X - (\overline{A} \cup \overline{B}) = (X - \overline{A}) \cap (X - \overline{B})$ can be expressed as a union of disjoint open intervals (vide Example 25, page 112),

$$G = \cup \{I_\nu = (a_\nu, b_\nu) : \nu \in K\},$$

where $K$ is an indexing set. Let $z_v$ be an arbitrary, but fixed, point in $I_v$, for each $v \in K$.

Let $p \in A$, then $p \in X - \overline{B}$. Hence $p$ is contained in some open interval in $X - \overline{B}$. We shall distinguish a special open interval $I_p$ in $X - \overline{B}$, containing $p$. We consider the following three different cases, for this purpose.

(*i*) There exists an element $x_p$, such that $p < x_p$ and $(p, x_p) = \Phi$; then we define $I_{pr} = [p, x_p)$. In this case,

$$I_{pr} = \{p\} \subset \overline{A} \cap (X - \overline{B}).$$

(*ii*) There exists an element $x_p$, such that $p < x_p$ and $[p, x_p) \subset \overline{A} \cap (X - \overline{B})$; then we define $I_{vr} = [p, x_p)$.

(*iii*) There exists an element $x_p \in G$ and an open interval $(a_v, b_v)$, such that $(a_v, b_v) \subset [p, x_p) \subset X - \overline{B}$; then we define $I_{pr} = [p, z_v)$.

Thus for every point $p \in A$, we can always find a right half open interval $I_{pr}$, of (at least) one of the types (*i*), (*ii*) and (*iii*), satisfying the conditions $p \in I_{pr} \subset X - \overline{B}$. In a similar way, we can find a left half open interval $I_{pl}$, such that $p \in I_{pl} \subset X - \overline{B}$, where $I_{pl} = (y_p, p]$ or $I_{pl} = (z_v, p]$ (for a suitable point $y_p$). Now, let $I_p = I_{pr} \cup I_{pl}$, then $p \in I_p \subset X - \overline{B}$. Let $U = \cup \{I_p : p \in A\}$; then $U$ is an open set in $(X, \tau)$, containing $A$.

Similarly, we can find an open set $V$ in $(X, \tau)$, containing $B$, where $V$ is the union of a family of open intervals,

$$V = \cup \{I_q = I_{qr} \cup I_{ql} : q \in B\},$$

$I_{qr}$ and $I_{ql}$ being defined as in the cases of $I_{pr}$ and $I_{pl}$.

We shall now show that $U$ and $V$ are disjoint sets. For this, it suffices to show that : (1) $I_{pr} \cap I_{qr} = \Phi$, $I_{pl} \cap I_{ql} = \Phi$, and (2) $I_{pr} \cap I_{ql} = \Phi$, for any $p \in A$ and $q \in B$. Without any loss of generality we may assume that $p < q$.

(1) Now, $I_{pr} \subset X - \overline{B}$ and $I_{ql} \subset X - \overline{A}$, imply $q \notin I_{pr}$ and $p \notin I_{ql}$, from which it follows that

$$I_{pr} \cap I_{qr} = I_{pl} \cap I_{ql} = \Phi.$$

(2) If $I_{pr}$ is of the type (*i*) or (*ii*), then $I_{pr} \subset \overline{A}$ and $I_{ql} \subset X - \overline{A}$, imply $I_{pr} \cap I_{ql} = \Phi$; and the same holds if $I_{ql}$ is of the type (*i*) or (*ii*). Next, let both $I_{pr}$ and $I_{ql}$ be of the type (*iii*), say $I_{pr} = [p, z_\mu)$ and $I_{ql} = (z_\nu, q]$. Now, if $x_p \le y_q$, then $z_\mu < z_\nu$, and if $y_q < x_q$, then $[p, x_p] \subset X - \overline{B}$ and $[y_q, q] \subset X - \overline{A}$, so that $[y_q, x_p] \subset G$, and therefore $z_\mu < z_\nu$. Hence in each case, $I_{pr} \cap I_{ql} = \Phi$.

Thus $U$ and $V$ are two disjoint open sets which separate $A$ and $B$ strongly. The axiom $T_5$, therefore, holds in $(X, \tau)$; and since it is also a $T_1$-space, it follows that $(X, \tau)$ is a $T_5$-space.

(*b*) **Ordinal spaces.** Let $\Gamma$ be an ordinal number, and $[0, \Gamma]$ consist of all ordinal numbers less than or equal to $\Gamma$. The set $[0, \Gamma]$, with the order (or interval) topology $\tau$ on it, is called the *closed ordinal space* $[0, \Gamma]$. This topology is generated by a base, consisting of all sets $(\alpha, \beta + 1) = (\alpha, \beta] = \{x : \alpha < x < \beta + 1\}$, where $\alpha, \beta + 1 \in [0, \Gamma]$. Thus $[\alpha, \beta)$ is an open set if, and only if, either $\alpha = 0$, or $\alpha$ has an immediate predecessor.

The *open ordinal space* $[0, \Gamma)$, consisting of all ordinal numbers which are strictly less than $\Gamma$, with the order (or interval) topology, is likewise defined.

The ordinal space (closed or open) with $\Gamma = \Omega$ (the first uncountable ordinal) is of particular importance.

It will be shown later that *both the open and the closed ordinal spaces are $T_5$-spaces.*

(*D*) *Let* $\mathbf{R}$ *denote the set of real numbers, and* $\tau$, $\tau_1$, $\tau_2$ *denote respectively the* (*i*) *usual (order) topology;* (*ii*) *the lower limit (or the upper limit) topology, and* (*iii*) *the left hand (or the right hand) topology on* $\mathbf{R}$. *Then* $(\mathbf{R}, \tau)$ *and* $(\mathbf{R}, \tau_1)$ *are* $\mathbf{T_5}$*-spaces; and the space* $(\mathbf{R}, \tau_2)$ *is* $T_0$, *normal and completely normal, but not regular,* $T_1$, $T_2$, $T_3$, $T_4$ *or* $T_5$.

(*i*) This property follows from (*C*) since $R$ is a linearly ordered set with the least upper bound property, and without any greatest or least element.

[It will be proved (in the theory of metric spaces) that any metric topology is always a $T_5$-topology, and, since the order (interval) topology $\tau$ on **R** is a metric topology (being induced by the Euclidean metric on **R**), it follows that (**R**, $\tau$) is a $T_5$ topology].

(*ii*) Let $\tau_1$ denote the lower limit topology on **R**. Then (**R**, $\tau_1$) is evidently a Hausdorff space, and hence also a $T_1$-space. In fact, any two distinct real numbers $p, q$, where $p < q$, are strongly separated in (**R**, $\tau_1$) by the two disjoint open sets $[p, p+t)$ and $[q, q+t)$, where $t = q - p$. Next, let $A$, $B$ be a pair of weakly separated subsets in (**R**, $\tau_1$). Let $a \in A$, then there exists a real number $u_\alpha$, such that $[a, u_a) \cap B = \Phi$ (since $a$ is not an accumulation point of $B$). Similarly, for each $b \in B$, there exists a real number $v_b$, such that $[b, v_b) \cap A = \Phi$. Then $[a, u_a) \cap [b, v_b) = \Phi$; otherwise, either $a \in [b, v_b)$ or $b \in [a, u_a)$, each of which is impossible. Let $U = \cup \{[a, u_a) : a \in A\}$, and $V = \cup \{[b, v_b) : b \in B\}$. Then $U$ and $V$ are two disjoint open sets, which separate $A$ and $B$ strongly in (**R**, $\tau_1$). Hence (**R**, $\tau_1$) is a $T_5$-space.

(*iii*) Let $\tau_2$ be the left hand topology on **R**. That is, $\tau_2$ consists of $\Phi$, **R** and the subsets $\{x : x < a\}$, for all $a \in$ **R**. The closed sets are therefore, given by $\Phi$, **R** and the subsets $\{x : a \leq x\}$, for all $a \in$ **R**. Hence, for any two distinct (non-empty) open sets, one must be contained in the other, and therefore no two (non-empty) subsets are weakly separated in (**R**, $\tau_2$). Also, no two (non-empty) closed sets are disjoint in (**R**, $\tau_2$). Consequently, the axioms $T_4$ and $T_5$ are vacuously, satisfied in (**R**, $\tau_2$). (**R**, $\tau_2$) is therefore a normal and a completely normal space.

For any two distinct points $p, q \in$ **R**, where $p < q$, the open set $\{x : x < q\}$ contains $p$, but not $q$. Thus the axiom $T_0$ holds in (**R**, $\tau_2$). Also, for every open set $\{x : x < r\}$, containing $q$, $q < r$, and therefore $p < q < r$; hence the open set also contains $p$. Thus the axiom $T_1$ does not hold in (**R**, $\tau_3$); consequently, the axiom $T_2$ also does not hold.

Since the axioms $T_0$ and $T_3$, taken together, imply the axiom $T_1$ (vide Theorem 17 (*ii*), page 139), it follows that the axiom $T_3$ does not hold in (**R**, $\tau_2$).

(*E*) **Half-disc topology.** Let $X$ denote the set of all points $(x, y)$ on the Euclidean plane $E^2$, for which $y \geq 0$. Let $L$ be the subset, consisting of all points on the $x$-axis, *i.e.*, all points for which $y = 0$. For each point $P$ in $X$ and each positive real number $r$, we define a set $U_r(P)$ as follows :

$$U_r(P) = S(P, r) \cap X \text{ if } P \in X - L \qquad \text{(Figure 1)}$$

and $$U_r(P) = [S(P, r) \cap (X - L)] \cup \{P\}, \text{ if } P \in L, \qquad \text{(Figure 2)}$$

where $S(P, r)$ denotes the open circle, consisting of all those points on the Euclidean plane whose distances from $P$ are less than $r$.

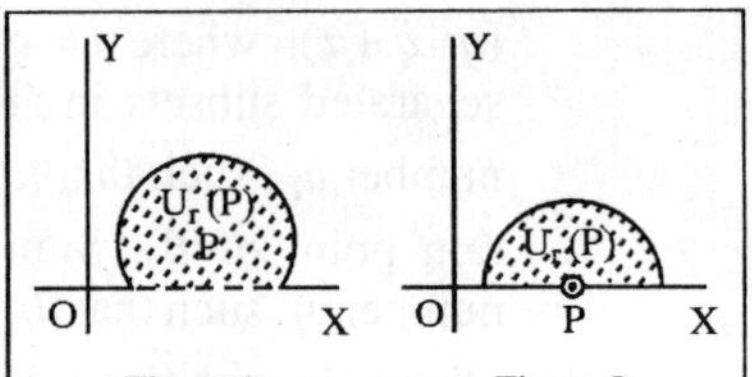

Figure 1 Figure 2

Then, for each point $P \in X$, the collection of all sets $U_r(P)$ constitutes a neighbourhood filter of $P$. Let $\tau$ be the resulting *nbd* topology on $X$; it is known as the *half-disc topology on X*. Then $(X, \tau)$ is a Hausdorff space, since any two distinct points $P, Q \in X$ are separated strongly by the two disjoint neighbourhoods $U_r(P)$ and $U_r(Q)$, where $r$ is any positive real number, less than or equal to half of the distance between the points $P$ and $Q$.

Again, no point $P$ lying outside $L$ is an accumulation point of $L$, since the *nbd* $U_r(P)$ of the point $P$, with $r$ less than the perpendicular distance of $P$ from $L$, does not intersect $L$. Also, if $T$ is any fixed point on $L$, then no *nbd* $U_r(T)$ of the point $T$ contains any point of $L$, other than $T$. Hence $L - T$ is a closed set, and $L - T$ and $T$ cannot be separated strongly in $(X, \tau)$. $(X, \tau)$ is therefore not a regular space.

Thus *the half-disc topology is* $T_0$, $T_1$ *and* $T_2$, *but it is not regular*, and hence *not normal, completely normal*, $T_3$, $T_4$ or $T_5$.

(*F*) **Niemytzki's tangent disc topology.** Let $X$ consist of all those points $(x, y)$ on the Euclidean plane $E^2$, for which $y \geq 0$. Let $L$ be the subset, consisting of all points on the $x$-axis, *i.e.*, all points for which $y = 0$. We now define :

$U_r(P) = S(P, r) \cap X$, where $S(P, r)$ denotes an open circle, with centre $P$ and radius $r > 0$, for each point $P \in X - L$ (Figure 1), and $U_r(P) =$ the union of $\{P\}$ and an open circle of radius $r > 0$, lying in $X$, the circumference of the circle touching $L$ at the point $P$, for each point $P \in L$ (Figure 2).

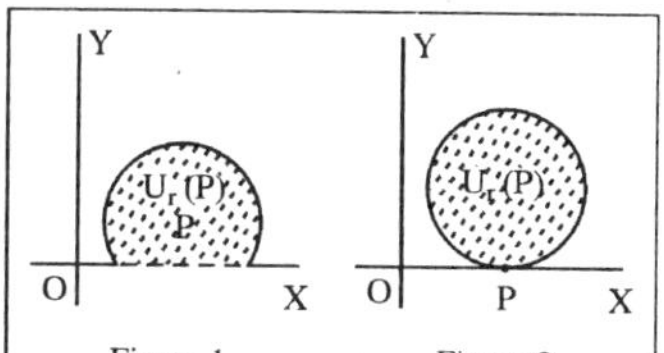


Figure 1 Figure 2

$U_r\,(P)$ consists of all points in the dotted region (inside the circumference or the arc of the circumference). It includes points on the $x$-axis intercepted by the arc in Figure 1. But it does not contain any point on the $x$-axis, excepting the point $P$ in Figure 2

It can be readily verified that, for each point $P \in X$, the family $\{U_r\,(P)\}$ forms a neighbourhood filter of the point $P$. Let $\tau$ be the resulting *nbd* topology on $X$; it is called Niemytzki's tangent disc topology on $X$. Then $(X,\ \tau)$ is a $T_2$-space (as in $(E)$).

$(X,\ \tau)$ is also a regular space, since for any *nbd* $U_r\,(P)$ of any point $P$, there exists a *nbd* $U_s\,(P)$ of $p$, such that $\overline{U_s\,(P)} \subset U_r\,(P)$. In fact, the condition is satisfied, if we take $s < r$.

Let $M = \{M_1, M_2, \ldots\}$ be the countable set of all rational points (*i.e.*, points for which $x$ is rational) on $L$. Then any open set $G$, containing $M$, is the union of a collection of *nbds* $U_{r_i}\,(M_i)$,

*i.e.*, $$G = \cup\ \{U_{r_i}\,(M_i) : i = 1, 2, \ldots\}.$$

Let $U_s\,(B)$ be a *nbd* of an irrational point $B$ on $L$. Let $M_i = (m_i,\ 0)$ and $B = (b,\ 0)$, then $U_{r_i}\,(M_i)$ and $U_s\,(B)$ are the open circles, having $(m_i,\ r_i)$ and $(b,\ s)$ as their respective centres and $r_i$ and $s$ as their radii, taken together with the points $M_i$ and $B$, respectively. Now, in order that $U_s\,(B)$ be disjoint with $G$, it must be disjoint with each $U_{r_i}\,(M_i)$,

*i.e.*, $$(m_i - b)^2 + (r_i - s)^2 \geq (r_i + s)^2,$$

or $(m_i - b)^2 \geq 4 r_i\, s$, *i.e.*, $(m_i - b)^2/2r_i \geq 2s$, for $i = 1, 2, \ldots$

Let us enclose each rational point $M_i$ in an open interval

$$(m_i - e_{in},\ m_i + e_{in}), \text{ where } (m_i - e_{in})^2 = 2r_i/n;$$

and let $U_n$ denote the union of these intervals. Then $L - U_n$ consists of irrational points only; and for any irrational point $B = (b,\ 0)$, belonging to $L - U_n$, we have

$$(m_i - b)^2/2r_i \geq (m_i - e_{in})^2/2r_i = 1/n, \text{ for } i = 1, 2, \ldots$$

so that $B$ possesses a *nbd* $U_s(B)$, disjoint with $G$, provided that $s \geq 1/2n$. In other words, $L - U_n$ is the set of all those irrational points $B$ which possess *nbds* $U_s(B)$, of radii $s \geq 1/2n$, disjoint with $G$. If it were possible to find a *nbd* of every irrational point disjoint with $G$, then $\cup \{L - U_n\}$ would be the entire set of irrational numbers. But each $U_n$ is a dense open subset of the real number space (with the order topology), and therefore, each $L - U_n$ is non-dense. But as the set of irrational numbers is of the second category (vide Example 7, page 69), *i.e.*, not expressible as a union of a countable aggregate of non-dense sets, it follows that the two sets $M$ and $L - M$ cannot be separated strongly in $(X, \tau)$. As $M$ and $L - M$ are disjoint closed sets, it follows that $(X, \tau)$ is not a normal space, and therefore also not a completely normal space.

*Thus Niemytzki's tangent disc topology is $T_0$, $T_1$, $T_2$, regular and $T_3$, but not normal, completely normal, $T_4$ or $T_5$.*

## 2.3 Finer and Coarser Topologies

**Theorem 6 :** *If $(X, \tau)$ is a $T_i$-space, and $\tau'$ is a topology on X, finer than $\tau$, then $(X, \tau')$ is also a $T_i$-space, for $i = 0$, 1 and 2; but the same is not true for regular, normal and completely normal spaces.*

**Proof :** As the topology $\tau'$ is finer than the topology $\tau$, every $\tau$-open set is also a $\tau'$-open set. Let $p$, $q$ be any two distinct points in $X$.

If $G$ is a $\tau$-open set, containing one only of the points $p$ and $q$, then $G$ is a $\tau'$-open set, with the same property. Hence if $(X, \tau)$ is a $T_0$-space, then $(X, \tau')$ is also a $T_0$-space.

If $U$ and $V$ are two $\tau$-open sets, which separate $p$ and $q$ weakly (respectively strongly), then $U$ and $V$ are two $\tau'$-open sets, which separate $p$ and $q$ weakly (respectively strongly). Hence, if $(X, \tau)$ is a $T_1$-space or a $T_2$-space, then $(X, \tau')$ is also respectively a $T_1$-space or a $T_2$-space.

Let $(X, \tau_i)$ be an indiscrete space, then it is regular, normal and completely normal by Example (*B*), page 122. Let $\tau'$ be a topology on $X$, which is not regular, normal or completely normal. For instance, $\tau'$ may be taken to be the half-disc topology (Example (*E*), page 125) on $X$ (where the set $X$ is defined accordingly). The topology $\tau'$ on $X$ is finer than $\tau_i$. Thus, whereas $(X, \tau_i)$ is regular, normal or completely normal, the space $(X, \tau')$, with a finer topology, is not regular, normal or completely normal.//

An example of a regular space $(X, \tau)$, such that $(X, \tau')$ is not regular, where $\tau'$ is finer than $\tau$, is given in Example 8, page 137.

**Coarser topology :** *The separation properties $T_i$ are not, in general, preserved under weakening of the topology.* Let $\tau_d$ denote the discrete topology on a set $X$ (with 2 or more elements). Then $(X, \tau_d)$ is a $T_5$-space (vide Example (*A*), page 122). Any topology $\tau$ on $X$ is coarser than $\tau_d$. Now, (*i*) if $\tau = \tau_i$ (the indiscrete topology), then $(X, \tau)$ is not a $T_i$-space, for $i = 0, 1, 2, 3, 4$ and 5 (vide Example (*B*), page 122). And (*ii*) if $\tau$ is the half-disc topology (and $X$ is appropriately defined), then $(X, \tau)$ is not regular, normal or completely normal (vide Example (*E*), page 125). Thus $(X\ \tau_d)$ is a $T_5$-space, the topology $\tau$ is coarser then $\tau_d$, and such that $(X, \tau)$ is not a $T_0$, $T_1$, $T_2$, $T_3$, $T_4$, or $T_5$-space in case (*i*), and not regular, normal or completely normal in case (*ii*).

## 2.4 $T_0$, $T_1$ and $T_2$-spaces

**$T_0$-spaces :** Most of the topological spaces are $T_0$-spaces; but there are some, which are not so, e.g.

**Example 3.** ***If X is any non-empty set, and $\tau$ the indiscrete topology on X, then $(X, \tau)$ is a topological space, which is not a $T_0$-space.***

A few characteristic properties of a $T_0$-space are given in the following Theorem.

**Theorem 7 :** *A topological space $(X, \tau)$ is a $T_0$-space if, and only if, any one of following conditions holds :*

(*i*) *For every pair of distinct points $p$, $q$ in $X$, either $p \notin \overline{\{q\}}$ or $q \notin \overline{\{p\}}$.*

(*ii*) *Closures of distinct points are distinct.*

(*iii*) *For every point $p \in X$, $\{p\}'$ is a union of closed sets.*

**Proof :** (*i*) Let $(X, \tau)$ be a $T_0$-space, and $p$, $q$ be two distinct points in $X$. By axiom $T_0$, there exists an open set $U$, which contains one only of the points $p$ and $q$; let $p \in U$ and $q \notin U$. Then the open set $U$, containing the point $p$, has an empty intersection with the (one-pointic) set $\{q\}$; hence $p \notin \{q\}'$. Also $p \neq q$; consequently, $p \notin \overline{\{q\}}$.

Conversely, let the condition (*i*) hold in a topological space $(X, \tau)$, and $p$, $q$ be a pair of distinct points in $X$. By hypothesis, either $p \notin \overline{\{q\}}$ or $q \notin \overline{\{p\}}$; say, $p \notin \overline{\{q\}}$ holds. Let $V = X - \overline{\{q\}}$, then $V$ is an open set, such that $p \in V$ and $q \notin V$ (since $q \in \overline{\{q\}}$). Hence $(X, \tau)$ is a $T_0$-space.

(*ii*) Let $(X, \tau)$ be a $T_0$-space, and $p$, $q$ be two distinct points in $X$. Then, by (*i*), either $p \notin \overline{\{q\}}$ or a $q \notin \overline{\{p\}}$; let $p \notin \overline{\{q\}}$. Also $p \in \overline{\{p\}}$; hence $\overline{\{p\}} \neq \overline{\{q\}}$.

On the other hand, if both $p \in \overline{\{q\}}$ and $q \in \overline{\{p\}}$ hold, then $\overline{\{p\}} = \overline{\overline{\{p\}}} \supset \overline{\{q\}} = \overline{\overline{\{q\}}} \supset \overline{\{p\}}$, so that $\overline{\{p\}} = \overline{\{q\}}$ holds. Hence, $\overline{\{p\}} \neq \overline{\{q\}}$ implies that either $p \notin \overline{\{q\}}$ or $q \in \overline{\{p\}}$ must hold. Consequently, $(X, \tau)$ is a $T_0$-space, by (*i*).

(*iii*) Let $(X, \tau)$ be a $T_0$-space, and $p \in X$. Let $q$ be a point in $X$, such that $q \in \overline{\{p\}}$ and $q \neq p$; then $\overline{\{q\}} \subset \overline{\overline{\{p\}}} = \overline{\{p\}}$. Again since $q \in \overline{\{p\}}$, it follows, by (*i*), that $p \notin \overline{\{q\}}$ must hold; consequently, $\overline{\{q\}} \subset \{p\}'$. Thus, for every point $q \in \{p\}'$ (*i.e.*, satisfying $q \in \overline{\{p\}}$, $q \neq p$), $\overline{\{q\}} \subset \{p\}'$. Hence $\{p\}' = \cup \overline{\{q\}} : q \in \{p\}'\}$; that is, $\{p\}'$ is a union of some closed sets.

Conversely let $\{p\}'$ be a union of some closed sets in $(X, \tau)$. Let $q \neq p$, then either $q \in \{p\}'$ or $q \notin \overline{\{p\}}$. If $q \in \{p\}'$ then $q \in F \subset \{p\}'$, where $F$ is one of the closed sets, whose union is the set $\{p\}'$; and then $X - F$ is an open set, which contains $p$, but not $q$. On the other hand, if $q \notin \overline{\{p\}}$, then $X - \overline{\{p\}}$ is an open set, which contains $q$, but not $p$. Hence $(X, \tau)$ is a $T_0$-space.//

**Theorem 8 :** *Any topological space can be made into a $T_0$-space by identifying points having the same closures.*

**Proof :** Let $(X, \tau)$ be a given topological space, and let a binary relation $\rho$ be defined on $X$, such that $(x, y) \in \rho$ iff $\overline{\{x\}} = \overline{\{y\}}$. Then $\rho$ is obviously an equivalence relation; and $X$ is accordingly partitioned into disjoint classes. It will be shown that if $Y$ denotes the quotient set $X/\rho$, and $\sigma$ be the quotient topology of $Y$, then $(Y, \sigma)$ is necessarily a $T_0$-space.

In fact, if $D_x$ denotes the class which contains the element $x$, *i.e.*, $D_x = \{u \in X : \overline{\{u\}} = \overline{\{x\}}\}$, then for any $x, y \in X$, either $D_x = D_y$, or $D_x \cap D_y = \Phi$. Again, it follows from the definition of the quotient topology $\sigma$ that a collection of classes $\{D_x\}$ forms a closed set in $(Y, \sigma)$ if, and only if, $\cup \{D_x\}$ is a closed set in $(X, \tau)$. Hence the closure of a set of classes $\{D_x\}$ in $(Y, \sigma)$ is the smallest collection of classes $\{D_y\}$, containing $D_x$, for which $\cup \{D_y\}$ is a closed set in $(X, \tau)$. Also, for any point $z \in D_x$, $\overline{\{z\}} = \overline{\{x\}}$ implies $z \in \overline{\{x\}}$, so that $D_x \subset \overline{\{x\}}$. And, therefore

$\overline{\{x\}} = \cup \{D_y : D_y$ belongs to the closure of $\{D_x\}$ in $(Y, \sigma)\}$.

The separation axiom $T_0$ may now be easily verified. In fact, if $D_x \neq D_y$, then $\overline{\{x\}} \neq \overline{\{y\}}$. Consequently, the closure of $D_x$ is different from the closure of $D_y$ in $(Y, \sigma)$. Hence $(Y, \sigma)$ is a $T_0$-space, by Theorem 7 (*ii*).//

**Example 4.** ***Let*** $(X, \tau)$ ***be an indiscrete space, and*** $(Y, \tau')$ ***a*** $T_0$***-space. If*** $f : (X, \tau) \to (Y, \tau')$ ***is continuous, then*** $f$ ***must be constant.***

**Solution :** If possible, let $f$ be not constant, and let $f(x_1) = a$ and $f(x_2) = b$, where $a \neq b$. As $(Y, \tau')$ is a $T_0$-space, there exists an open set $V \in \tau'$, such that $V$ contains one only of the points $a$ and $b$; let $V$ contain $a$, but not $b$. Since $x_1 \in f^{-1}(V), f^{-1}(V)$ is a proper subset of $X$; and, as $f$ is continuous, $f^{-1}(V) \in \tau$. Hence $f^{-1}(V)$ must be the null set $\phi$, since the topology $\tau$ is indiscrete. But then $x_1 \in f^{-1}(V) = \phi$, which is impossible. Hence $f$ must be constant.

$T_1$**-spaces :** Every $T_1$-space is $T_0$-space, by Theorem 3 (*i*). But the converse is not true; e.g.

**Example 5.** ***Let*** $X = \{a, b, c\}$, ***and*** $\tau = \{\phi, \{a\}, \{a, b\}, X\}$. ***Then*** $(X, \tau)$ ***is a*** $T_0$***-space, but not a*** $T_1$***-space.***

**Hints :** Verify that $(X, \tau)$ is a topological space, in which the axiom $T_0$ holds. The axiom $T_1$ does not hold in $(X, \tau)$. In fact, all the open sets, containing the point $b$ (viz. $\{a, b\}$ and $X$), also contain the point $a$; thus $a$ and $b$ cannot be weakly separated in $(X, \tau)$.

It is, however, true that :

**Theorem 9 :** *A* $T_0$*-space, in which* $p \in \overline{\{q\}}$ *implies* $q \in \overline{\{p\}}$ *is a* $T_1$*-space.*

**Proof :** Let $(X, \tau)$ be a $T_0$-space, in which the given condition holds. Let $a, b$ be a pair of distinct points in $X$. Since $(X, \tau)$ is a $T_0$-space, either $a \notin \overline{\{b\}}$ or $b \notin \overline{\{a\}}$ must hold, by Theorem 6 (*i*); let $a \notin \overline{\{b\}}$ hold. Again, since $b \in \overline{\{a\}}$ would imply $a \in \overline{\{b\}}$ (by hypothesis), it follows that $b \notin \overline{\{a\}}$ must also hold. Consequently, the points $a$ and $b$ are weakly separated by the open sets $X - \overline{\{b\}}$ and $X - \overline{\{a\}}$. Hence $(X, \tau)$ is a $T_1$-space.//

A $T_1$-space can be characterised in several ways, as given in the following Theorem.

**Theorem 10 :** *The following conditions are equivalent for a topological space* $(X, \tau)$ :

(*i*) *$(X, \tau)$ is a $T_1$-space.*

(*ii*) *Every one-pointic set is closed in $(X, \tau)$.*

(*iii*) *Every finite set is closed in $(X, \tau)$.*

(*iv*) *For every point $x \in X$, $\{x\}' = \Phi$.*

(*v*) *For every subset $A \subset X$, the intersection of all open sets, containing $A$, is the set $A$.*

(*vi*) *Any accumulation point $p$ of a subset $A$ in $(X, \tau)$ is an $\omega$-accumulation point of $A$ (i.e., every open set, containing $p$, intersects $A$ in a denumerably infinite collection of points).*

**Proof :** *Equivalence of* the conditions (*i*) and (*ii*) has already been established in the Lemma, page 120.

*Equivalence of* (*ii*) *and* (*iii*) : Since finite unions of closed sets are closed sets, (*iii*) follows from (*ii*). Again, since every single point forms a finite set, (*ii*) follows from (*iii*).

*Equivalence of* (*ii*) *and* (*iv*) : Let (*ii*) hold. Then the set $\{x\}$ is closed; hence $\{x\} = \overline{\{x\}} = \{x\} \cup \{x\}'$, which implies that $\{x\}' \subset \{x\}$. But $x \notin \{x\}'$, since the set $\{x\}$ does not contain any point other than $x$. Hence $\{x\}' = \Phi$, so that (*iv*) holds. Conversely, let $\{x\}' = \Phi$; then $\overline{\{x\}} = \{x\} \cup \{x\}'$ $= \{x\}$, that is, $\{x\}$ is a closed set. Thus (*iv*) implies (*ii*).

(*ii*) *implies* (*v*), *and* (*v*) *implies* (*i*) : Let (*ii*) hold. Then, for any point $b \notin A$, there exists an open set $X - \{b\}$, such that $A \subset X - \{b\}$, and $b \notin X - \{b\}$. Hence $A = \cap \{X - \{b\} : b \in X - A\}$, so that the intersection of all open sets, containing $A$, is the set $A$. Thus (*v*) holds. Next, let (*v*) hold. Let $a$, $b$ be a pair of distinct points in $X$. By taking $A = \{a\}$, it follows from (*v*) that the intersection of all open sets, containing the point $a$, is the set $\{a\}$. Hence there exists an open set $U$, such that $a \in U$ and $b \notin U$. Similarly (by taking $A = \{b\}$), there exists an open set $V$, such that $b \in V$ and $a \notin V$. Thus the points $a$ and $b$ are separated weakly by the open sets $U$ and $V$ in $(X, \tau)$. Hence $(X, \tau)$ is a $T_1$-space; that is, (*i*) holds.

*Equivalence of* (*i*) *and* (*vi*) : Let $(X, \tau)$ be a $T_1$-space, and let $p$ be an accumulation point of a subset $A$ in $(X, \tau)$. Let $U$ be an open set containing $p$. Then $U$ intersects $A$ in some point $x_1 \neq p$. By (*ii*), $\{x_1\}$ is a closed set in $(X, \tau)$. Hence $U - \{x_1\} = U \cap [X - \{x_1\}]$ is an open set containing $p$, and it intersects $A$ in some point $x_2 \neq p$. Also $x_2 \neq x_1$. Now, $\{x_1, x_2\}$ is a closed set, by (*iii*). Hence, $U - \{x_1, x_2\}$ is an open set containing $p$; and, it must intersect $A$ in some point $x_3 \neq p$. Also, $x_3 \neq x_1$ and $x_3 \neq x_2$. Continuing this process, one obtains an infinite sequence of distinct points, $x_1$, $x_2$, ... in $U \cap A$. Consequently $p$ is an ω-accumulation point of $A$.

On the other hand, let $(X, \tau)$ be not a $T_1$-space. Then by (*ii*), some one-pointic set must be *not* closed in $(X, \tau)$. Let, $\{x\} \neq \overline{\{x\}}$; and let $y \in \overline{\{x\}}$, where $y \neq x$. Then $y$ is an accumulation point of $\{x\}$; but it cannot be an ω-accumulation point of $\{x\}$, since $\{x\}$ consists of only one point. Thus, if $(X, \tau)$ is not a $T_1$-space, an accumulation point need not be an ω-accumulation point.//

**Theorem 11 :** (*i*) *The coarsest $T_1$-topology, that can be defined on a set X, is the cofinite topology on X.*

(*ii*) *The intersection of any collection of $T_1$-topologies on a set X is also a $T_1$-topology on X.*

**Proof :** (*i*) Let $\tau$ denote the cofinite topology on $X$ (vide Example 3, page 45). Then $\tau$ is a $T_1$-topology, since finite sets are closed in $(X, \tau)$. Again, for any $T_1$-topology $\tau'$ on $X$, every finite set is $\tau'$-closed, so that complements of finite sets are $\tau'$-open. Consequently $\tau \subset \tau'$. Hence, $\tau$ is the coarsest among the $T_1$-topologies that can be defined on $X$.

(*ii*) Let $\{\tau_\alpha\}$ be a given collection of $T_1$-topologies on $X$, and let $\sigma$ be their intersection. Then $\sigma$ is also a topology on $X$ (vide page 54). Since each $\tau_\alpha$ contains the cofinite topology $\tau$ on $X$ (by (*i*)), $\sigma$ also contains $\tau$; that is $\sigma$ is finer than $\tau$. As $\tau$ is a $T_1$-topology on $X$, $\sigma$ is also a $T_1$-topology (by Theorem 6).//

## EXERCISES

**1.** Show that there is only one $T_1$-topology on a finite set, viz. the discrete topology.

**2.** If $f$ is a continuous mapping of a topological space $(X, \tau)$ into a $T_1$-space $(Y, \tau')$, then the inverse images of points (belonging to $Y$) are closed sets in $(X, \tau)$.

**3.** Show that, for any subset $A$, in a $T_1$-space,

(*i*) $A'' \subset A'$, (*ii*) $\overline{A}' = A' = (\overline{A})'$, and

(*iii*) every derived set is closed.

Give an example to show that the above properties may not hold in a $T_0$-space.

**Hausdorff spaces :** Every Hausdorff space is always a $T_1$-space (by Theorem 3 (*ii*)); hence a Hausdorff space is the same as a $T_2$-space. The following example shows that a $T_1$-space need not be a $T_2$-space.

**Example 6.** ***Let X be an uncountable set, and τ the topology of countable complements on X (vide Example 4, page 46); Then (X, τ) is a $T_1$-space since every finite subset (being countable) is closed in (X, τ). But it is not a $T_2$-space, since no two (non-null) open sets in (X, τ) are disjoint. In fact, if G and H be two (non-null) disjoint open sets in (X, τ), then $(X-G)\cup(X-H)=X-(G\cap H)=X-\phi=X$, which is impossible, since each of the closed sets X – G and X – H being countable, (X – G) ∪ (X – H) is also countable, whereas the set X is uncountable.***

The following properties characterise a Hausdorff space :

**Theorem 12 :** *A topological space (X, τ) is a Hausdorff space if, and only if, any one of the following conditions holds :*

(*i*) *For every point $x \in X$, $\{x\} = \cap\{\overline{N}_x : N_x \in \mathbf{N}_x\}$, where $\mathbf{N}_x$ denotes the neighbourhood filter of the point x.*

(*ii*) *The diagonal set $\Delta = \{(x, x) : x \in X\}$ is a closed set in the product space $(X\times X, \tau\times\tau)$.*

**Proof :** (*i*) Let (X, τ) be a Hausdorff space, and let $x \in X$. Then, for any point $y \neq x$, there exist two disjoint neighbourhood $N_x$ and $N_y$ of $x$ and $y$ respectively. Then $N_x \cap N_y = \Phi$ implies $N_x \subset X - N_y$, so that

$$\overline{N}_x \subset \overline{X - N_y} \subset \overline{X - V} = X - V,$$

where $V$ is an open set satisfying $y \in V \subset N_y$; and one obtains $\overline{N}_x \cap V = \Phi$, which implies that $y \notin \overline{N}_x$. Hence,

$$y \notin \cap[\overline{N}_x : N_x \in \mathbf{N}_x], \text{ that is, } \cap[\overline{N}_x : N_x \,\varepsilon\, \mathbf{N}_x] = \{x\}.$$

Conversely, let the condition (*i*) hold in (X, τ). Then for any $y \neq x$, there exists $N_x$, such that $y \notin \overline{N}_x$. Then $X - \overline{N}_x$ is an open neighbourhood of $y$, and $x$ and $y$ are strongly separated by the two disjoint neighbourhoods $N_x$ and $X - \overline{N}_x$ of $x$ and $y$ respectively.

(*ii*) Let (X, τ) be a Hausdorff space, and (Z, σ) denote the product space $(X\times X, \tau\times\tau)$. Let $(x, y) \,\varepsilon\, Z - \Delta$, then $x \neq y$. And as $\{x\} = \cap[\overline{N}_x : N_x \,\varepsilon\, \mathbf{N}_x]$ (by (*i*)), there exists a neighbourhood $N_x$ of $x$, such that $y \notin \overline{N}_x$. Then $X - \overline{N}_x$ is an open neighbourhood of $y$; and since $N_x$ and

$X - \overline{N}_x$ are disjoint, $N_x \times (X - \overline{N}_x)$ is a neighbourhood of $(x, y)$, lying wholly inside $Z - \Delta$. Consequently, $Z - \Delta$ is an open set, and therefore $\Delta$ is a closed set in the product space $(Z, \sigma)$.

Conversely, let the condition (*ii*) hold in $(X, \tau)$. Let $x$, $y$ be two distinct points in $X$. Then $(x, y) \in Z - \Delta$. As $Z - \Delta$ is an open set in $(Z, \sigma)$, and $\{N_x \times N_y : N_x \in \mathbf{N}_x \text{ and } N_y \in \mathbf{N}_y\}$ forms a base of the product topology $\sigma$, there exists a suitable $N_x \times N_y$ satisfying $(x, y) \in N_x \times N_y \subset Z - \Delta$. Now, $N_x \times N_y \subset Z - \Delta$ implies that $N_x$ and $N_y$ are disjoint in $X$. Consequently, $x$ and $y$ are separated strongly by the disjoint neighbourhoods $N_x$ and $N_y$ of $x$ and $y$ respectively. Hence $(X, \tau)$ is a Hausdorff space.//

**Corollary :** *Let f and g be two continuous functions on a topological space* $(X, \tau)$, *with values in a Hausdorff space* $(Y, \tau')$. *Then*

1. $F = \{x \in X : f(x) = g(x)\}$ *is a closed subset in* $(X, \tau)$.
2. *If* $f(x) = g(x)$ *for all points x in a subset A, which is every-where dense in X, then f and g are identical.*

**Proof :** 1. Let $h$ be a mapping of $(X, \tau)$ into $(Y, \tau') \times (Y, \tau')$, defined by $h(x) = (f(x), g(x))$, then $h$ is continuous, by Theorem 64, page 94. Also, $F$ is the inverse image of $\Delta$ under $h$. And as $\Delta$ is closed in $(Y, \tau') \times (Y, \tau')$, it follows, by Theorem 46 (*i*), page 78, that $F$ is a closed set in $(X, \tau)$.

2. Let $F = \{x \in X : f(x) = g(x)\}$. Then $F \supset A$, and so $\overline{F} \supset \overline{A} = X$, as $A$ is dense on $X$. Also, by (1), $F$ is a closed set. Hence $F = \overline{F} \supset X$, so that $F = X$. Thus $f(x) = g(x)$ holds for all points $x \in X$; hence $f$ and $g$ are identical.//

**Theorem 13 :** *Any convergent sequence in a Hausdorff space possesses a unique limit.*

**Proof :** Let $\{x\} = \{x_1, x_2, ..., x_n, \dots\}$ be a convergent sequence, having a point $p$ as a limit, in a Hausdorff space $(X, \tau)$. Let $q$ be any point in $X$ different from $p$. By the axiom $T_2$, there exist two open sets $U$ and $V$, such that $p \in U$, $q \in V$ and $U \cap V = \Phi$. Now, since $p$ is a limit of the sequence $\{x\}$, and $U$ an open set containing $p$, there exists a positive integer $m$, such that $x_i \in U$ for all $i > m$. Then, as $V$ is disjoint with $U$, $V$ can contain at most $m$ elements, viz., $x_1, x_2, ..., x_m$ of the sequence $\{x\}$. Consequently, the point $q$ cannot be a limit of the sequence $\{x\}$. Hence $p$ is the unique limit of the given sequence $\{x\}$ in $(X, \tau)$.//

The converse of Theorem 13 is, however, not true, as can be seen from the following example :

**Example 7.** ***Let X be an uncountable set, and τ be the topology of countable complements on X. Then (X, τ) is $T_1$-space, but not a $T_2$-space (vide Example 6, page 134). Let $\{x\} = \{x_1, x_2, \ldots\}$ be a convergent sequence in (X, τ), having a point $p \in X$ as a limit. Let q be any point in X, other than p. As (X, τ) is a $T_1$-space, there exists an open set U, containing p but not q. As the open set U contains p, and p is a limit of the sequence {x}, there exists a positive integer m, such that $x_i \in U$, for all $i > m$. Let $V = X - \{x_{m+j} : j = 1, 2, \ldots\}$, then V (being the complement of a countable set) is an open set, containing the point q. But there does not exist any positive integer k, such that $x_i \in V$, for all $i > k$. Hence, q cannot be a limit of the sequence {x}; in other words, the sequence {x} has a unique limit, viz. p.***

***Thus every convergent sequence has a unique limit in the space (X, τ), although (X, τ) is not a Hausdorff space.***

## EXERCISES

1. Show that the cofinite topology on an infinite set is $T_1$ but not $T_2$.

2. If $f$ is a continuous mapping of a Hausdorff space into itself, show that the set of fixed points, *i.e.*, the set $\{x : f(x) = x\}$, is closed.

3. Let $(X, \tau)$ be a Hausdorff space; show that, for any $p \in X$,

   (*i*) $\cap \{F \subset X : p \in F \text{ and } F \text{ is closed}\} = \{p\}$;

   (*ii*) $\cap \{G \subset X : p \in G \text{ and } G \text{ is open}\} = \{p\}$.

   Show by suitable examples that none of the above two properties is sufficient for a Hausdorff topology.

4. Show that convergent sequences in Hausdorff spaces are homeomorphic.

   **Hints :** Let $(X, \tau)$ and $(Y, \tau')$ be two Hausdorff spaces. Let $\{x\} = \{x_1, x_2, \ldots\}$, and $\{y\} = \{y_1, y_2, \ldots\}$ be convergent sequences (with $x_i \neq x_j$ and $y_i \neq y_j$) in the spaces $(X, \tau)$ and $(Y, \tau')$ respectively. Show that the mapping, $f : \{x\} \to \{y\}$, defined by $f(x_i) = y_i$, for $i = 1, 2, \ldots$, is a homeomorphism.

5. Show that it is possible to define two different Hausdorff topologies $\tau$ and $\tau'$ on a given set $X$, such that $\tau_A = \tau_A'$ and $\tau_B = \tau_B'$, where $B = X - A$, for a suitable subset $A \subset X$.

   **Hints :** Let $\tau$ be a Hausdorff topology on $X$, and $A$ a non-open subset in $(X, \tau)$. Then $\tau' = \{G \subset X : G \cap A \in \tau_A \text{ and } G \cap B \in \tau_B\}$ is a Hausdorff topology (since $\tau \subset \tau'$) on $X$, different from $\tau$ (since $A$ is $\tau'$-open, but not $\tau$-open), and such that $\tau_A' = \tau_A$ and $\tau_B' = \tau_B$.

## 2.5 Regular and $T_3$-spaces

A regular space need not be a $T_0$-space and also neither a $T_1$-space nor a $T_2$-space (vide Example 1, page 120). A $T_3$-space is always a $T_2$-space and, hence, also a $T_1$-space and a $T_0$-space (by Theorem 4). The converse is, however, not true; that is, a $T_2$-space (which is necessarily a $T_1$-space and also a $T_0$-space) need not be a regular space, as can be seen from the following example.

**Example 8.** ***Let R denote the set of real numbers, and Q the subset, formed by the rational numbers. Let S be the collection of all open intervals and the subset Q; then S is a subbase of a topology $\sigma$ on R. Since the usual topology $\tau$ on R is a Hausdorff topology, and the topology $\sigma$ is finer than $\tau$, it follows, by Theorem 6, that (R, $\sigma$) is a Hausdorff space. But (R, $\sigma$) is not a regular space, since the closed set $R - Q$ (i.e., the set of all irrational numbers) and the point 1, lying outside $R - Q$, cannot be separated strongly in (R, $\sigma$). Thus***

(*i*) ***(R, $\sigma$) is a Hausdorff space, but not a regular space, and***

(*ii*) ***the usual topology $\tau$ on R is regular, but the topology $\sigma$ (on R), which is finer than $\tau$, is not regular.***

A regular space need not be a $T_3$-space. In fact, *any indiscrete space* (with more than 2 elements) *is* (vacuously) *a regular space, which is not a $T_3$-space* (Example (*A*), page 122).

Some characteristic properties of a regular space are given in the following Theorems :

**Theorem 14 :** *(Regularity criterion of Tychonov) : A topological space $(X, \tau)$ is regular if, and only if, for every point $p \in X$, and every nbd U of p, there exists a nbd V of p, such that $\overline{V} \subset U$.*

**Proof :** Let $(X, \tau)$ be a regular space, and $U$ a *nbd* of a point $p \in X$. There exists an open set $G$, satisfying the condition $p \in G \subset U$. Then $X - G$ is a closed set, which does not contain $p$. Hence, $X - G$ and $p$ are strongly separated in $(X, \tau)$; that is, there exist open sets $H$ and $K$, such that $X - G \subset H$, $p \in K$, and $H \cap K = \Phi$. It follows from $H \cap K = \Phi$ that $K \subset X - H$, so that $\overline{K} \subset \overline{X - H} = X - H \subset G \subset U$. Then $K = V$ is an (open) *nbd* of the point $p$, such that $\overline{V} \subset U$.

Conversely, let the given criterion hold in $(X, \tau)$. Let $F$ be a closed set, and $p$ a point lying outside $F$. Then $U = X - F$ is an (open) *nbd* of $p$. Hence, by the hypothesis, there exists a *nbd* $V$ of $p$, such that $\overline{V} \subset U = X - F$. Let $G$ be an open set, satisfying $p \in G \subset V$. Then $F$ and $p$ are separated strongly by the two disjoint open sets $X - \overline{V}$ and $G$. Hence $(X, \tau)$ is a regular space.//

The regularity criterion of Tychonov may also be formulated in the following equivalent forms:

**Theorem 14 (A) :** (*i*) *A topological space $(X, \tau)$ is regular if, and only if, for every point $p \in X$, and every open set $U$ containing $p$, there exists an open set $V$, such that $p \in V \subset \overline{V} \subset U$.*

(*ii*) *A topological space $(X, \tau)$ is regular if, and only if, every nbd of a point $p$ always encloses a closed nbd of $p$.*

As an immediate consequence of the regularity criterion, we have the following :

**Theorem 15 :** *A topological space $(X, \tau)$ is regular if, and only if, any one of the following conditions holds:*

(*i*) *For any closed set $F$ and a point $p \notin F$, there exists a nbd $V$ of $p$, such that $F \cap \overline{V} = \Phi$.*

(*ii*) *If $A$ and $B$ are a pair of subsets of $X$, satisfying $A \cap B = \neq \Phi$, then there exists an open set $G$, such that $A \cap G \neq \Phi$ and $\overline{G} \subset B$.*

(*iii*) *For every non-empty subset $A$ and a closed subset $B$, satisfying $A \cap B = \Phi$, there exists a pair of disjoint open sets $U$, $V$, such that $A \cap U \neq \Phi$ and $B \subset V$.*

**Proof :** The proof is left as an exercise.

**Theorem 16 :** *A topological space $(X, \tau)$ is regular if, and only if, any one of the following conditions holds :*

(*i*) *The closed nbds of any point $p \in X$ form a base about the point $p$ in $(X, \tau)$.*

(*ii*) *For every closed subset $A$, the intersection of all closed nbds of $A$ in $(X, \tau)$ is the set $A$.*

[$V$ is called a *nbd of a set A*, if $A \subset \text{Int}(V)$].

**Proof :** (*i*) Let $(X, \tau)$ be a regular space, and $G$ be an open set, containing the point $p$. Then $X - G$ is a closed set, and $p \notin X - G$. Hence there exist two open sets $U$, $V$, such that $p \in U$, $X - G \subset V$, and $U \cap V = \Phi$. Then $p \in U \subset X - V \subset G$; thus there exists a closed *nbd* $X - V$ of $p$, contained in $G$. Hence, the closed *nbds* of $p$ form a local base at the point $p$.

Conversely, let the closed *nbds* of a point $p \in X$ form a local base at the point $p$, for every point $p$ in $(X, \tau)$. Let $F$ be a closed set, and $p$ any point, lying outside $F$. Then $X - F$ is an open set, containing $p$. Hence, by the hypothesis, there exists a closed *nbd* $N$ of $p$, such that $N \subset X - F$. As $N$ is a (closed) *nbd* of $p$, there exists an open set $G$, such that $p \in G \subset N$. Then $F$ and $p$ are separated strongly by the two disjoint open sets $X - N$ and $G$. Hence, $(X, \tau)$ is a regular space.

(*ii*) Let $(X, \tau)$ be a regular space, and $A$ a closed subset in $(X, \tau)$. Let $p$ be any point in $X$, lying outside the set $A$. Then $X - A$ is an (open) *nbd* of $p$. Hence, by the regularity criterion (Theorem 14 ($A$)), there exists an open set $G$ satisfying $p \in G \subset \overline{G} \subset X - A$; that is, $A \subset X - \overline{G} \subset X - G$. Thus $X - G$ is a closed *nbd* of the set $A$, and such that $p \notin X - G$. Hence, the intersection of all closed *nbds* of $A$ is the set $A$.

Conversely, let the given condition be satisfied in $(X, \tau)$. Let $F$ be a closed set and $p$ a point lying outside $F$. It follows from the hypothesis that there exists a closed *nbd* $V$ of $A$, such that $p \notin V$. Then $F$ and $p$ are separated strongly by the two disjoint open sets Int $(V)$ and $X - V$. Hence, $(X, \tau)$ is a regular space.//

**Theorem 17 :** (*i*) *For any two distinct points $p$, $q$ in a regular space, either $\overline{\{p\}} = \overline{\{q\}}$, or $\overline{\{p\}} \cap \overline{\{q\}} = \Phi$.*

(*ii*) *Every regular $T_0$-space is a $T_1$-space, and is therefore a $T_3$-space.*

**Proof :** (*i*) Let $p$, $q$ be any two distinct points in a regular space $(X, \tau)$. If both $p \in \overline{\{q\}}$ and $q \in \overline{\{p\}}$ hold, then $\overline{\{p\}} \subset \overline{\overline{\{q\}}} = \overline{\{q\}} \subset \overline{\overline{\{p\}}} \subset \overline{\{p\}}$ implies $\overline{\{p\}} = \overline{\{q\}}$. On the other hand, if (at least) one of the two relations be not true, say $q \notin \overline{\{p\}}$, then the closed set $\overline{\{p\}}$ and the point $q$, lying outside it, are separated strongly in $(X, \tau)$. Hence there exists an open set $G$, such that $\overline{\{p\}} \subset G$ and $q \in X - G$, so that $\overline{\{q\}} \subset \overline{X - G} = X - G$; consequently, $\overline{\{p\}} \cap \overline{\{q\}} = \Phi$. Thus, either $\overline{\{p\}} = \overline{\{q\}}$, or $\overline{\{p\}} \cap \overline{\{q\}} = \Phi$ must hold.

(*ii*) Let $(X, \tau)$ be a regular space, and $p$ be any point in $X$. Let $q$ be a point in $X$, different from $p$. As $(X, \tau)$ is a $T_0$-space, either $p \notin \overline{\{q\}}$ or $q \notin \overline{\{p\}}$ holds (by theorem 7 (*i*)), from which it follows that $p \neq q$ implies $\overline{\{p\}} \neq \overline{\{q\}}$. Also, as $(X, \tau)$ is a regular space, it now follows, by (*i*), that $\overline{\{p\}} \cap \overline{\{q\}} = \Phi$. Consequently, $q \notin \overline{\{p\}}$, which means that $\overline{\{p\}}$ does not contain any point $q$, different from $p$; that is, $\overline{\{p\}} = \{p\}$. Thus every one-pointic set is closed in $(X, \tau)$, so that

$(X, \tau)$ is necessarily a $T_1$-space. And, as it is also regular, it follows that $(X, \tau)$ is a $T_3$-space.//

**Theorem 18 :** *The following properties hold in a $T_3$-space $(X, \tau)$:*

(*i*) *Each pair of distinct points in X have nbds, whose closures are disjoint.*

(*ii*) *If A is an infinite subset of X, then there exists an infinite sequence $\{V_i : i = 0, 1, \ldots\}$ of open subsets of X, such that $\overline{V}_m \cap \overline{V}_n = \Phi$, for $m \neq n$, and $A \cap V_n = \Phi$, for each $n \geq 1$.*

**Proof :** (*i*) Let $p$, $q$ be any two distinct points in $X$. Since $(X, \tau)$ is a $T_3$-space, and therefore also a $T_1$-space, $\{q\}$ is a closed set. By the regularity criterion, there exists an open set $U$, satisfying $p \in U \subset \overline{U} \subset X - \{q\}$. Then $q \in X - \overline{U}$, and there exists an open set $V$, such that $q \in V \subset \overline{V} \subset X - \overline{U}$. Now, $U$ and $V$ are (open) *nbds* of $p$ and $q$, such that their closures $\overline{U}$ and $\overline{V}$ are disjoint.

(*ii*) The proposition will be proved by induction. Take $V_0 = \Phi$; and assume that the open sets $V_0, V_1, \ldots, V_m$ have been obtained, such that $\overline{V}_0, \overline{V}_1, \ldots, \overline{V}_m$ are pairwise disjoint, $A \cap V_k \neq \Phi$ for $k = 1, 2, \ldots, m$, and

$$A_m = A - [\overline{V}_0 \cap \overline{V}_1 \cap \ldots \cap \overline{V}_m]$$

is an infinite set. Let $p$, $q$ be a two distinct points in $A_m$. By the regularity criterion, there exists an open set $V$, satisfying

$$p \in V \subset \overline{V} \subset X - [\overline{V}_0 \cup \overline{V}_1 \cup \ldots \cup \overline{V}_m \cup \{q\}],$$

and also an open set $W$, such that

$$q \in W \subset \overline{W} \subset X - [\overline{V}_0 \cup \overline{V}_1 \cup \ldots \cup \overline{V}_m \cup \overline{V}].$$

Let, now $V_{m+1}$ be defined by $V_{m+1} = V$ if $A \cap V$ is finite, and $V_{m+1} = W$ if $A \cap V$ is infinite.

Then $V_{m+1}$ is an open set, such that $\overline{V}_0, \overline{V}_1, \ldots, \overline{V}_m, \overline{V}_{m+1}$ are pairwise disjoint, and $A \cap V_k \neq \Phi$, for $k = 1, 2, \ldots, m+1$. Also the set $A - [\overline{V}_0 \cup \overline{V}_1 \cup \ldots \cup \overline{V}_{m+1}]$ is infinite. This completes the proof, by the principle of mathematical induction.//

The converse of Theorem 18 (*i*) is, however, not true, in general (vide Example 3, page 141).

## EXERCISES

**1.** Prove the following in a regular space $(X, \tau)$ :

(*a*) For arbitrary points $p, q \in X$, $p \in \overline{\{q\}}$ implies $q \in \overline{\{p\}}$.

(*b*) If $N(A)$ denotes the intersection of all *nbds* of a subset $A$ of $X$, then :

(*i*) $N(A) \subset \overline{A}$;

(*ii*) $N(\{x\})$ is closed, and the subspace topology of $N(\{x\})$ is indiscrete.

**2.** Prove the following in a $T_3$-space $(X, \tau)$ :

(*a*) If $A$ is a closed set in $(X, \tau)$, then the intersection of all open sets, containing $A$, is the set $A$.

(*b*) Every pair of weakly separated *countable* subsets can always be strongly separated.

**3.** Let $\mathbf{R}$ denote the set of real numbers. For any $x \in \mathbf{R}$, let

$$V_n(x) = \{x\} \cup \{y : y \text{ is rational, and } y \in (x - 1/n,\ x + 1/n)\}$$

Show that $N_x = \{V_n(x) : n = 1, 2, \ldots\}$ forms a neighbourhood filter of the point $x$. Also, if $\sigma$ is the neighbourhood topology on $\mathbf{R}$, determined by the family of neighbourhood filters $\{\mathbf{N}_x : x \in \mathbf{R}\}$, then $(\mathbf{R}, \sigma)$ *is not a regular space, but each pair of distinct points have nbds, whose closures are disjoint.* (This shows that the converse of Theorem 18 (*i*) is not true).

## 2.6 Normal and Completely Normal Spaces

**Normal and $T_4$-spaces :** A normal space need not be a regular space, nor a Hausdorff space, nor a $T_1$-space, nor even a $T_0$-space (vide Example 2, page 120). But a $T_4$-space is always a $T_3$-space, and hence also a $T_2$-space, a $T_1$-space and a $T_0$-space (by Theorem 4). But the converses are not true; in fact *Niemytzki's tangent disc topology* (Example (*F*), page 126) is $T_0$, $T_1$, $T_2$, regular and $T_3$, but not normal or $T_4$.

There are several necessary and sufficient conditions for a topological space to be normal. Two such conditions, Urysohn's lemma and Tietze's theorem, formulated in terms of real-valued continuous functions, will be considered in the next chapter. We now prove :

**Theorem 19 :** *(Normality criterion of Urysohn) : A topological space $(X, \tau)$ is normal if, and only if, for every closed set $F$, and any open set $U$ containing $F$, there exists an open set $V$, such that $F \subset V$, $\overline{V} \subset U$.*

**Proof :** Let $(X, \tau)$ be a normal space, and $U$ any open set containing a closed set $F$. Then the pair of disjoint closed sets $F$ and $X - U$ are strongly separated in $(X, \tau)$. That is, there exist two open sets $V$ and $G$, such that $F \subset V$, $X - U \subset G$ and $V \cap G = \Phi$. Now, $V \cap G = \Phi$ implies $V \subset X - G$, so that $\overline{V} \subset \overline{X - G} = X - G \subset U$. Thus $F \subset V$ and $\overline{V} \subset U$. This proves the necessity of the condition.

Converselly, let the given condition hold in $(X, \tau)$. Let $F$, $K$ be a pair of disjoint closed sets in $(X, \tau)$. Then the closed set $F$ is contained in the open set $X - K$. Hence, by hypothesis, there exists an open set $V$, such that $F \subset V$ and $\overline{V} \subset X - K$. Then $F$ and $K$ are separated strongly by the two disjoint open sets $V$ and $X - \overline{V}$. Hence, $(X, \tau)$ is a normal space.//

It now follows immediately from the above Theorem that :

**Corollary :** *A topological space $(X, \tau)$ is normal if, and only if, any one of the following conditions holds :*

1. *For each pair of disjoint closed sets $P$, $Q$, there exists an open set $U$, such that $P \subset U$ and $Q \cap \overline{U} = \Phi$.*
2. *Each pair of disjoint closed sets have nbds, whose closures are disjoint.*

**Proof :** Proof is left as an exercise.//

**Coverings.** Normality can be characterised in terms of coverings, as shown below.

A collection $\mathbf{A}$ of subsets of a set $X$, such that $\cup \{A : A \in \mathbf{A}\} = X$, is called a covering (or cover) of the set $X$. If a sub-collection $\mathbf{B}$ of $\mathbf{A}$ is also a covering of $X$, then $\mathbf{B}$ is called a sub-covering of the covering $\mathbf{A}$.

Let $(X, \tau)$ be a topological space, then a covering $\mathbf{A}$ (respectively a subcovering $\mathbf{B}$) is called an *open covering* (respectively an *open sub-covering*) of $X$, if the members of $\mathbf{A}$ are open sets in $(X, \tau)$; like-wise we define *closed coverings* and *closed sub-covering* of $X$. If $Y$ is a subset of $X$, then by a *covering* of $Y$ *in* $X$, we mean a collection of subsets of $X$, whose union contains $Y$.

Let $\mathbf{G} = \{G_\alpha : \alpha \in I\}$, where $I$ is an indexing set, be an open covering of $X$. Let $\mathbf{H} = \{H_\alpha : \alpha \in I\}$ be also an open covering of $X$, such that $\overline{H_\alpha} \subset G_\alpha$, for each $\alpha \in I$. Then the open covering $\mathbf{G}$ is said to be *shrinkable*, and the open covering $\mathbf{H}$ is called a *shrinking* of the open covering $\mathbf{G}$.

**Theorem 20 :** *A topological space $(X, \tau)$ is normal if, and only if, every finite open covering of $X$ is shrinkable.*

**Proof :** Let $(X, \tau)$ be a normal space, and $\mathbf{G} = \{G_1, G_2, ..., G_n\}$ be a finite covering of $X$. Then $X = G_1 \cup G_2 \cup ... \cup G_n$; that is,

$$\Phi = X - (G_1 \cup G_2 \cup ... \cup G_n) = (X - G_1) \cap (X - G_2) \cap ... \cap (X - G_n),$$

so that $X - G_1$ and $(X - G_2) \cap ... \cap (X - G_n)$ form a pair of disjoint closed sets in $(X, \tau)$. As $(X, \tau)$ is a normal space, there exist two open sets $V_1$ and $H_1$, such that :

$$X - G_1 \subset V_1,\ (X - G_2) \cap ... \cap (X - G_n) \subset H_1, \text{ and } V_1 \cap H_1 = \Phi.$$

Then $H_1 \subset X - V_1$, so that $\overline{H_1} \subset \overline{X - V_1} = X - V_1 \subset G_1$. Again, since $X - (G_2 \cup ... \cup G_n) = (X - G_2) \cap ... \cap (X - G_n) \subset H_1$, it follows that

$$X - (H_1 \cup G_2 \cup ... \cup G_n) = (X - H_1) \cap [X - (G_2 \cup ... \cup G_n)] = \Phi,$$

so that $H_1 \cup G_2 \cup ... \cup G_n = X$. Thus $\{H_1, G_2, ..., G_n\}$ forms an open covering of $X$, where $\overline{H_1} \subset G_1$. By continuing the above process, we can replace $G_2, ... G_n$, successively by open sets $H_2, ..., H_n$, and thus obtain a shrinking $\mathbf{H} = \{H_1, H_2, ..., H_n\}$ of the given open covering $\mathbf{G}$.

Conversely, let every finite open covering of $X$ be shrinkable, for a topological space $(X, \tau)$. Let $P$, $Q$ be a pair of disjoint closed sets in $(X, \tau)$. Then $\{X - P, X - Q\}$ forms a (finite) open covering of $X$. Hence, by hypothesis, there exists a shrinking $\{G, H\}$ of the covering $\{X - P, X - Q\}$. That is, $G$, $H$ are two open sets, satisfying the conditions $G \cup H = X$, $\overline{G} \subset X - P$ and $\overline{H} \subset X - Q$. Then $P \subset X - \overline{G}$, $Q \subset X - \overline{H}$, and

$$(X - \overline{G}) \cap (X - \overline{H}) = X - (\overline{G} \cup \overline{H}) = \Phi$$

(since $\overline{G} \cup \overline{H} \supset G \cup H = X$). Thus $P$ and $Q$ are separated strongly by the two disjoint open sets $X - \overline{G}$ and $X - \overline{H}$; hence $(X, \tau)$ is a normal space.//

Theorem 20 can be generalised for a particular type of infinite open coverings as shown in Theorem 21.

A covering $\mathbf{A}$ of a set $X$ is said to be point-finite if each $x \in X$ belongs to only finitely many members of $\mathbf{A}$.

**Theorem 21 :** *A topological space $(X, \tau)$ is normal if, and only if, every point-finite open covering of $X$ is shrinkable.*

**Proof :** Let $(X, \tau)$ be a normal space, and let $\mathbf{G} = \{G_\nu : \nu \in I\}$ be a point-finite open covering of $X$. Let the indexing set $I$ be well-ordered; and for convenience, let us denote $I = \{1, 2, ..., \nu, ...\}$. We now construct a collection of subsets $\mathbf{H} = \{H_\nu;\ \nu \in I\}$ of $X$, by transfinite induction, as follows.

Let $F_1 = X - \cup \{G_\lambda : \lambda > 1\}$; then $F_1$ is a closed set, contained in the open set $G_1$. Hence, by the normality criterion there exists an open set $H_1$, such that $F_1 \subset H_1$ and $\overline{H}_1 \subset G_1$.

Let us assume that sets $H_\mu$ has been defined, as above, for each $\mu < v$. Let, now, $F_v = X - [(\cup \{H_\mu : \mu < v\}) \cup (\cup \{G_a : \alpha > v)\}]$. Then $F_v$ is a closed set, contained in the open set $G_v$. Hence, by the normality criterion, there exists an open set $H_v$, such that $F_v \subset H_v$ and $\overline{H}_v \subset G_v$. Then the family $\mathbf{H} = \{H_v : v \in I\}$ forms a shrinking of the given (point-finite) open covering $\mathbf{G}$, provided that $\mathbf{H}$ forms a covering of $X$. Let $x \in X$, then $x$ belongs to only finitely many members of $\mathbf{G}$, say $G_{\alpha 1}, G_{\alpha 2} \dots G_{\alpha n}$.

Let $\beta = \max\ \{\alpha_1, \alpha_2, \dots, \alpha_n\}$; then $x \notin H_\mu$ for any $\mu > \alpha$. Now there are two possibilities, viz. either, (*i*) $x \in H_\beta$ for some $\beta < \alpha$ or, (*ii*) also $x \in H_\beta$ for any $\beta < \alpha$, and in that case, $x \in F_\beta \subset H_\beta$. Thus in either case, $x \in H_\beta$, for some $\beta \le \alpha$. Hence $\mathbf{H}$ forms an open covering of $X$, and since $\overline{H}_v \subset G_v$, for each $v \in I$ (as shown above), it follows that $\mathbf{H}$ is a shrinking of $\mathbf{G}$; in other words, the given (point-finite) open covering $G$ is shrinkable.

The proof of the converse part is the very same as for Theorem 20.//

The following property is often used to show that a topological space is *not* normal.

**Theorem 22 :** *(Jones' lemma) : If a topological space* $(X, \tau)$ *contains a dense subset* $D$*, and a closed (relatively) discrete subspace* $Y$*, such that card* $(Y) \ge 2^\alpha$*, where* $\alpha = card\,(D)$*, then* $(X, \tau)$ *is not normal.*

**Proof :** If possible, let $(X, \tau)$ be a normal space, containing subsets $D$ and $Y$, with the above properties. Let $Z_i$ be a subset $Y$; then the subsets $Z_i$ and $Y - Z_i$ are disjoint and closed in $(X, \tau)$. Hence, there exist two disjoint open sets $U_i$, $V_i$, such that $Z_i \subset U_i$ and $Y - Z_i \subset V_i$. Now, for two subsets $Z_1$ and $Z_2$, such that $Z_1 - Z_2 \ne \Phi$, $U_1 \cap V_2$ is a non-empty open set in $(X, \tau)$. Hence $U_1 \cap V_2 \cap D \ne \Phi$, and $U_1 \cap V_2 \cap D$ is a subset of $U_1 \cap D$, but not a subset of $U_2 \cap D$. Thus if $Z_1$ and $Z_2$ are different subsets of $Y$, then $U_1 \cap D$ and $U_2 \cap D$ are different subsets of $D$; consequently

$$\text{card}\,[\mathbf{P}\,(Y)] \le \text{card}\,[\mathbf{P}\,(D)].$$

But this is impossible if card $(Y) \ge 2^\alpha$, where $\alpha = \text{card}\,(D)$.//

## EXERCISES

1. A normal space, in which $p \in \overline{\{q\}} \Rightarrow q \in \overline{\{p\}}$, is regular.
2. A topological space, which contains exactly one isolated point, is normal.
3. If a topological space $(X, \tau)$ is the union of subsets, each of which is open, closed and normal, then $(X, \tau)$ is a normal space.

**Complėtely normal and $T_5$-spaces :** Every completely normal space is normal (by Theorem 3 (*iii*)), but it need not be regular or a $T_i$-space, for $i = 0, 1, 2, 3, 4$ and 5. In fact,

(*a*) an indiscrete space is completely normal, normal and regular, but not a $T_i$-space, for $i = 0, 1, 2, 3, 4$ and 5 (vide Example (*B*) page 122);

(*b*) the set of real numbers **R**, with the left hand topology, is completely normal, and $T_0$, but not regular and not a $T_i$-space, for $i = 1, 2, 3$, 4 and 5 (vide Example (*C*), page 122).

On the other hand, since a topology, which is regular and $T_i$, for $i = 0, 1, 2$ and 3, may not be normal and $T_4$, (viz. Niemytzki's tangent disc topology), such a topology is not completely normal and $T_5$. Also, a normal space may not be completely normal, as shown below in Example 9, page 146.

A completely normal space is characterised by the following property:

**Theorem 23 :** *A topological space is completely normal if and only if, every subspace of it is normal.*

**Proof :** Let $(X, \tau)$ be a completely normal space, and $Y$ a subset of $X$. Let $P$, $Q$ be a pair of disjoint closed sets in $(Y, \tau_Y)$. Then there exist two sets $P_1$ and $Q_1$ (nor necessarily disjoint), closed in $(X, \tau)$, such that $P = P_1 \cap Y$ and $Q = Q_1 \cap Y$. Then

$$P_1 \cap Q = [(P_1 - P) \cup P] \cap Q = [(P_1 - P) \cap Q] \cup [P \cap Q] = \Phi,$$

since $P \cap Q = \Phi$ and $(P_1 - P) \cap Q = \Phi$ (as $P_1 - P \subset X - Y$ and $P \subset Y$). Similarly, $P \cap Q_1 = \Phi$. Consequently, $P$ and $Q$ are weakly separated in $(X, \tau)$ by the two open sets $X - Q_1$ and $X - P_1$. As $(X, \tau)$ is completely normal, it follows that $P$ and $Q$ are also strongly separated in $(X, \tau)$ by a suitable pair of disjoint open sets $U$, $V$. Then $P$ and $Q$ are strongly separated in $(Y, \tau_Y)$ by the pair of disjoint open sets $U \cap Y$ and $V \cap Y$. Hence $(Y, \tau_Y)$ is normal.

Conversely, let every subspace of a topological space $(X, \tau)$ be normal. Let $A$, $B$ be two weakly separated sets in $(X, \tau)$. Then (by Hausdorff-Lennes condition) $\overline{A} \cap B = \Phi$ and $A \cap \overline{B} = \Phi$. Let $G = X - (\overline{A} \cap \overline{B}) = \text{Ext}(A) \cup \text{Ext}(B)$. Then $G \cap \overline{A}$ and $G \cap \overline{B}$ are two disjoint subsets, closed in $(G, \tau_G)$. Since, by hypothesis, $(G, \tau_G)$ is a normal space, these can be strongly separated by a pair of disjoint sets $U$, $V$, open in $(G, \tau_G)$. Since $G$ is open in $(X, \tau)$, it follows (by Example 1 (*ii*), page 91 that $U$ and $V$, which are open in $(G, \tau_G)$, are also open in $(X, \tau)$. Also

$$G \cap \overline{A} = [\text{Ext}(A) \cup \text{Ext}(B)] \cap \overline{A} = [\text{Ext}(A) \cap \overline{A}] \cup [\text{Ext}(B) \cap \overline{A}]$$
$$= \text{Ext}(B) \cap \overline{A} \supset A, \text{ since } A \subset X - \overline{B} = \text{Ext}(B).$$

Similarly, $G \cap \overline{B} \supset B$. Thus $A \subset G \cap \overline{A} \subset U$ and $B \subset G \cap \overline{B} \subset V$. That is, $A$ and $B$ are strongly separated by $U$ and $V$ in $(X, \tau)$. Hence, $(X, \tau)$ is a completely normal space.//

**Example 9.** ***Example of a normal space which is not completely normal.***

Let $X = \{a, b, c, d\}$, and $\tau = [\phi, \{d\}, \{b, d\}, \{c, d\}, \{b, c, d\}, X]$. Then $(X, \tau)$ is a topological space possessing the requisite properties. The closed sets in $(X, \tau)$ are $X$, $\{a\}$, $\{a, b\}$, $\{a, c\}$, $\{a, b, c\}$, and $\phi$. As no pair of (non-null) closed sets is disjoint, it follows that $(X, \tau)$ is (vacuously) a normal space. The subset $Y = \{b, c, d\}$ with the relative topology $\tau_Y$ is however not normal. In fact, the open sets in $(Y, \tau_Y)$ are $\phi$, $\{d\}$, $\{b, d\}$, $\{c, d\}$ and $Y$; and the closed sets are $Y$, $\{b, c\}$, $\{c\}$, $\{b\}$, and $\phi$. The pair of disjoint closed sets $\{b\}$, $\{c\}$ cannot be strongly separated in $(Y, \tau_Y)$. Thus

(*i*) A subspace $(Y, \tau_Y)$ of the normal space $(X, \tau)$ is not normal, and

(*ii*) $(X, \tau)$ is a normal space, but not completely normal.

## EXERCISE

1. Prove that a topological space $(X, \tau)$ is completely normal if, and only if, every *open* subspace of it is normal.

# Art. 3 SUBSPACE, SUM, PRODUCT AND QUOTIENT SPACES

## 3.1 Subspaces of $T_i$-spaces

**Theorem 24 :** *Every subspace of a $T_i$-space is also a $T_i$-space, for $i = 0$, 1, 2, 3 and 5, but not so for $i = 4$.*

**Proof :** Let $Y$ be a non-empty subset of a $T_i$-space $(X, \tau)$, and $\tau_Y$ be the subspace topology on $Y$, so that $(Y, \tau_Y)$ is a subspace of $(X, \tau)$.

1. Let $(X, \tau)$ be a $T_i$-space, for $i = 0, 1$ or 2, and let $p$, $q$ be a pair of distinct points in $Y$.

   (*i*) If $(X, \tau)$ is a $T_0$-space, then there exists an open set $G \in \tau$, which contains one only of the points $p$ and $q$; say, $p \in G$ and $q \notin G$. Then $Y \cap G \in \tau_Y$, and $p \in Y \cap G$ and $q \notin Y \cap G$. Hence, $(Y, \tau_Y)$ is also $T_0$-space.

   (*ii*) If $(X, \tau)$ is a $T_1$-space, then there exist $U, V \in \tau$, such that $p \in U$, $p \notin V$ and $q \in V$, $q \notin U$. Then $Y \cap U$, $Y \cap V \in \tau_Y$, and $p \in Y \cap U$, $p \notin Y \cap V$, $q \in Y \cap V$ and $q \notin Y \cap U$. Thus $p$ and $q$ are weakly separated by the open sets $Y \cap U$ and $Y \cap V$ in $(Y, \tau_Y)$. Hence, $(Y, \tau_Y)$ is also a $T_1$-space.

   (*iii*) If $(X, \tau)$ is a $T_2$-space, then $p$ and $q$ are strongly separated by a pair of disjoint open sets $G$ and $H$ in $(X, \tau)$, say $p \in G$ and $q \in H$. Then $Y \cap G$ and $Y \cap H$ are two open sets, which separate $p$ and $q$ strongly in $(Y, \tau)$, since $p \in Y \cap G$, $q \in Y \cap H$, and $(Y \cap G) \cap (Y \cap H) = Y \cap (G \cap H) = \Phi$. Hence, $(Y, \tau_Y)$ is also a $T_2$-space.

2. Let $(X, \tau)$ be a regular space. Let $F$ be a closed set in $(Y, \tau_Y)$, and $p$ a point in $Y$, lying outside $F$. Then $F = Y \cap K$ for some closed set $K$ in $(X, \tau)$, by Theorem 58 (*iii*), page 89; also $p \notin F \subset K$. Since the axiom $T_3$ holds in $(X, \tau)$, there exist two disjoint open sets $G$, $H \in \tau$, such that $K \subset G$ and $p \in H$. Then $Y \cap G$, $Y \cap H \in \tau_Y$, and $F \subset Y \cap K \subset Y \cap G, p \in Y \cap H$, and $(Y \cap G) \cap (Y \cap H) = Y \cap (G \cap H) = \Phi$. Hence $(Y, \tau_Y)$ is also a regular space.

   If $(X, \tau)$ be in particular, a $T_3$-space, *i.e.*, a regular $T_1$-space, then $(Y, \tau_Y)$ is also a regular $T_1$-space, that is, a $T_3$-space.

3. Let $(X, \tau)$ is a normal space. It follows, by Example 9, page 146, that a subspace of $(X, \tau)$ may *not* be a normal space. Hence, a subspace of a $T_4$-space need *not* be $T_4$-space.

4. Let $(X, \tau)$ be a completely normal space. Then every subspace of $(X, \tau)$ is a normal space, by Theorem 23. Every subspace of $(Y, \tau_Y)$ is also a subspace of $(X, \tau)$, by Theorem 60, page 91. Hence, every subspace of $(Y, \tau_Y)$ is also a normal space; consequently, $(Y, \tau_Y)$ is a completely normal space (by Theorem 23).

If $(X, \tau)$ is in particular, a $T_5$-space, that is, a completely normal $T_1$-space, then $(Y, \tau_Y)$ is also a completely normal $T_1$-space, that is, a $T_5$-space.//

It has been shown, in course of the above proofs, that :

**Corollary 1 :** *Any subspace of a regular or a completely normal space is, correspondingly, regular or completely normal.*

2. *A subspace of a normal space may not be normal.*

It is however true that :

**Theorem 25 :** *Any closed subspace of a normal space is normal.*

**Proof :** Let $Y$ be a closed subset of a normal space $(X, \tau)$. Let $P, Q$ be any two disjoint sets, closed in $(Y, \tau_Y)$. Then $P, Q$ are two disjoint sets, also closed in $(X, \tau)$, by Example 1 (*i*), page 91. As the axiom $T_4$ holds in $(X, \tau)$, $P$ and $Q$ are strongly separated in $(X, \tau)$ by two disjoint open sets $U, V$. Then $P, Q$ are also strongly separated by the two disjoint open sets $Y \cap U$ and $Y \cap V$ in $(Y, \tau_Y)$. Thus the axiom $T_4$ holds in $(Y, \tau_Y)$; in other words, $(Y, \tau_Y)$ is a normal space.//

Theorem 25 can be generalised to the following :

**Theorem 26 :** *(P. Urysohn) : Let $(X, \tau)$ be a normal space, and $Y$ any $F_\sigma$-set in $(X, \tau)$, then the subspace $(Y, \tau_Y)$ is also normal.*

**Proof :** Let $Y = \cup \{F_i : i = 1, 2, \ldots\}$, where each $F_i$ is closed in $(X, \tau)$. Let $P, Q$ be any two disjoint closed sets in $(Y, \tau_Y)$. We define, inductively, two sequences $U_1, U_2, \ldots$ and $V_1, V_2, \ldots$ of open subsets of $(X, \tau)$, such that

$$\overline{U_{i-1}} \cup (P \cap F_i) \subset U_i,\ \overline{U_i} \cap (Q \cup \overline{V_{i-1}}) = \Phi, \quad \ldots(1)$$

and

$$\overline{V_{i-i}} \cup (Q \cap F_i) \subset V_i,\ \overline{V_i} \cap (P \cup \overline{U_i}) = \Phi \quad \ldots(2)$$

taking $U_0 = V_0 = \Phi$.

In fact, the existence of an open set $U_i$, satisfying (1), is assured by corollary (1) of Theorem 19, since $\overline{U_{i-1}} \cup (P \cap F_i)$ and $Q \cup \overline{V_{i-1}}$ form a pair of disjoint closed sets in $(X, \tau)$, for $i = 1, 2, \ldots$ Likewise, since $\overline{V_{i-1}} \cup (Q \cap F_i)$ and $P \cup \overline{U_i}$ form a pair of disjoint closed sets in $(X, \tau)$, there exists an open set $V_i$ in $(X, \tau)$, satisfying the condition (2), for $i = 1, 2, \ldots$

Let $G = \cup \{Y \cap U_i : i = 1, 2, \ldots\}$ and $H = \cup \{Y \cap V_i : i = 1, 2, \ldots\}$, then $P$ and $Q$ are separated strongly by the two disjoint open sets $G$ and $H$ in $(Y, \tau_Y)$ (verify). Hence $(Y, \tau_Y)$ is a normal space.//

We sometimes express the results, proved in Theorem 24, by saying that *the property "that a topological space is a $T_i$-space" is hereditary* (*or cogradient*), for $i = 0, 1, 2, 3$ and 5, *but not for* $i = 4$.

## 3.2 Sum of $T_i$-spaces

A property $\pi$ of topological spaces is said to be *additive*, if for every family of disjoint topological spaces $\{(X_\alpha, \tau_\alpha) : \alpha \in I\}$, each of which has the property $\pi$, their topological sum $\Sigma\,\{(X_\alpha, \tau_\alpha : \alpha \in I\}$ also possesses this property.

**Theorem 27 :** *The property that a topological space is a $T_i$-space is additive, for $i = 0, 1, 2, 3$ and 4.*

**Proof :** We shall prove only one case, viz that normality is an additive property. The proofs for other cases are similar.

Let $\{(X_\alpha, \tau_\alpha) : \alpha \in I\}$, $I$ being an arbitrary indexing set, be a family of disjoint normal spaces, and let $(X, \tau)$ denote their topological sum. Let $A$ and $B$ be two disjoint closed sets in $(X, \tau)$, then $A \cap X_\alpha$ and $B \cap X_a$ are two disjoint closed sets in $(X_\alpha, \tau_\alpha)$, for each $\alpha \in I$. And, since $(X_\alpha, \tau_\alpha)$ is a normal space, there exist sets $U_\alpha$ and $V_\alpha$, open in $(X_\alpha, \tau_\alpha)$, such that

$$A \cap X_\alpha \subset U_\alpha,\; B \cap X_\alpha \subset V_\alpha \text{ and } U_\alpha \cap V_\alpha = \Phi.$$

Then $A$ and $B$ are separated strongly by the two disjoint open sets $U$ and $V$ in $(X, \tau)$, where $U = \cup\{U_\alpha : \alpha \in I\}$ and $V = \cup\{V_\alpha : \alpha \in I\}$. $(X, \tau)$ is, therefore, a normal space.//

## 3.3 Product of $T_i$-spaces

In considering the topological products of $T_i$-spaces, we first show that *the topological product of two $T_5$-spaces may not even be a normal space.*

**Theorem 28 :** *Let $(X, \sigma)$ denote the topological product $(\mathbf{R}, \tau_l) \times (\mathbf{R}, \tau_l)$, where $(\mathbf{R}, \tau_l)$ is the Sorgenfrey line, i.e., the real numbers space, with its lower limit topology. Then $(X, \sigma)$ is not a normal space (and therefore also not a $T_4$-space), whereas $(\mathbf{R}, \tau_l)$ is a $T_5$-space.*

**Proof :** The product topology $\sigma$ is generated by a base **B**, consisting of all rectangles (with the lower and left sides included );

$$\{(x, y) : a \le x < b,\; c \le y < d, \text{ where } a, b, c, d \in \mathbf{R}\}.$$

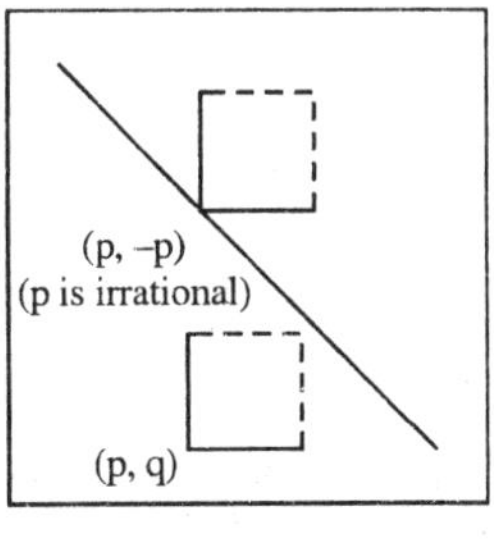


Let $F = \{(x, -x) : x$ is rational$\}$ and $K = \{(x, -x) : x$ is irrational$\}$. Then $F$ and $K$ form a pair of disjoint closed sets in $(X, \sigma)$. In fact, for any point $(p, q) \in X - F$, there exists a rectangle

$$\{(x, y) : p \le x < p + t,\; q \le y < q + t\},$$

which contains $(p, q)$ as its left-lower corner, but does not intersect $F$. Thus $(p, q)$ cannot be an accumulation point of $F$, and $F$ is therefore a closed set in $(X, \sigma)$; and similarly for $K$.

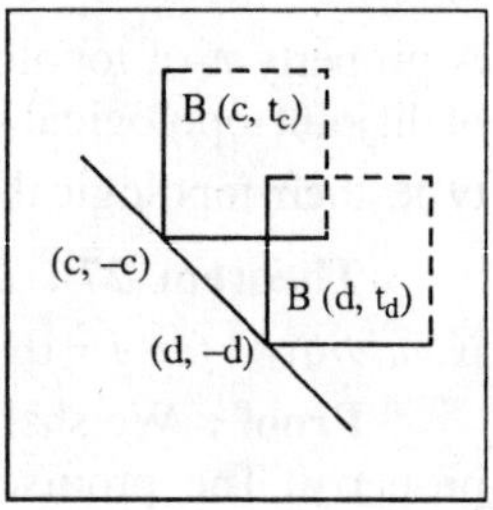


Let $U$ and $V$ be two open sets in $(X, \sigma)$, such that $F \subset U$ and $K \subset V$. As **B** forms a base for the topology $\sigma$, it follows that, for each $(r, -r) \in U$, there exists a $t_r > 0$, such that $B(r, t_r) \subset U$, where

$$B(r, t_r) = \{(x, y):\ r \le y < r + t_r\},\ r \le x < r + t_r,$$

for every real number $r$.

We shall now show that there exists a $t > 0$ and $a, b \in \mathbf{R}$ (with $a < b$), such that the set $\{x : x \text{ is irrational, for which } t_x \ge t\}$ is dense in the open interval $(a, b)$, with respect to the usual (order) topology on $\mathbf{R}$. In fact, the set of all irrational numbers $T$ can be expressed as $T = \cup\{T_n : n = 1, 2, \ldots\}$, where $T_n = \{r \in T : t_r \ge 1/n\}$. Also $T$ is a set of the second category (vide Example 7, page 71). Hence there exists a positive integer $m$, such that $T_m$ contains an open set, and therefore $T_m$ contains an open interval $(a, b)$ for some $a, b \in \mathbf{R}$, with $a < b$. Taking $t = 1/m$, we have $T_m = \{x : x \text{ is irrational and } t_x \ge t\}$.

Let $c$ be a rational number with $a < c < b$, and let $t_c$ be such that $B(c, t_c) \subset V$ and $t_c < t$. Then there exists a $d$ with $c < d < b$, such that $t_d \ge t$ and $d - c < t_c$. Then $B(c, t_c) \cap B(d, t_d) \ne \Phi$, which implies $U \cap V \ne \Phi$. Thus $(X, \sigma)$ is not a normal space and, therefore, also not a $T_4$-space. But $(\mathbf{R}, \tau_1)$ is a $T_5$-space, by Example (*D* (*ii*)), page 125.//

**Theorem 29 :** *Let $(X, \tau)$ be the topological product of a family of topological spaces $\{(X_\alpha, \tau_\alpha) : \alpha \in I\}$, where $I$ is an indexing set. Then $(X, \tau)$ is a $T_i$-space if, and only if, each $(X_\alpha, \tau_\alpha)$ is a non-empty $T_i$-space, for $i = 0, 1, 2,$ and 3.*

**Proof :** We shall prove the theorem for the values $i = 2$ and 3 only. The cases for $i = 0$ and 1 can be proved in exactly the same way as in the case of $i = 2$.

Let $i = 2$. Let $(X_\alpha, \tau_\alpha)$ be a non-empty Husdorff space, for each $\alpha \in I$. Let $x = \{x_\beta\}$, and $y = \{y_\beta\}$ be two distinct points in $X$, then $x_\nu \ne y_\nu$, for some $\nu \in I$. Since $(x_\nu, \tau_\nu)$ is a Hausdorff space, $x_\nu$ and $y_\nu$ can be separated strongly by two open sets $U_\nu$ and $V_\nu$ in $(X_\nu, \tau_\nu)$. Then $p_\nu^{-1}(U_\nu)$ and $p_\nu^{-1}(V_\nu)$ are two disjoint open sets in $(X, \tau)$, which separate $x$ and $y$ strongly. Hence $(X, \tau)$ is necessarily a Hausdorff space.

Conversely, let the product space $(X, \tau)$ be a Hausdorff space. Each coordinate space is homeomorphic with a subspace of the product space $(X, \tau)$ (vide Example 3, page 98). And, as every subspace of a Hausdorff space is also a Hausdorff space (by Theorem 24), it follows that each coordinate space is homeomorphic to a Hausdorff space, and is therefore itself a Hausdorff space. Thus the coordinate space $(X_\alpha, \tau_\alpha)$ is a Hausdorff space, for each $\alpha \in I$.

Next, let $i = 3$. Let $(X_\alpha, \tau_\alpha)$ be a $T_3$-space, *i.e.*, a regular $T_1$-space, for each $\alpha \in I$. Then the product space $(X, \tau)$ is a $T_1$-space, as the proposition holds for $i = 1$. Next, to show that $(X, \tau)$ is a regular space, we shall apply the regularity criterion. So, let $y = \{y_\alpha\}$ be a point in $(X, \tau)$, and $U$ be an open set, containing $y$ in $(X, \tau)$. Let **S** denote the subbase which generates the product topology $\tau$ on $X$, and **B** be the base generated by **S**. Then there exists a $V \in \mathbf{B}$, such that $y \in V \subset U$. Also $V$ is the intersection of finitely many members of **S**; consequently, $V = \cap \{p^{-1}(W_\beta) : W_\beta \in \tau_\beta,\ \beta \in J\}$, where $J$ is a finite subset of $I$. Thus $V = \pi\{V_\alpha : V_\alpha = W_\alpha$ for $\alpha \in J$ and $V_\alpha = X_\alpha$ for $\alpha \notin J\}$. Now, $y_\alpha = p_\alpha(y) \in p_\alpha(V) = V_q$ in $(X_\alpha, \tau_\alpha)$. And since $(X_\alpha, \tau_\alpha)$ is a regular space, there exists (by the regularity criterion) a set $G_\alpha \in \tau_a$, such that $y_\alpha \in G_a \subset \overline{G_a} \subset V$. Let $G = \cap \{p_\beta^{-1}(G_\beta) : \beta \in J\}$. Then $\overline{G} = \cap \{p_\beta^{-1}(\overline{G_\beta}) : \beta \in J\}$, and so $y \in G \subset \overline{G} \subset V \subset U$. Hence, by the regularity criterion, $(X, \tau)$ is a regular space, and therefore also a $T_3$-space.

Conversely, let the product space $(X, \tau)$ be a $T_3$-space, *i.e.*, a regular $T_1$-space. Since the converse part of the proposition holds for $i = 1$, each co-ordinate space $(X_a, \tau_a)$ is a $T_1$-space. Again, each co-ordinate space is homeomorphic to a subspace of the product space $(X, \tau)$ (vide Example 3, page 98). And, as every subspace of a regular space is regular (by Theorem 24), it follows that each coordinate space $(X_a, \tau_a)$ is homeomorphic to a regular space, and is therefore itself a regular space. Thus each co-ordinate space $(X_a, \tau_\alpha)$ is a regular $T_1$-space, *i.e.*, a $T_3$-space.//

It has been shown, while proving Theorem 28, that :

(*i*) *The topological product of two $T_5$-spaces may not even be a $T_4$-space, and hence not a $T_5$-space; and, therefore,*

(*ii*) *The topological product of two $T_4$-spaces need not be a $T_4$-space.//*

It is however true that :

**Theorem 30 :** *If $(X, \tau)$ is the topological product of a family of $T_1$-spaces $\{(X_\nu, \tau_\nu) : \nu \in I\}$, I being an indexing set, and the product space $(X, \tau)$ is a $T_4$-space, then each coordinate space $(X_\nu, \tau_\nu)$ is also a $T_4$-space.*

**Proof :** As the product space $(X, \tau)$ is a $T_1$-space, each coordinate space $(X_\nu, \tau_\nu)$ is homeomorphic to a closed subspace of $(X, \tau)$. Since every closed subspace of a $T_4$-space is also a $T_4$-space, it follows that the coordinate space $(X_\nu, \tau_\nu)$, for each $\nu \in I$, is homeomorphic to a $T_4$-space, and is, therefore, a $T_4$-space.//

## 3.4 Functions on $T_i$-spaces

Let $f$ be a single-valued function on a topological space $(X, \tau)$ with values in a topological space $(Y, \tau')$. We shall now obtain certain properties of the space $(X, \tau)$ which remain invariant under different types of functions $f$ on it.

**Theorem 31 :** *If $f$ is a single-valued function on a topological space $(X, \tau)$ with values in a topological space $(Y, \tau')$, and*

(*i*) *if $f$ is a closed surjection, and $(X, \tau)$ is a $T_1$-space, then $(Y, \tau')$ is also a $T_1$-space;*

(*ii*) *if $f$ is a closed bijection, and $(X, \tau)$ is a $T_2$-space, then $(Y, \tau')$ is also a $T_2$-space;*

(*iii*) *if $f$ is a closed continuous surjection, and $(X, \tau)$ is a $T_i$-space, then $(Y, \tau')$ is also a $T_i$-space, for $i = 4$ and 5.*

**Proof :** (*i*) Let $(X, \tau)$ be a $T_1$-space, and $p$ be any point in $Y$. As $f$ is surjective, there exists a point $x$ in $X$, such that $f(x) = p$. The one-pointic set $\{x\}$ is closed in $(X, \tau)$. As $f$ is a closed mapping, the one pointic set $\{p\} = \{f(x)\}$ is closed in $(Y, \tau')$. Consequently, $(Y, \tau')$ is a $T_1$-space.

(*ii*) Let $(X, \tau)$ be a $T_2$-space, and $f$ a closed bijection on $(X, \tau)$ onto $(Y, \tau')$. Then $f$ is also an open mapping (by Theorem 52 (*i*), page 84). Let $p, q$ be a pair of distinct points in $Y$. Then $f^{-1}(p), f^{-1}(q)$ are two distinct points in $(X, \tau)$, since $f$ is bijective. As $(X, \tau)$ is a $T_2$-space, there exist two disjoint open sets $G, H \in \tau$, such that $f^{-1}(p) \in G$ and $f^{-1}(q) \in H$. Then $f(G)$ and $f(H)$ are two disjoint open sets (since $f$ is a bijective open mapping) which separate $p$ and $q$ strongly in $(Y, \tau')$. Hence $(Y, \tau')$ is a $T_2$-space.

(*iii*) Let $f$ be a closed, continuous surjection of a $T_4$-space (or a $T_5$-space) $(X, \tau)$ onto $(Y, \tau')$. As $(X, \tau)$ is a $T_1$-space and $f$ is a closed surjection, $(Y, \tau')$ is also a $T_1$-space, by case (*i*). Next, let $P$, $Q$ be two disjoint subsets (respectively weakly separated subsets) in $(Y, \tau')$. Then $f^{-1}(P)$ and $f^{-1}(Q)$ are two disjoint closed (respectively, weakly separated) subsets in $(X, \tau)$, since $f$ is a continuous surjection. Since $(X, \tau)$ is normal (respectively completely normal), there exist two disjoint open sets $G, H \in \tau$, such that $f^{-1}(P) \subset G$ and $f^{-1}(Q) \subset H$. Let $U = Y - f(X - G)$ and $V = Y - f(X - H)$. Then $U$ and $V$ are two disjoint open sets in $(Y, \tau')$, since $f$ is a closed surjection; also $P \subset U$ and $Q \subset V$. Thus $P$ and $Q$ are strongly separated by $U$ and $V$ in $(Y, \tau')$. Hence, $(Y, \tau')$ is a normal (respectively completely normal) space, and hence also a $T_4$-space (respectively $T_5$-space).//

**Theorem 32 :** *A topological space $(Y, \tau')$, homeomorphic to a $T_i$-space $(X, \tau)$, is also a $T_i$-space, for $i = 0, 1, 2, 3, 4$ and 5.*

In other words, *the property "that a topological space $(X, \tau)$ is a $T_i$-space" is a topological property, for $i = 0, 1, 2, 3, 4$ and 5.*

**Proof :** Let $f$ be a homeomorphism of $(X, \tau)$ onto $(Y, \tau')$. Let $P$, $Q$ be (*i*) two distinct points, or (*ii*) $P$ is a point and $Q$ a closed set, not containing the point $P$, or (*iii*) two disjoint closed sets or (*iv*) a pair of weakly separated sets in $(Y, \tau')$. Since $f$ is a homeomorphism, $f^{-1}$ is also a homeomorphism and, therefore, bijective and closed. Hence $f^{-1}(P)$ and $f^{-1}(Q)$ are (*i*) two distinct points, or (*ii*) $f^{-1}(P)$ is a point and $f^{-1}(Q)$ a closed set, not containing the point $f^{-1}(P)$, or (*iii*) two disjoint closed sets, or (*iv*) a pair of weakly separated sets (respectively) in $(X, \tau)$. If $G$ and $H$ be two open sets, which separate $f^{-1}(P)$ and $f^{-1}(Q)$ weakly or strongly in $(X, \tau)$, then, correspondingly, $f(G)$ and $f(H)$ are two open sets (since $f$ is an open mapping), which separate $P$ and $Q$ weakly or strongly in $(Y, \tau')$.

Also, if $G$ is an open set which contains one only of the points $f^{-1}(P)$ and $f^{-1}(Q)$ in $(X, \tau)$ then $f(G)$ is an open set, which contains one only of the points $P$ and $Q$ in $(Y, \tau')$.

Thus, if $(X, \tau)$ is a $T_i$-space, then $(Y, \tau')$ is also a $T_i$-space, for $i = 0, 1, 2, 3, 4$ and 5.//

It may be noted that, since a homeomorphism $f$ is a closed, continuous, bijection, the cases for $i = 1, 2, 4$ and 5 of the above Theorem follow directly from Theorem 31 (*i*), (*ii*) and (*iii*).

It will be shown in the next chapter how some $T_i$-spaces can be completely characterised by real-valued continuous functions, defined on them.

## 3.5 Quotient Spaces of $T_i$-spaces

A quotient space of a $T_i$-space need not be a $T_i$-space, for $i = 0, 1, 2, 3, 4$ and 5. Likewise, a quotient space of a completely normal, normal, regular or Hausdorff space may not respectively be completely normal, normal, regular or Hausdorff. These follow from Example 10 and Example 11, given below. First we prove two lemmas.

**Lemma 1 :** *If every element (i.e., one-pointic subset) is dense in a topological space $(X, \tau)$, then $(X, \tau)$ is an indiscrete spaces.*

**Proof :** Let $(X, \tau)$ be not an indiscrete space. If $U$ is any open set in $(X, \tau)$, other than $\Phi$ and $X$, then for any element $p \in X - U$, $\{p\}$ cannot be dense in $(X, \tau)$, since for any point $q \in U$, $q \notin \{p\}'$, so that $q \notin \overline{\{p\}}$.//

**Lemma 2 :** *If $\mu$ is an equivalence relation, defined on a topological space $(X, \tau)$, such that each equivalence class (i.e., each member of $X/\mu$) is dense in $(X, \tau)$, then the quotient space $(X/\mu, \tau_\mu)$ is indiscrete.*

**Proof :** Let each equivalence class be dense in $(X, \tau)$, and $h$ denote the canonical mapping of $X$ onto $X/\mu$. Let $D$ be an equivalence class, *i.e.*, $D \in X/\mu$. Then $\overline{\{D\}} = \overline{h(D)} \supset h(\overline{D})$ (since $h$ is continuous) $= h(X) = X/\mu$. Thus every member of $X/\mu$ is a dense subset in the quotient space $(X/\mu, \tau_\mu)$. Hence $(X/\mu, \tau_\mu)$ is an indiscrete space, by lemma 1.//

**Example 10.** ***Let $Q$ and $T$ denote respectively the set of all rational numbers and the set of all irrational numbers. Then $Q \cup T = R$ and $Q \cap T = \phi$, where $R$ denotes the set of all real numbers. Let $\mu$ be the equivalence relation on $R$, generated by the partition $\{Q, T\}$. Then each of the two equivalence classes $Q$ and $T$ is dense in the real number space $(R, \tau)$. Hence the quotient space $(R/\mu, \tau_\mu)$ is indiscrete (by lemma 2), and is therefore not a $T_0$-space (vide Example (B), page 122). Thus :***

***The real number space $(R, \tau)$ is a $T_5$-space (vide Example (D (i)), page 124) and is therefore a $T_i$-space, for $i = 0, 1, 2, 3, 4$ and 5, but the quotient space $(R/\mu, \tau_\mu)$ is not a $T_0$-space, and hence also not a $T_i$-space, for $i = 0, 1, 2, 3, 4$ and 5.***

**Example 11.** ***Let $X$ be the union of the two lines $y = 0$ and $y = 1$ on the Euclidean plane $E^2$, and $\tau$ be the subspace topology on $X$, induced by the Euclidean topology on $E^2$. Since $E^2$ is (a metric space, and hence) a $T_5$-space, $(X, \tau)$ is also a $T_5$-space (by Theorem 24). Let $Y$ be the quotient***

***space of $(X, \tau)$, obtained by identifying the point $(a, 0)$ with the point $(a, 1)$, for each real number $a \neq 0$. Then $Y$ is a $T_1$-space, but not a Hausdorff space, since $h(0, 0)$ and $h(0, 1)$ are two distinct points, which cannot be strongly separated in $Y$ ($h$ being the canonical mapping of $X$ onto $Y$). Thus :***

***$(X, \tau)$ is a $T_5$-space (and therefore completely normal, normal, regular and Hausdorff) a quotient space $Y$ of which is $T_1$, but not Hausdorff (and therefore neither regular, nor normal, nor completely normal).***

**Theorem 33 :** *Let $(X, \tau)$ be a topological space, $\mu$ an equivalence relation on $X$, and $h$ the canonical mapping of $X$ onto $X/\mu$. Then*

(*i*) *the quotient space $(X/\mu, \tau_\mu)$ is a $T_1$-space if, and only if, each equivalence class, i.e., each member of $X/\mu$, is a closed set in $(X, \tau)$;*

(*ii*) *if $\mu$ is closed in $X \times X$, and $h : (X, \tau) \to (X/\mu, \tau_\mu)$ is an open mapping, then the quotient space $(X/\mu, \tau_\mu)$ is Hausdorff;*

(*iii*) *if $(X, \tau)$ is a normal space, and $h : (X, \tau) \to (X/\mu, \tau_\mu)$ is an open, closed mapping, then the quotient space $(X/\mu, \tau_\mu)$ is also a normal space.*

**Proof :** (*i*) Let the quotient space $(X/\mu, \tau_\mu)$ be a $T_1$-space, and $D \in X/\mu$, then the one member set $\{D\}$ is closed in $(X/\mu, \tau_\mu)$. Consequently $h^{-1}[\{D\}] = D$ is a closed subset in $(X, \tau)$, since $h$ is continuous. Conversely, let each member of $X/\mu$ be a closed subset of $(X, \tau)$. Let $D \in X/\mu$, then $h^{-1}[\{D\}] = D$ is a closed subset of $(X, \tau)$, by hypothesis. Hence the one member subset $\{D\}$ is closed in $(X/\mu, \tau_\mu)$; $(X/\mu, \tau_\mu)$ is therefore a $T_1$-space.

(*ii*) Let $h(p)$, $h(q)$ be two distinct members of $X/\mu$; then $(p, q) \notin \mu$, *i.e.*, $(p, q) \in X \times X - \mu$. Since $\mu$ is closed in $X \times X$, and $X \times X - \mu$ is therefore open, there exists a *nbd* $U \times V$ of $(p, q)$, such that $U \times V \in X \times X - \mu$. Hence $h(U)$, $H(V)$ are disjoint; and since $h$ is an open mapping (by hypothesis), $H(U)$ and $h'(V)$ are open in $(X/\mu, \tau_\mu)$. Thus $h(p)$ and $h(q)$ are separated strongly by the two disjoint open sets $h(U)$ and $h(V)$; and $(X/\mu, \tau_\mu)$ is therefore a Hausdorff space.

(*iii*) Let $P, Q$ be two disjoint closed subsets of the quotient space $(X/\mu, \tau_\mu)$. Then $h^{-1}(P)$ and $h^{-1}(Q)$ are two disjoint subsets of $X$, which are also closed in $(X, \tau)$, since $h$ is closed and continuous. Hence there exist two disjoint open sets $U, V \in \tau$, which separate $h^{-1}(P)$

and $h^{-1}(Q)$ strongly in $(X, \tau)$. Then $h(U)$ and $h(V)$ are two disjoint open sets (since $h$ is an open mapping) which separate $P$ and $Q$ strongly in $(X/\mu, \tau_\mu)$. Consequently, $(X/\mu, \tau_\mu)$ is a normal space.

## EXERCISES

**1.** Prove the following :

(*a*) The property of a space being a $T_1$-space is preserved by open bijections (and is therefore is a topological property).

(*b*) Hausdorff topologies are not preserved under continuous surjections, nor even under continuous open surjections.

**2.** Let $(X, \tau)$ be a topological space, $\mu$ an equivalence relation on $X$, and $h$ the canonical mapping of $X$ onto $X/\mu$. Prove that :

(*a*) If $(X/\mu, \tau_\mu)$ is Hausdorff, then $\mu$ is closed in $X \times X$ (with product topology).

(*b*) If $(X, \tau)$ is, in particular, a $T_3$-space, and

(*i*) if $h$ is an open mapping, then $\mu$ need not be open in $X \times X$;

(*ii*) if $h$ is a closed mapping, then $\mu$ is closed in $X \times X$;

(*iii*) if $h$ is an open, closed mapping, then $(X/\mu, r_\mu)$ is Hausdorff.

**3.** Prove that :

(*i*) the image of a $T_3$-space, under an open, continuous mapping, need not be regular;

(*ii*) a closed, continuous image of a $T_3$-space may not be a $T_2$-space; and, even if it is a $T_2$-space, it need not be regular;

(*iii*) a closed open continuous image of a $T_3$-space is a $T_2$-space.

**4.** Let $(X, \tau)$ be a regular space, but not normal, and $P$, $Q$ be two disjoint closed sets, which cannot be strongly separated in $(X, \tau)$. Let $\mu$ be the equivalence relation on $X$, determined by the partition of $X$ into the sets $P$, $Q$ and one-pointic sets $\{x\}$, for all $x \in X - (P \cup Q)$. Show that the quotient space $(X/\mu, \tau_\mu)$ is $T_1$, but not Hausdorff.

## Art. 4 $T_D$, Urysohn and Semi-Regular Spaces

### 4.1 $T_D$-spaces

A topological space $(X, \tau)$ is called a $T_D$-space, if $\{x\}'$ is a closed set, for every $x \in X$.

**Theorem 34 :** (*i*) *Every $T_1$-space is a $T_D$-space, and*

(*ii*) *every $T_D$-space is a $T_0$-space.*

**Proof :** (*i*) In a $T_1$-space, $\{x\}' = \Phi$, for every $x \in X$ (by Theorem 10 (*v*)); and $\Phi$ is a closed set. Hence every $T_1$-space is necessarily a $T_D$-space.

(*ii*) Since $\{x\}'$ is a closed set, for every $x \in X$, in a $T_D$-space, it follows by Theorem 7 (*iii*) that every $T_D$-space is necessarily a $T_0$-space.//

The converses of (*i*) and (*ii*) in Theorem 34 are, in general, not true, as can be seen from the following examples.

**Example 12.** ***Let R denote the set of all real numbers, and $\tau$ consist of $\phi$, X, all left open improper intervals $(a, \infty)$, and all left closed improper intervals $[a, \infty)$. Then $\tau$ is a topology on R, and $(R, \tau)$ is a $T_D$-space, but not a $T_1$-space (verify).***

**Example 13.** ***Let $\sigma$ denote the left hand topology on the set of real numbers R. Then $(R, \sigma)$ is a $T_0$-space, but not a $T_D$-space (verify).***

**Theorem 35 :** *A topological space $(X, \tau)$ is a $T_D$-space if, and only if, any one of the following conditions holds :*

(*i*) *For every point $x \in X$, there exists an open set G and a closed set F, such that $\{x\} = G \cap F$.*

(*ii*) *All derived sets are closed in $(X, \tau)$.*

**Proof :** (*i*) Let $(X, \tau)$ be a $T_D$-space, and $x \in X$. Then $G = X - \{x\}'$ is an open set, and $F = \overline{\{x\}}$ is a closed set, such that $G \cap F = \{x\}$.

Conversely, let the condition (*i*) be satisfied for a topological space $(X, \tau)$. Since $F$ is a closed set containing $x$, it follows that $\overline{\{x\}} \subset F$. And, then

$$\{x\}' = \overline{\{x\}} - \{x\} = \overline{\{x\}} - (G \cap F) = \overline{\{x\}} - (G \cap \overline{\{x\}}) = \overline{\{x\}} \cap (X - G)$$

is a closed set (being intersection of two closed sets. Hence $(X, \tau)$ is a $T_D$-space.

(*ii*) If every derived set is closed in $(X, \tau)$, then $\{x\}'$ is also closed, for every $x \in X$. Hence $(X, \tau)$ is a $T_D$-space.

Conversely, let $(X, \tau)$ be a $T_D$-space. Let $A$ be an arbitrary subset of $X$. We shall show that the derived set $A'$ is closed in $(X, \tau)$. If $A' = \Phi$, or $A'$ has no limiting point, then $A'$ is obviously a closed set. So, let $A' \neq \Phi$ and $A'' \neq \Phi$, and let $p \in A''$ (*i.e.*, $p$ is a limiting point of $A'$). Then, either $p \in A'$, or else $p \in A$, since $A'' \subset A \cup A'$ (vide Example 1 (*v*), page 57). If possible, let $p \notin A'$, and then $p \in A$. Then there exists an open set $G$, containing $p$, such that $G \cap A = \{p\}$. As $p$ is a limiting point of $A'$, there exists a point $q \in A' \cap G$, $q \neq p$. If $q \notin \{p\}'$, then there exists an open set $H$, containing $q$, such that $p \notin H$. Then the open set $G \cap H$ contains $q$ but does not contain any point of $A$, which is impossible, since $p \in A'$; hence $q \in \{p\}'$. Again, since $\{p\}'$ is a closed set (as $(X, \tau)$ is a $T_D$-space), and $p \notin \{p\}'$, $G \cap (X - \{p\}')$ is an open set, containing $p$, but not containing any point of $A'$. This is also impossible, since $p$ is a limiting point of $A'$. Hence, our initial assumption that $p \notin A'$ has not been valid, so that we must have $p \in A'$, and $A'$ is therefore a closed set.//

## EXERCISES

**1.** Prove the following :

(*a*) A subspace of a $T_D$-space is a $T_D$-space.

(*b*) The topological product of a finite family of $T_D$-spaces is a $T_D$-space.

(*c*) Any topology, finer than a $T_D$-topology, is also a $T_D$-topology.

(*d*) The property "that a topological space is a $T_D$-space" is a topological property.

## 4.2 Urysohn Spaces

A topological space $(X, \tau)$ is called a *Urysohn space*, if every pair of distinct points in $x$ have *nbds*, whose closures are disjoint.

It follows from Theorem 18 (*i*), and the definitions of Urysohn and Hausdorff spaces that :

**Theorem 36 :** (*i*) *Every $T_3$-space is a Urysohn space, and*

(*ii*) *every Urysohn space is a Hausdorff space.*

By virtue of the properties (*i*) and (*ii*) of Theorem 36, a Urysohn space is usually called a $T_2\frac{1}{2}$*-space*. Some authors also use the term *completely Hausdorff* in place of Urysohn; but we shall define the term completely Hausdorff in a different sense.

Since a regular space may not be a Hausdorff space (vide Example 1, page 120), it follows that *a regular space need not be a Urysohn space.*

The converses of the results (*i*) and (*ii*) of Theorem 36 are, in general, not true. In fact, *an example of a Urysohn space that is not a $T_3$-space,* is given in Example 3, page 141, and *an example of a Hausdorff space, that is not a Urysohn space*, is given in Example 14 in Miscellaneous Exercises II.

## EXERCISES

**1.** Prove the following :

(*a*) A subspace of a Urysohn space is a Urysohn space.

(*b*) The topological product of a family of topological spaces $\{(X_\nu, \tau_\nu) : \nu \in I\}$, where $I$ is an indexing set, is a Urysohn space if, and only if, $(X_\nu, \tau_\nu)$ is a Urysohn space, for each $\nu \in I$.

(*c*) Any topology, finer than a Urysohn topology, is also a Urysohn topology.

(*d*) The property "that a topological space is a Urysohn space" is a topological property.

## 4.3 Semi-Regular Spaces

A topological space $(X, \tau)$ is called a semi-regular space, if the family of regular open sets forms an open base of the topology $\tau$.

**Theorem 37 :** *Every regular space is semi-regular.*

**Proof :** Let $(X, \tau)$ be a regular space, $p$ a point, and $G$ any open set, containing $p$, in $(X, \tau)$. By the regularity criterion, there exists an open set $U$, such that $p \in U \subset \overline{U} \subset G$. Let $V = \text{Int}(\overline{U})$; then $V$ is a regular open set, satisfying the condition $p \in V \subset G$. Hence, the regular open sets form an open base of the topology $\tau$. $(X, \tau)$ is therefore a semi-regular space.//

The converse of Theorem 37 is, in general, not true. In fact, an example of a semi-regular space, that is not Urysohn, and therefore also not regular, is given in Example 14 in Miscellaneous Exercises II. Also Urysohn space may not be semi-regular (the topological space, defined in Example 3, page 141, is Urysohn but not semi-regular).

**Example 14.** **(*i*)** ***Any topological space can be embedded in a semi-regular space, and*** **(*ii*)** ***a subspace of a semi-regular space may not be semi-regular.***

**Solution :** (*i*) Let $(X, \tau)$ be a given topological space, and $\mathbf{I}$ denote the closed unit interval [0, 1]. On the product set $Y = X \times \mathbf{I}$, we define a topology $\tau'$, by choosing *nbd* filters $\mathbf{N}_p$ of the points $p = (x, y) \in Y$, as follows:

(*a*) For $y \neq 0$, $\mathbf{N}_p = \{(x, z) : y - \varepsilon < z < y + \varepsilon\}$, for small positive $\varepsilon$; and (*b*) For $y = 0$, $\mathbf{N}_p = \{(x', z) : x' \varepsilon V,\ 0 < z < \varepsilon(x')\}$, where $V$ is a *nbd* of $x$ in $(X, \tau)$, and for each $x' \varepsilon V$, $\varepsilon(x')$ is some small positive number.

The resulting space $(Y, \tau')$ is semiregular (verify), and $(X, \tau)$ is homeomorphic with the subspace $\{(x, 0) : x \varepsilon X\}$ of $(Y, \tau')$.

(*ii*) Taking $(X, \tau)$ to be a topological space, that is not semi-regular, it follows from the result (*i*) that whereas $(Y, \tau')$ is semi-regular, the subspaces $\{(x\ 0) : x \varepsilon X\}$ of it is not semi-regular.

## EXERCISES

1. Show that the cofinite topology on an infinite set is not semi-regular.
2. Let $\tau$ be the topology of countable complements on the set of real numbers. Examine whether $\tau$ is semi-regular.

*Chapter 4*

# REAL FUNCTIONS

## Art. 1 REAL-VALUED CONTINUOUS FUNCTIONS

Let ($\mathbf{R}$, $\sigma$) denote the real number space, and ($I$, $\sigma_I$) the subspace formed by the closed unit interval [0, 1]. A function $f$ on a topological space ($X$, $\tau$) into ($\mathbf{R}$, $\sigma$) (or into ($I$, $\sigma_I$)) will be called a *real-valued function* or, simply, a *real function* on $X$.

Since the open intervals (together with $\Phi$) form a base of the topology $\sigma$, and since any open *nbd* $V$ of a point $p \in \mathbf{R}$ always contains an open interval $(p-\varepsilon,\ p+\varepsilon)$, containing $p$, it follows that :

**Theorem 1 :** *A real function f on a topological space (X, τ) is continuous at a point $p \in X$ if, corresponding to a pre-assigned positive number ε, there always exists an open nbd G of p, such that $|f(x)-f(p)|<\varepsilon$, for all $x \in G$.*

**Example 1.** ***If f is a continuous real function on a topological space (X, τ), then :***

(*i*) ***the sets A, B, C, defined below, are closed in (X, τ) :***

$$A=\{x \in X : f(x) \le a\},\ B=\{x \in X : f(x) \ge a\},$$

***and*** $$C=\{x \in X : a \le f(x) \le b\};$$

(*ii*) ***the sets E, F, G, defined below, are open sets in (X, τ) :***

$$E=\{x \in X : f(x) < a\},\ F=\{x \in X : f(x) > a\},$$

***and*** $$G=\{x \in X : a < f(x) < b\},$$

***where a, b (a < b) are any two real numbers.***

**Hints :** The improper intervals $(-\infty,\ a]$, $[a,\ \infty)$ and the closed interval $[a,\ b]$ are closed subsets of ($\mathbf{R}$, $\sigma$). Hence the subsets $A=f^{-1}(-\infty,\ a]$, $B=f^{-1}[a,\ \infty)$ and $C=f^{-1}[a,\ b]$ are closed in ($X$, $\tau$).

Similarly, as the intervals $(-\infty,\ a)$, $(a,\ \infty)$ and $(a,\ b)$ are open subsets of ($\mathbf{R}$, $\sigma$), the subsets $E=f^{-1}(-\infty,\ a)$, $F=f^{-1}(a,\ \infty)$ and $G=f^{-1}(a, b)$ are open in ($X$, $\tau$).

**Theorem 2 :** *If f and g are two continuous real functions on a topological space (X, τ) and c any real number, then c.f., $f+g$, $f-g$, f.g and f/g are* *also continuous real functions on* $(X, \tau)$, *provided that in case of* $f/g$, $g(x) \neq 0$ *for all* $x \in X$, *where these functions are defined point-wise as follows:*

$$(c.f)(x) = c.f(x),\ (f+g)(x) = f(x) + g(x),\ (f-g)(x) = f(x) - g(x),$$

$$(f.g)(x) = f(x).g(x), \text{ and } (f/g)(x) = f(x)/g(x),$$

*for each* $x \in X$.

**Proof :** The proofs can be obtained in the same way as in elementary real analysis (and are therefore left out).//

**Theorem 3 :** *If f is a continuous real function on a topological space* $(X, \tau)$, *and* $g : (\mathbf{R}, \sigma) \to (\mathbf{R}, \sigma)$ *is continuous, then the composite function gof, defined point-wise by*

$$(gof)(x) = g[f(x)], \text{ for each } x \in X,$$

*is a continuous real function on* $(X, \tau)$.

**Proof :** Follows directly from Theorem 47, page 80.//

**Theorem 4 :** *If f and g are two continuous real functions on a topological space* $(X, \tau)$, *then* $|f|$, *max* $(f, g)$ *and min* $(f, g)$ *are also continuous real functions on* $(X, \tau)$, *where these functions are defined, point-wise, by*

(*i*) $(|f|)(x) = |f(x)|$,

(*ii*) $[max(f, g)](x) = max[f(x), g(x)]$, *and*

(*iii*) $[min(f, g)](x) = min[f(x), g(x)]$, *for each* $x \in X$.

**Proof :** (*i*) $|f| : x \to |f(x)|$ is continuous (by Theorem 3), since it is the composite of two continuous functions, $x \to f(x)$ and $f(x) \to |f(x)|$.

(*ii*) and (*iii*) : As :

$$\max[f(x), g(x)] = \frac{1}{2}[f(x) + g(x) + |f(x) - g(x)|],$$

and $$\min[f(x), g(x)] = \frac{1}{2}[f(x) + g(x) - |f(x) - g(x)|],$$

the results (*ii*) and (*iii*) follow by Theorems 2, 3 and 4 (*i*).//

## 1.1 Uniform Convergence

An infinite sequence of real functions $\{f_n : n = 1, 2, \ldots\}$ on a topological space $(X, \tau)$ is said to *converge uniformly* to a real function $f$ on $(X, \tau)$ if, for every $\varepsilon > 0$, there exists a positive integer $k$, such that

$$|f_n(x) - f(x)| < \varepsilon, \text{ for all } n > k, \text{ and for all } x \in X.$$

**Theorem 5 :** *An infinite sequence of real functions* $\{f_n : n = 1, 2, \ldots\}$, *on a topological space* $(X, \tau)$, *converge uniformly on* $(X, \tau)$ *if, and only* *if, for every* $\varepsilon > 0$, *there exists a positive integer* $k$, *such that, for all* $m > k$ *and* $n > k$,

$$|f_n(x) - f_m(x)| < \varepsilon, \text{ for all } x \in X. \quad \text{...(1)}$$

(*Cauchy's criterion for uniform convergence*)

**Proof :** Let the sequence $\{f_n : n = 1, 2, \dots\}$ converge uniformly on $(X, \tau)$, and let $f$ be the limit function. Let $\varepsilon > 0$ be pre-assigned, then there exists a positive integer $k$, such that, for all $n > k$ and for all $x \in X$,

$$|f_n(x) - f(x)| < \varepsilon/2.$$

Then for any $m > k$ and $n > k$, we have

$$|f_n(x) - f_m(x)| \leq |f_n(x) - f(x)| + |f(x) - f_m(x)| < \varepsilon/2 + \varepsilon/2 = \varepsilon,$$

for all $x \in X$. Thus the condition (1) holds.

Conversely, let the condition (1) hold. Then, for any $x \in X$, $\{f_n(x) : n = 1, 2, \dots\}$ is a Cauchy sequence in the real number space $(\mathbf{R}, \sigma)$. As $(\mathbf{R}, \sigma)$ is a complete metric space, $\{f_n(x) : n = 1, 2, \dots\}$ is necessarily a convergent sequence; and let, $f(x)$ be the limit. Thus the sequence $\{f_n : n = 1, 2, \dots\}$ converges, on $(X, \tau)$, to $f$. We shall show that the convergence is uniform.

Let $\varepsilon > 0$ be pre-assigned. By hypothesis, there exists a positive integer $k$, such that the condition (1) holds, for all $m > k$, $n > k$ and all $x \in X$. Let $n$ be kept fixed, and let $m \to \infty$ in (1). Since $f_m(x) \to f(x)$ as $m \to \infty$, we have $|f_n(x) - f(x)| < \varepsilon$ for all $n > k$ and all $x \in X$. Thus $\{f_n : n = 1, 2, \dots\}$ converges uniformly on $(X, \tau)$ to $f$.//

**Theorem 6 :** *The limit of a uniformly convergent sequence of continuous real functions on a topological space* $(X, \tau)$ *is a continuous real function on* $(X, \tau)$.

**Proof :** Let $\{f_n : n = 1, 2, \dots\}$ be a uniformly convergent sequence of continuous real functions on $(X, \tau)$, and let $f$ be the limit of the sequence. Then $f$ is a real function on $(X, \tau)$.

Let $\varepsilon > 0$ be a pre-assigned positive number, and let $x'$ be a given point of $X$. As the given sequence $\{f_n : n = 1, 2, \dots\}$ is uniformly convergent, having $f$ as the limit, it follows that there exists a positive integer $k$, such that

$$|f_k(x) - f(x)| < \varepsilon/3, \text{ for all } x \in X. \quad \text{...(2)}$$

Hence, substituting $x'$ for $x$, we get

$$|f_k(x') - f(x')| < \varepsilon/3. \quad \text{...(3)}$$

Again, since the function $f_k$ is continuous at the point $x'$, there exists an open set $G$, containing $x'$, such that

$$|f_k(x) - f_k(x')| < \varepsilon/3, \text{ for all } x \in G. \quad ...(4)$$

Now, by (2), (3) and (4), we get, for all $x \in G$,

$$|f(x) - f(x')| \le |f(x) - f_k(x)| + |f_k(x) - f_k(x')| + |f_k(x') - f(x')|$$
$$< \varepsilon/3 + \varepsilon/3 + \varepsilon/3 = \varepsilon$$

Consequently, the real function $f$ is continuous at the point $x'$ in $(X, \tau)$, and is therefore continuous on $(X, \tau)$.//

Let $\{f_n : n = 1, 2, \ldots\}$ be an infinite sequence of real functions on a topological space $(X, \tau)$. Then the infinite series $\sum_{n=1}^{\infty} f_n$ or, simply, $\Sigma f_n$ is said to converge uniformly to the sum $s$ on $(X, \tau)$, if the sequence of partial sums $\{s_n = f_1 + f_2 + \ldots + f_n : n = 1, 2, \ldots\}$, converges uniformly to $s$ in $(X, \tau)$.

**Theorem 7 :** *(Weierstrass' M-test) : If $\{f_n : n = 1, 2, \ldots\}$ be an infinite sequence of real functions on a topological space $(X, \tau)$, such that*

$$|f_n(x)| \le M_n, \text{ for } n = 1, 2, .., \text{ and for all } x \in X,$$

*where $M_n$ are positive constants, and the series $\Sigma M_n$ is convergent, then the infinite series $\Sigma f_n$ converges uniformly on $(X, \tau)$.*

**Proof :** As the series (of positive constants) $\Sigma M_n$ is convergent, corresponding to a pre-assigned $\varepsilon > 0$, there exists a positive integer $k$, such that $\sum_{r=m}^{n} M_r < \varepsilon$ holds, for any $n > m > k$. Hence,

$$|s_n(x) - s_{m-1}(x)| = |f_m(x) + f_{m+1}(x) + \ldots + f_n(x)|$$
$$\le \sum_{r=m}^{n} |f_r(x)| \le \sum_{r=m}^{n} M_r < \varepsilon, \text{ for all } x \in X.$$

Consequently, the sequence of partial sums $\{s_n\}$ is uniformly convergent on $(X, \tau)$, by Cauchy's criterion (Theorem 5); hence the infinite series $\Sigma f_n$ is uniformly convergent on $(X, \tau)$.//

If (in the above Theorem) every $f_n$ is continuous on $(X, \tau)$, then every partial sum, $s_n = f_1 + f_2 + \ldots + f_n$, for $n = 1, 2, \ldots$, is also continuous on $(X, \tau)$. Also the infinite series $\Sigma f_n$ converges uniformly on $(X, \tau)$ if, and only if, the infinite sequence $\{s_n\}$ converges uniformly on $(X, \tau)$. Hence, it follows, by Theorem 6, that :

**Theorem 8 :** *If* $\{f_n : n = 1, 2, \ldots\}$ *is an infinite sequence of continuous real functions on a topological space* $(X, \tau)$, *and if the infinite series* $\Sigma f_n$ *converges uniformly on* $(X, \tau)$, *then the sum of the series is also a continuous real function on* $(X, \tau)$.

## Art. 2 SEPARATION BY REAL FUNCTIONS

Let $(X, \tau)$ be a topological space, and $C(X)$ the set of all continuous real functions on $X$. Two disjoint subsets $A$, $B$ are said to be separated by a continuous real function (or *completely separated*), if there exists a function $f \in C(X)$, such that

$$f(x) = 0 \text{ for all } x \in A,\ f(x) = 1 \text{ for all } x \in B,$$

and $$0 \leq f(x) \leq 1, \text{ for all } x \in X.$$

The function $f$ is called the *Urysohn function* on $X$, corresponding to the pair of (disjoint) subsets $A, B$.

**Theorem 9 :** *In a topological space* $(X, \tau)$, *if two subsets* $P$, $Q$ *are separated by a continuous real function, then* $P$ *and* $Q$ *are also strongly separated (by a pair of disjoint open sets).*

**Proof :** Let $f$ denote the Urysohn function on $(X, \tau)$, corresponding to the pair of (disjoint) subsets $P$, $Q$.

Let $I$ denote the closed unit interval $[0, 1]$, and $\sigma_I$ denote the topology on $I$, relative to the usual topology $\sigma$ of the real number space **R**.

The half-open intervals $[0, 1/4)$ and $(3/4, 1]$ are open sets in $(I, \sigma_I)$. And, since $f$ is a continuous function on $(X, \tau)$ with values in $I$. It follows that $U = f^{-1}[0, 1/4)$ and $V = f^{-1}(3/4, 1]$ are two open sets in $(X, \tau)$. $U$ and $V$ are obviously disjoint; also, $P \subset U$ and $Q \subset V$. Thus $P$ and $Q$ are strongly separated by $U$ and $V$ in $(X, \tau)$.//

The converse of Theorem 9 is not true (vide Example 2, page 172).

In the previous chapter we introduced the concept of separation by open sets, and defined different types of spaces (viz. $T_i$-spaces, regular, normal and completely normal spaces) in terms of different separation axioms. In a similar manner, we shall now characterise some types of topological spaces in terms of separation by continuous real functions.

Novak has given an example of a $T_3$-space, on which every continuous real function is constant. Hewitt modified Novak's example, and gave an example of a $T_1$-space, on which every continuous real function is a constant

(Hewitt—On two problems of Urysohn, Annals of Maths. Volume 47 (1946), pp 503-509). Thus $T_1$, $T_2$ and $T_3$ spaces cannot be characterised in terms of separation by continuous real functions. So we now consider normal spaces.

## 2.1 Normal Spaces

We shall show that a topological space $(X, \tau)$ is normal if, and only if, every pair of disjoint closed sets in it can always be separated by a continuous real function. This is proved in Urysohn's lemma, given below :

**Theorem 10 :** *(Urysohn's lemma) : A topological space $(X, \tau)$ is normal if, and only if, every pair of disjoint closed sets $P$, $Q$ in $(X, \tau)$ are separated by a continuous real function. That is, there exists a continuous real function $f$ on $(X, \tau)$, such that*

$$\left.\begin{array}{ll} & f(x) = 0 \text{ for all } x \in P,\ f(x) = 1 \text{ for all } x \in Q, \\ \text{and} & 0 \leq f(x) \leq 1 \text{ for all } x \in X. \end{array}\right\} \quad \ldots(1)$$

**Proof :** Let $(X, \tau)$ be a topological space, such that there exists a continuous real function $f$ on $(X, \tau)$, satisfying the condition (1), for every pair of disjoint closed sets $P$, $Q$ in $(X, \tau)$. That is $P$, $Q$ are separated by the continuous function $f$. Then $P$ and $Q$ are strongly separated by a pair of disjoint open sets in $(X, \tau)$, by Theorem 9. Hence $(X, \tau)$ is a normal space.

Conversely, let $(X, \tau)$ be a normal space, and let $P$, $Q$ be a pair of disjoint closed sets in the $(X, \tau)$. Then the closed set $P$ is contained in the open set $X - Q$. Hence, by the normality criterion (Theorem 19, page 141), there exists an open set, which we shall denote by $U(1/2)$, such that

$$P \subset U(1/2) \text{ and } \overline{U(1/2)} \subset X - Q. \quad \ldots(2)$$

Similarly, corresponding to the two relations in (2) (in each of which a closed set is contained in an open set), there exist open sets $U(1/4)$ and $U(3/4)$, such that

$$P \subset U(1/4) \text{ and } \overline{U(1/4)} \subset U(1/2),$$

$$\text{and} \quad \overline{U(1/2)} \subset U(3/4) \text{ and } \overline{U(3/4)} \subset X - Q.$$

Suppose, by repeating the above process, we have defined the sets $U(r/2^n)$, for a fixed $n$, and $r = 1, 2, \ldots, 2^n - 1$, such that

$$P \subset U(1/2^n) \text{ and } \overline{U(r/2^n)} \subset U(r+1)/2^n), \quad \ldots(3)$$

where $U(1) = X - Q$. Then for each value of $r$, corresponding to the relation (3), there exists, by the normality criterion, an open set $U(2r+1)/2^{n+1})$, such that

$\overline{U(r/2^n)} \subset U((2r+1)/2^{n+1})$ and $\overline{U((2r+1)/2^{n+1})} \subset U(r+1)/2^n)$.

Thus, for every dyadic proper fraction $r/2^n$, there exists an open set $U(r/2^n)$ with the following properties :

(*a*) $P$ is contained in each $U(r/2^n)$;

(*b*) each $\overline{U(r/2^n)}$ is contained in $X - Q$; and

(*c*) if $r/2^n < r'/2^m$ then $\overline{U(r/2^n)} \subset U(r'/2^m)$.

Let now, a real function $f$ be defined on $X$, with values in $I$, as follows:

$f(x)$ = the greatest lower bound of all those numbers $r/2^n$, such that $x \in U(r/2^n)$,

and $f(x) = 1$, if $x$ does not lie in any $U(r/2^n)$. Then

(*i*) for any point $x$ in $X$, $0 \le f(x) \le 1$, since $0 < r/2^n < 1$;

(*ii*) for any point $x$ in $P$, $x$ belongs to every $U(r/2^n)$, so that $f(x) = 0$; and

(*iii*) for any point $x$ in $Q$, $x$ does not lie in any $U(r/2^n)$, so that $f(x) = 1$.

Thus all the conditions in (1) are satisfied for the function $f$ as defined above. Next we shall prove the continuity of $f$.

Let $p$ be any point in $X$, and let $\varepsilon$ be a pre-assigned positive number. The following three cases may arise:

**Case 1 :** When $f(p) = 0$ : Let $m$ be a positive integer such that $1/2^m < \varepsilon$, and let $G = U(1/2^m)$. Then $G$ is an open set containing $p$. Also, for any point $x$ in $G$, $f(x) \le 1/2^m < \varepsilon$. And, since $0 \le f(x)$, it follows that $|f(x) - f(p)| = f(x) < \varepsilon$. Hence $f$ is continuous at the point $p$.

**Case 2 :** When $f(p) = 1$ : Let $m$ be a positive integer such that $1/2^m < \varepsilon$, and let $G = X - \overline{U((2^m - 1)/2^m)}$. Then $G$ is an open set containing $p$, since $p \in \overline{U((2^m - 1)/2^m)} \subset U((2^{m+1} - 1)/2^{m+1})$ would imply $f(p) \le (2^{m+1} - 1)/2^{m+1} < 1$. Now, for any point $x$ in $G$, we must have

$$x \notin \overline{U((2^m - 1)/2^m)},$$

so that $x \notin U(2^m - 1)/2^m)$. Hence $x \notin U(r/2^n)$, whenever $r/2^n < (2^m - 1)/2^m$. Consequently,

$$f(x) \ge (2^m - 1)/2^m = 1 - 1/2^m > 1 - \varepsilon.$$

But $f(x) \leq 1$. Hence $|f(p) - f(x)| = 1 - f(x) < \varepsilon$. The function $f$ is therefore continuous at the point $p$.

**Case 3 :** When $0 < f(p) < 1$ : let $n$ be a positive integer such that $1/2^{n-1} < \varepsilon$ and $r/2^n < f(p) < (r+1)/2^n < 1$, for a suitable positive integer $r$. Then $p \notin U(r/2^n)$, and therefore $p \notin \overline{U(r-1)/2^n)}$, but $p \in U(r+1)/2^n$. Let now $G = U((r+1)/2^n) - \overline{U((r-1)/2^n)}$, then $G$ is an open set containing $p$. Also, for any point $x$ in $G$, $x \in U((r+1)/2^n)$ and $x \notin \overline{U((r-1)/2^n)}$, and so $x \notin U((r-1)/2^n)$. Hence, it follows (from the definition of the function $f(x)$) that

$$(r-1)/2^n \leq f(x) \leq (r+1)/2^n,$$

so that
$$|f(x) - f(p)| \leq 1/2^{n-1} < \varepsilon.$$

The function $f$ is therefore continuous at the point $p$.

Thus the function $f$ is continuous at every point $p$ in $X$, and is therefore continuous on $X$.//

Urysohn's lemma finds a significant application in the following Theorem, which is another important characterisation for a normal space.

**Theorem 11 :** *(Tietze's extension theorem) : A topological space* $(X, \tau)$ *is a normal space if, and only if, for every closed set* $F$ *in* $(X, \tau)$ *and every continuous real function* $g$ *on* $(F, \tau_F)$, *there exists a continuous real function* $h$ *on* $(X, \tau)$ *such that* $h(x) = g(x)$ *for all* $x \in F$ *(i.e.,* $g$ *has an extension* $h$ *on* $X$*).*

*Also, if, in particular,* $|g(x)| \leq \alpha$ *(a positive real number) for all* $x \in F$, *then* $|h(x)| \leq \alpha$ *for all* $x \in X$.

**Proof :** Let the condition hold in $(X, \tau)$, and let $P$, $Q$ be any pair of disjoint closed sets in $(X, \tau)$. Then $F = P \cup Q$ is a closed set in $(X, \tau)$. Let a function $g$ be defined on $(F, \tau_F)$ by $g(x) = 0$ if $x \in P$, and $g(x) = 1$ if $x \in Q$. Then $g$ is a continuous real function on $(F, \tau_F)$, since $P$ and $Q$ are disjoint. Hence, by hypothesis, there exists a continuous real function $h$ on $(X, \tau)$, such that $h(x) = g(x)$ for all $x \in F$. Let

$$U = \{x \in X : h(x) < 1/2\}, \text{ and } V = \{x \in X : h(x) > 1/2\}.$$

Then $U$ and $V$ are open sets in $(X, \tau)$, since $h$ is continuous (vide Example 1 (*ii*), page 161). Also, $U$ and $V$ are disjoint, and $P \subset U$ and $Q \subset V$. Thus $P$ and $Q$ are strongly separated by $U$ and $V$ in $(X, \tau)$. Hence $(X, \tau)$ is a normal space.

Conversely, let $(X, \tau)$ be a normal space, and let $g$ be a continuous real function on $(F, \tau_F)$, where $F$ is a closed subset in $(X, \tau)$.

Firstly, we assume that $g$ is bounded, *i.e.*, $|g(x)| \leq \alpha$ for all $x \in F$, where $\alpha$ is some fixed positive real number. Let

$$P_0 = \{x \in F : g(x) \leq (1/3)\,\alpha\},\ Q_0 = \{x \in F : g(x) \geq (2/3)\,\alpha\}.$$

Then $P_0$, $Q_0$ are disjoint, and they are also closed subsets of $(F, \tau_F)$, (vide Example 1 (*i*), page 161). As $F$ is itself a closed subset in $(X, \tau)$, it follows (by Example 1 (*i*), page 91) that $P_0$, $Q_0$ are also closed in $(X, \tau)$. As $(X, \tau)$ is a normal space, there exists, by Urysohn's lemma, a continuous real function on $(X, \tau)$, $f_0$ say, such that $f_0(x) = 0$ if $x \in P_0$ and $f_0(x) = 1$ if $x \in Q$, and $0 \leq f_0(x) \leq 1$ for all $x \in X$. Let

$$h_0(x) = (2\alpha/3)\,[f_0(x) - 1/2], \text{ for all } x \in X.$$

Then $h_0$ is a continuous real function on $(X, \tau)$, such that

$$|h_0(x)| \leq (1/3)\,\alpha,\ x \in X, \text{ and } |g(x) - h_0(x)| \leq (2/3)\,\alpha,\ x \in F.$$

Next, let $g_1(x) = g_0(x) - h_0(x)$ where $g_0(x) = g(x)$. Then $g_1$ is a continuous real function on $(F, \tau_F)$, where $|g_1(x)| \leq (2/3)\,\alpha$. By applying an argument similar to that in case of $g\ (= g_0)$, we obtain, corresponding to the function $g_1$, a continuous real function $h_1$ on $(X, \tau)$, and a continuous real function $g_2$ on $(F, \tau_F)$, satisfying the conditions :

$$|h_1(x)| \leq (2/3^2)\,\alpha,\ x \in X,$$

and

$$|g_2(x)| = |g_1(x) - h_1(x)| \leq (2/3)^2\,\alpha,\ x \in F.$$

By repeating this process, we obtain continuous real functions $h_m$ on $(X, \tau)$, and continuous real functions $g_m$ on $(F, \tau_F)$, for $m = 0, 1, 2, \ldots,$ such that

$$\left.\begin{aligned} &|h_m(x)| \leq (2^m/3^{m+1})\,\alpha,\ x \in X, \\ \text{and}\quad &|g_{m+1}(x)| \leq (2/3)^{m+1}\,\alpha,\ x \in F, \end{aligned}\right\} \qquad \ldots(4)$$

where $g_{m+1}(x) = g_m(x) - h_m(x)$, for all $x \in F$. ...(5)

It now follows, by putting $m = 0, 1, \ldots, n$ in (5), that

$$g_1 = g_0 - h_0,\ g_2 = g_1 - h_1,\ \ldots,\ g_{n+1} = g_n - h_n,$$

and therefore $g_{n+1} = g_0 - (h_0 + h_1 + \ldots + h_n) = g - \sum_{i=0}^{n} h_i$. Hence

$$|g(x) - \sum_{i=0}^{n} h_i(x)| = |g_{n+1}(x)| \leq (2/3)^{n+1}\,\alpha,\ x \in F. \qquad \ldots(6)$$

It follows from (4), by Weierstrass M-test (Theorem 7), that the infinite series $\Sigma h_i(x)$ converges uniformly on $(X, \tau)$; and the sum function $h(x) = \Sigma h_i(x)$ is a continuous real function on $(X, \tau)$, by Theorem 8. Again, it follows from (6) that $g(x) = \Sigma h_i(x) = h(x)$ for all $x \in F$. Thus $h(x)$ is a continuous real function on $(X, \tau)$, such that $h(x) = g(x)$ for all $x \in F$. Also, by (1),

$$|h(x)| \leq (1/3 + 2/3^2 + \ldots + 2^n/3^{n+1})\,\alpha < \alpha,\; x \in X.$$

Finally, let $g$ be unbounded on $(F, \tau)$. Then $\tan^{-1}(g)$ is a continuous real function, which is bounded on $(F, \tau_F)$, since $\tan^{-1}(g) < \pi/2$. Hence, by the first part of this proof, there exists a continuous real function $u$ (say) on $(X, \tau)$, such that $u(x) = \tan^{-1}(g(x))$ for all $x \in F$, $|u(x)| < \pi/2$.

Let $P = \{x \in X : u(x) = \pi/2 \text{ or } -\pi/2\}$; then $P$ is a closed subset in $(X, \tau)$, disjoint with $F$. Hence, by Urysohn's lemma, there exists a continuous real function $v$ on $(X, \tau)$, such that $v(x) = 0$ if $x \in P$, $v(x) = 1$ if $x \in F$, and $0 \leq v(x) \leq 1$ for all $x \in X$. Now putting $vu = w$, we obtain a continuous real function $w$ on $(X, \tau)$, such that $|w(x)| = |v(x) \,.\, u(x)| < \pi/2$ for all $x \in X$, and $w(x) = v(x) \,.\, u(x) = 1 \,.\, \tan^{-1}(g(x))$, *i.e.*, $\tan(w(x)) = g(x)$ for all $x \in F$. Hence $h = \tan(w)$ is a continuous real function on $(X, \tau)$, such that $h(x) = \tan(w(x)) = g(x)$ for $x \in F$. Thus $h(x)$ is the required extension of $g(x)$ on $(X, \tau)$.//

**Note :** The condition that $F$ is closed in $(X, \tau)$ is indispensable; the Theorem does not hold if this condition is dropped. In fact, in the real number space $(\mathbf{R}, \sigma)$, the function $g(x) = \sin(1/x)$ is continuous on the subspace $(F, \tau_F)$, where $F = (0, \infty)$. But $g(x)$ cannot be extended continuously on $(\mathbf{R}, \sigma)$, since $\lim_{x \to +0} g(x)$ is indeterminate.

## 2.2 Completely Regular Spaces

A topological space $(X, \tau)$ is called a *completely regular space* if, for any closed set $F$ and any point $p$, lying outside $F$, there always exists a continuous real function $f$ on $(X, \tau)$, such that

$$f(p) = 0,\; f(x) = 1 \text{ if } x \in F, \text{ and } 0 \leq f(x) \leq 1 \text{ for all } x \in X;$$

that is, $\{p\}$ and $F$ are separated by a continuous real function $f$ on $(X, \tau)$.

A completely regular $T_1$-space is called a *Tychonoff space.*

**Theorem 12 :** (*i*) *Every $T_4$-space is a Tychonoff space.*

(*ii*) *Every Tychonoff space is completely regular.*

(*iii*) *Every completely regular space is regular.*

(*iv*) *Every Tychonoff space is a $T_3$-space.*

**Proof :** (*i*) Let $(X, \tau)$ be a $T_4$-space, and $F$ be a closed set and $p$ any point, lying outside $F$, in $(X, \tau)$. As $(X, \tau)$ is a $T_1$-space, the one-pointic set $\{p\}$ and $F$ form a pair of disjoint closed sets in $(X, \tau)$. Also since $(X, \tau)$ is a $T_4$-space, there exists, by Urysohn's lemma, a continuous real function $f$ on $(X, \tau)$, such that $f(p) = 0$, $f(x) = 1$ if $x \in F$, and $0 \leq f(x) \leq 1$ for all $x \in X$. Hence $(X, \tau)$ is a completely regular space, and hence also a Tychonoff space, as it is a $T_1$-space.

(*ii*) This follows immediately from the definitions.

(*iii*) Let $(X, \tau)$ be a completely regular space, and let $F$ be a closed set and $p$ any point in $(X, \tau)$, where $p \notin F$. As $(X, \tau)$ is a completely regular space, $\{p\}$ and $F$ are separated by a continuous real function. Hence, by Theorem 9, $\{p\}$ and $F$ are strongly separated by a pair of disjoint open sets in $(X, \tau)$. Hence $(X, \tau)$ is a regular space.

(*iv*) It is an immediate consequence of the property (*iii*).//

By virtue of the properties (*i*) and (*iv*) in Theorem 12, a Tychonoff space is also called a $T_3\frac{1}{2}$-space.

It should however be noted that, inspite of the property (*i*) above, *a normal space may not be completely regular.* e.g., the space $(X, \tau)$, defined in Example 2, page 120, is a normal space but not a regular space; and hence it is not completely regular (by property (*iii*) above). In fact,

**Theorem 13 :** *A normal space is completely regular if, and only if, it is regular.*

**Proof :** Let $(X, \tau)$ be a normal space. If it is completely regular, then it is necessarily regular (by Theorem 12 (*iii*)).

Conversely, let $(X, \tau)$ be both regular and normal. Let $F$ be a closed subset, and $p$ a point in $(X, \tau)$, such that $p \notin F$. Then $X - F$ is an open set, containing the point $p$. It follows by the regularity criterion (Theorem 14, page 137) that there exists an open set $V$, such that $p \in V, \overline{V} \subset X - F$. Since $(X, \tau)$ is a normal space, there exists a Urysohn function $f$, corresponding to the pair of disjoint closed sets $\overline{V}$ and $F$. Thus $f(p) = 0$, since $p \in \overline{V}, f(x) = 1$ if $x \in F$, and $0 \leq f(x) \leq 1$ for all $x \in X$. Hence $(X, \tau)$ is completely regular.//

The converses of Theorem 9 and properties (*i*)-(*iv*) in Theorem 12 are not true, as can be seen from the following cases.

**Example 2.** ***Niemytzki's tangent disc topology (vide Example (F), page 126) is Tychonoff but not normal (and hence not $T_4$).***

**Solution :** Let $\tau$ be the Niemytzki's tangent disc topology on the set $X$, as defined in Example ($F$), page 126. It has been shown there that $(X, \tau)$ is a $T_2$-space, but not normal. To show that $(X, \tau)$ is completely regular, let $F$ be a closed set and $p$ a point in $(X, \tau)$, not contained in $F$.

Let $p \in X - L$, then there exists a *nbd* $U$ of $p$, which is contained in $X - F$, and is open relative to both $\tau$ and the Euclidean topology on $X$. So, $X - U$ is closed relative to the Euclidean topology $\sigma$ on $X$; and since the topology $\sigma$ is completely regular, there exists a Urysohn function $f$, corresponding to the pair $\{p, X - U\}$. Since $f$ is also a continuous (real) function, relative to $\tau$, and $F \subset X - U$, $f$ is also a Urysohn function corresponding to the pair $\{p, F\}$ in $(X, \tau)$.

Next, let $p \in L$, and let $D$ be a disc, tangent to $L$ at $p$, such that $D$ does not intersect $F$. Let the radius of $D$ be $r$. We define a function $f: X \to [0, 1]$, such that $f(p) = 0$, $f(x) = 1$ if $x$ is not in $D \cup \{p\}$, and $f(x, y) = [(x - p)^2 + y^2]/2ry$, for every point $(x, y) \in D$. Then $f$ is continuous, since $f^{-1}(c, 1]$ is the open set $X - \overline{D}_c$ and $f^{-1}[0, c)$ is the open set $\{p\} \cup D_c$, where $0 < c < 1$, and $D_c$ is the open disc of radius $cr$, tangent to $L$ at the point $p$. Thus $f$ is a Urysohn function for the pair $\{p, F\}$ in $(X, \tau)$.

Hence, $(X, \tau)$ is a completely regular space; and as it is $T_2$, it is also a Tychonoff space.

An example of *a completely regular space, which is not a Tychonoff space*, is given in Example 15 in Miscellaneous Exercises II.

We have alluded to *a topological space*, due to Novak, *which is $T_3$ (and therefore also regular)*, on which every continuous real function is constant; this space is, evidently, *not completely regular (and therefore also not a Tychonoff space).*

**Theorem 14 :** *A topological space $(X, \tau)$ is completely regular if, and only if, for each point $p$ in $X$ and each member $U$ belonging to a subbase* **S** *of $\tau$, such that $p \in U$, there exists a continuous real function $f$ on $(X, \tau)$, such that*

$$f(p) = 0,\ f(x) = 1 \text{ if } x \in X - U, \text{ and } 0 \le f(x) \le 1 \text{ for all } x \in X.$$

**Proof :** Let $F$ be any closed set in $(X, \tau)$ and let $p \notin F$. Then $p \in X - F$, and there exists an open set $V$, belonging to the base **B**, generated by **S**, such that $p \in V \subset X - F$. Now $V$ is an intersection of finitely many members of the subbase **S**, say $V = U_1 \cap U_2 \cap \ldots \cap U_n$, where $U_i \in$ **S** (for $i = 1, 2, ..., n$). Then $p \in U_i$, and there exists, by the hypothesis, a continuous real function $f_i$ on $(X, \tau)$, such that $f_i(p) = 0$, $f_i(x) = 1$ if $x \in X - U_i$, and $0 \leq f_i(x) \leq 1$ for all $x \in X$; and this holds for $i = 1, 2, ..., n$. Let, now

$$g(x) = \text{sup.} \{f_i(x) : i = 1, 2, ..., n\}.$$

Then $g$ is a continuous real function on $(X, \tau)$, such that $0 \leq g(x) \leq 1$, for all $x \in X$. Also $g(p) = 0$. And for any $x \in F$, $x \in F \subset X - V$ (since $V \subset X - F$), so that $x \in X - U_j$ for some $j$ (since $X - V = (X - U_1) \cup (X - U_2) \cup \ldots \cup (X - U_n)$); consequently, $g(x) = 1$ for all $x \in F$. Hence $(X, \tau)$ is a completely regular space.

Conversely, let $(X, \tau)$ be a completely regular space, and $p \in U$, where $p \in X$ and $U \in$ **S** (**S** being a subbase of $\tau$). Then $X - U$ is a closed set and $p \notin X - U$. Hence there exists a continuous real function $f$ on $(X, \tau)$, such that $f(p) = 0$, $f(x) = 1$ if $x \in X - U$, and $0 \leq f(x) \leq 1$ for all $x \in X$.//

**Theorem 15 :** *Every subspace of a completely regular space is a completely regular space.*

**Proof :** Let $(X, \tau)$ be a completely regular space, and $A$ any non-empty subset of $X$. Let $F$ be a closed set in $(A, \tau_A)$, and $p$ any point in $A$, where $p \notin F$. There exists a set $K$, closed in $(X, \tau)$, such that $F = A \cap K$, and then $p \notin K$. As $(X, \tau)$ is a completely regular space (by hypothesis), there exists a continuous real function $f$ on $(X, \tau)$, such that $f(p) = 0$, $f(x) = 1$ if $x \in K$, and $0 \leq f(x) \leq 1$ for all $x \in X$. Let $g = f|_A$ (the restriction of $f$ on $A$), then $g$ is also a continuous function on $(A, \tau_A)$ (vide theorem 59, page 91), such that $g(p) = 0$, $g(x) = 1$ if $x \in F = A \cap K$, and $0 \leq g(x) \leq 1$ for all $x \in A$. Hence the subspace $(A, \tau_A)$ is also completely regular.//

Since any subspace of a $T_1$-space is always a $T_1$-space (by Theorem 24, page 147), it follows immediately from the above Theorem that :

**Corollary :** *Any subspace of a Tychonoff space is also a Tychonoff space.*

**Theorem 16 :** *The topological product $(X, \tau)$ of any collection of topological spaces $\{(X_\alpha, \tau_\alpha)\}$ is completely regular if, and only if, each of the spaces $\{X_\alpha, \tau_\alpha\}$ is completely regular.*

**Proof :** Let each of the spaces $\{(X_\alpha, \tau_\alpha) : \alpha \in A\}$, where $A$ is an indexing set, be completely regular. Let $x$ be a point in $X$, and $U$ a member of subbase **S**, which generates the product topology $\tau$ on $X$, such that $p \in U$. Then $U = p_\beta^{-1}(U_\beta)$, for some $U_\beta \in \tau_\beta$, and for some $\beta \in A$, where

$p_\beta$ is the projection mapping of $X$ onto $X_\beta$. Also $x_\beta = p_\beta(x) \in U_\beta$. As $(X_\beta, \tau_\beta)$ is completely regular, there exists a continuous real function $f$ on $(X_\beta, \tau_\beta)$, such that $f(x_\beta) = 0$, $f(y) = 1$ if $y \in X_\beta - U_\beta$, and $0 \le f(y) \le 1$ for all $y \in X$. Then $g = f o p_\beta$ is a continuous real function on $(X, \tau)$, such that $g(x) = f(p_\beta(x)) = f(x_\beta) = 0$, and $g(y) = f(p_\beta(y)) = 1$, for all $y \in X - U$, since $p_\beta(y) \in X_\beta - U_\beta$, and also $0 \le g(y) \le 1$ for all $y \in X$. Hence, by Theorem 14, $(X, \tau)$ is a completely regular space.

Conversely, let $(X, \tau)$ be completely regular. Since each $(X_\alpha, \tau_\alpha)$ is homeomorphic with a subspace of $(X, \tau)$, it follows by Theorem 15 that each $(X_\alpha, \tau_\alpha)$ is also completely regular.//

Since the topological product of any collection of $T_1$-spaces is a $T_1$-space, and a subspace of a $T_1$-space is $T_1$, it follows immediately from the above Theorem that :

**Corollary :** *The topological product of any collection of topological spaces is a Tychonoff space if, and only if, each of the spaces is a Tychonoff space.*

*The quotient space of a Tychonoff space need not be even completely regular or $T_2$.* The space $(X, \tau)$, described in Example 11, page 154, is Tychonoff (since it is $T_4$), and its quotient space $Y$ is neither regular nor $T_2$.

The real number space $(\mathbf{R}, \sigma)$, with its usual topology, is a $T_5$-space, hence a normal space, and therefore also a Tychonoff space. Hence its subspace $(\mathbf{I}, \sigma_I)$, where $\mathbf{I}$ denotes the closed unit interval $[0, 1]$, is also a Tychonoff space (by Corollary to Theorem 15). The topological product of a family $\{(\mathbf{I}_\alpha, (\sigma_\mathbf{I})_\alpha) : (\mathbf{I}_\alpha, (\sigma_\mathbf{I})_\alpha = (\mathbf{I}, \sigma_\mathbf{I}),\ \alpha \in A\}$, where $A$ is an indexing set, is called a *unit cube* and denoted by $\mathbf{I}^A$. *A unit cube* $\mathbf{I}^A$ *is*, by Corollary to Theorem 16, always a Tychonoff space. On the other hand, we have the following:

**Theorem 17 :** *A topological space $(X, \tau)$ is a Tychonoff space if, and only if, it is homeomorphic with a subspace of a unit cube.*

**Proof :** It has been shown above that a unit cube is a Tychonoff space, and hence any subspace of a unit cube is also a Tychonoff space. Consequently, if $(X, \tau)$ is homeomorphic with a subspace of a unit cube, then $(X, \tau)$ is also a Tychonoff space.

Conversely, let $(X, \tau)$ be a Tychonoff space. Let $\{f_\alpha : \alpha \in A\}$ denote the family of all those continuous real functions on $(X, \tau)$, for which

$0 \le f_\alpha(x) \le 1$ for all $x \in X$. For the indexing set $A$, we consider the unit cube $I^A$. We now define a mapping $f$ of $(X, \tau)$ into $\mathbf{I}^A$, by

$$f(x) = \{ f_\alpha(x) : \alpha \in A \}.$$

We shall show that $f$ is a homeomorphism of $(X, \tau)$ onto the subspace $f(X) \subset I^A$. $f$ is obviously a surjective mapping on $X$ onto $f(X) \subset \mathbf{I}^A$.

Let $x$, $y$ be two distinct points in $X$. As $(X, \tau)$ is completely regular, there exists a continuous real function $g$ on $(X, \tau)$, such that $g(x) = 0$, $g(y) = 1$, and $0 \le g(u) \le 1$ for all $u \in X$, the one-pointic set $\{y\}$ being closed (as $(X, \tau)$ is a $T_1$-space); then $g = f_\beta$, for some $\beta \in A$, so that $f_\beta(x) = 0$ and $f_\beta(y) = 1$. Consequently, $f(x)$ and $f(y)$ are distinct, as they differ in their $\beta$th coordinates. Hence $f$ is injective, and therefore also bijective.

Again, $f_\alpha$, being the projection mapping of $(X, \tau)$ onto the co-ordinate space $(I, \sigma_I)$, is an open and continuous mapping, for each $\alpha \in A$. Consequently, the mapping $f$ is found to be both open and continuous (as in the case of an evaluation mapping). Thus $(X, \tau)$ is homeomorphic with the subspace $F(X)$ of the unit cube $\mathbf{I}^A$.//

By virtue of the property, proved in the Theorem 17, a unit cube is known as the *universal Tychonoff space.*

## 2.3 Perfectly Normal Spaces

A normal space has been characterised, in the Urysohn's lemma, by the existence of a Urysohn function on the space, corresponding to every pair of disjoint closed sets, $P$, $Q$ in the space. The Urysohn function $f$ is a continuous real function on the space $(X, \tau)$, such that $f(x) = 0$ if $x \in P$, $f(x) = 1$ if $x \in Q$, $0 \le f(x) \le 1$ for all $x \in X$. But there may exist points $x \notin P$ such that $f(x) = 0$, and points $y \notin Q$, for which $f(y) = 1$. That is, although $f^{-1}(0) \supset P$ and $f^{-1}(1) \supset Q$, but $f^{-1}(0) = P$ and $f^{-1}(1) = Q$ may not hold. Now, in order that a Urysohn function $f$ may satisfy these additional conditions, it is necessary to impose stronger conditions on the space, and this leads to the concept of a perfectly normal space.

**Lemma:** *Let $\{P, Q\}$ be a pair of disjoint closed sets in a normal space $(X, \tau)$. Then a necessary and sufficient condition for the existence of a Urysohn function $f$ on $X$, corresponding to the pair $\{P, Q\}$, and satisfying $f^{-1}(0) = P$, is that $P$ is a $G_\delta$-set.*

**Proof:** Let there exist a Urysohn function $f$ on $X$, corresponding to the pair $\{P, Q\}$, satisfying the additional condition $f^{-1}(0) = P$. Then

$$P = \cap \{ U_n : n = 1, 2, \dots \}$$

is a $G_\delta$-set, where $U_n = \{x \in X : f(x) < 1/n\}$ is an open set (by Example 1, page 161), for $n = 1, 2, \ldots$

Conversely, let $P$ be a $G_\delta$-set, say $P = \cap \{U_n : n = 1, 2, \ldots\}$, where each $U_n$ is an open set. Let $V_n = U_1 \cap U_2 \cap \ldots \cap U_n$, then $V_n$ is also an open set, for $n = 1, 2, \ldots$, and $P = \cap \{V_n : n = 1, 2, \ldots\}$. Also, $V_1 \supset V_2 \supset \ldots$; and we can assume, without any loss of generality, that $V_1 \cap Q = \Phi$. Let $f_n$ be a Urysohn function on $X$, corresponding to the pair of disjoint closed sets $\{P, X - V_n\}$, where $f(P) = 0$. Let $f = \sum_{n=1}^{\infty} (1/2^n) f_n$, then $f$ is a continuous function on $X$, by Theorems 7 and 8. It can readily be seen that $f$ is a Urysohn function on $X$, corresponding to the pair $\{P, Q\}$. Also, if $y \notin P$, then $y \notin V_r$, *i.e.*, $y \notin X - V_r$, for some $r$. Hence $f_r(y) = 1$, and therefore $f(y) \geq 1/2^r$; consequently, $f^{-1}(0) = P$.//

**Theorem 18 :** *Let $P$, $Q$ be a pair of disjoint closed sets in a normal space $(X, \tau)$. Then a necessary and sufficient condition that there exists a Urysohn function $f$ on $X$, corresponding to the pair $\{P, Q\}$, satisfying the additional conditions $f^{-1}(0) = P$ and $f^{-1}(1) = Q$, is that both $P$ and $Q$ be $G_\delta$-sets.*

**Proof :** Let both $P$ and $Q$ be $G_\delta$-sets. Then there exists, by the above lemma, a Urysohn function $g$ on $X$, corresponding to the pair $\{P, Q\}$, satisfying $g^{-1}(0) = P$, and a Urysohn function $h$ on $X$, corresponding to the pair $\{Q, P\}$, satisfying $h^{-1}(0) = Q$.

Let $f(x) = g(x)/[g(x) + h(x)]$, for all $x \in X$. Then

(*i*) $f(x)$ is a continuous function on $X$, by Theorem 2, since $g(x) + h(x) \neq 0$ for all $x \in X$ (as $P$ and $Q$ are disjoint),

(*ii*) $0 \leq f(x) \leq 1$, for all $x \in X$,

(*iii*) $f(x) = 0$ if, and only if, $g(x) = 0$, so that $f^{-1}(0) = P$, and

(*iv*) $f(x) = 1$ if, and only if, $h(x) = 0$, so that $f^{-1}(1) = Q$. Thus $f$ satisfies all the requisite conditions of the Theorem.

Conversely, let there exist a Urysohn function $f$ on $X$, corresponding to the pair $\{P, Q\}$, and satisfying the additional conditions $f^{-1}(0) = P$ and $f^{-1}(1) = Q$. Then $P = \cap \{U_n : n = 1, 2, \ldots\}$, and $Q = \cap \{V_n : n = 1, 2, \ldots\}$ are $G_\delta$-sets, where $U_n = \{x \in X : f(x) < 1/n\}$ and $V_n = \{x \in X : f(x) > n/(n+1)\}$ are open sets (vide Example 1, page 161), for $n = 1, 2, \ldots$//

A topological space $(X, \tau)$ is called a *perfectly normal space* if, for each pair of disjoint closed sets $P$, $Q$ in $(X, \tau)$, there exists a continuous real function $f$ on $(X, \tau)$, such that

$$0 \leq f(x) \leq 1 \text{ for all } x \in X, \text{ and } f^{-1}(0) = P \text{ and } f^{-1}(1) = Q.$$

It then follows immediately from the definition, by Urysohn's lemma, that *every perfectly normal space is a normal space.*

As an immediate consequence of Theorem 18, we obtain :

**Theorem 19 :** *A topological space is perfectly normal if, and only if, it is a normal space, in which every closed set is a $G_\delta$-set.*

Another necessary and sufficient condition for a perfectly normal space is the following :

**Theorem 20 :** *A topological space $(X, \tau)$ is perfectly normal if, and only if, for any non-empty closed set $F$ and any point $p \notin F$ in $(X, \tau)$, there exists a continuous real function $f : (X, \tau) \to \mathbf{I}$, (where $\mathbf{I}$ is the closed unit interval $[0, 1]$), such that $f^{-1}(0) = F$ and $f(p) = 1$.*

**Proof :** Let the condition be satisfied in $(X, \tau)$, and let $P$, $Q$ be a pair of disjoint closed set in $(X, \tau)$. Let $a \in P$ and $b \in Q$. Then there exist, by hypothesis, two continuous real functions $f$ and $g$ from $(X, \tau)$ to $\mathbf{I}$, satisfying the conditions $f^{-1}(0) = P$, $f(b) = 1$, and $g^{-1}(0) = Q$ and $g(a) = 1$. Then $h = f - g$ is a continuous real function from $(X, \tau)$ to the closed interval $[-1, 1]$, and $h(x) > 0$ for all $x \in P$, and $h(x) < 0$ for all $x \in Q$. Hence the two disjoint open sets $U = \{h^{-1}(r) : r < 0\}$ and $V = \{h^{-1}(r) : r > 0\}$ separate $P$ and $Q$ strongly; and $(X, \tau)$ is therefore a normal space.

Let $F$ be a closed set in $(X, \tau)$. If $F = X$ or $\Phi$, then $F$ is obviously a $G_\delta$-set. Next, let $F \neq X, \Phi$; then there exists a point $p \notin F$ in $X$. Then, by hypothesis, there exists a continuous real function $f$ from $(X, \tau)$ to $I$, such that $f^{-1}(0) = F$ and $f(p) = 1$. Hence $F = \cap \{f^{-1}(r) : r < 1/n, n = 1, 2, \ldots\}$ is a $G_\delta$-set. Thus $(X, \tau)$ is a normal space, in which every closed set is a $G_\delta$-set; it is therefore perfectly normal, by Theorem 19.

Conversely, let $(X, \tau)$ be a perfectly normal space. Let $F$ be a non-empty closed set and $p \notin F$ be a point in $(X, \tau)$. Then $F$ is a $G_\delta$-set; and let $F = \cap \{G_n : n = 1, 2, \ldots\}$, where each $G_n$ is an open set in $(X, \tau)$. As $p \notin F$, there exists a positive integer $r$, such that $p \notin G_r$. Let $H_1 = G_1 \cap G_2 \cap \ldots \cap G_r$, then $F \subset H_1$; and by the normality criterion, there exists an open set $H^*$, such that $F \subset H^*$, $\overline{H}^* \subset H_1$. Next, let $H_2 = G_{r+1} \cap H^*$. Then $F \subset H_2$ and $\overline{H}_2 \subset H_1$. Assume that open sets $H_1, H_2, \ldots, H_k$ have been obtained, such that $F \subset H_m$, $\overline{H}_m \subset H_{m-1}$, and $H_m \subset G_{r+m-1}$, for $m = 2, 3, \ldots k$. Now, we put $H_{k+1} = H_k^* \cap G_{r+k}$, where $H_k^*$ is an open set, satisfying $F \subset H_k^*$, $\overline{H}_k^* \subset H_k$. Thus, we construct inductively a sequence of open sets $\{H_n : n = 1, 2, \ldots\}$, such that $F = \cap \{H_n : n = 1, 2, \ldots\}$ (since $F \subset H_n \subset G_{r+n-1}$), and which satisfies the condition $\overline{H}_{n+1} \subset H_n \subset X - \{p\}$.

Now, putting $H_{n+1} = U(1/2^n)$, $n = 0, 1, 2, \ldots$, and taking $P = F$, $Q = X - H_1$, and proceeding as in the proof of Urysohn's lemma, we find that the sets $U(r/2^n)$ possess the properties (*a*), (*b*) and (*c*), in page 167, and we obtain a continuous real function $f$ on $X$, having the properties (*i*), (*ii*) and (*iii*).

Then $F \subset f^{-1}(0) = \cap \{f^{-1}(y) : y < 1/2^n,\ n = 1, 2, \ldots\}$

$\subset \cap \{H_n : n = 1, 2, \ldots\} = F.$

Thus $f^{-1}(0) = F$ and $f(p) = 1$. Also the continuity of $f$ follows in exactly the same way as in the case of Urysohn's lemma.//

**Theorem 21 :** *Every perfectly normal space is* (*i*) *completely regular, and* (*ii*) *completely normal.*

**Proof :** (*i*) Follows immediately from Theorem 20.

(*ii*) Let $\{A, B\}$, be a pair of weakly separated sets in a perfectly normal space $(X, \tau)$, so that $\overline{A} \cap B = \Phi$ and $A \cap \overline{B} = \Phi$. also every closed set in $(X, \tau)$ is a $G\delta$-set. Hence, for any closed set $F$, disjoint with $\overline{A}$, there exists, by the lemma (page 175), a Urysohn function $f$ on $X$, corresponding to the pair $\{\overline{A}, F\}$ and satisfying the condition $f^{-1}(0) = \overline{A}$. Likewise, for any closed set $K$, disjoint with $\overline{B}$, there exists a Urysohn function $g$ on $X$, corresponding to the pair $\{\overline{B}, K\}$, and satisfying the condition $g^{-1}(0) = \overline{B}$. Let

$$U = \{x \in X : f(x) < g(x)\} \text{ and } V = \{x \in X : g(x) < f(x)\}.$$

Then $U$ and $V$ are open sets, and $U \cap V = \Phi$. Also, $A \subset U$, since if $x \in A$, then $f(x) = 0$, and as $\overline{B} \cap A = \Phi$, we have $g(x) > 0$; similarly, $B \subset V$. Thus $A$ and $B$ are separated strongly by the pair of disjoint open sets $U$ and $V$ in $(X, \tau)$. $(X, \tau)$ is therefore a completely normal space.//

The converse of Theorem 21 is not true, as shown below :

**Example 3.** ***Let Z denote the set of all integers, and τ consist of Z and the set of all even integers. Then (Z, τ) forms a topological space,***

***which is completely normal but not perfectly normal. (In fact, the set of all odd integers is closed, but not a $G_\delta$-set in $(Z, \tau)$). This space is also not regular.***

## 2.4 Completely Hausdorff Spaces

A topological space $(X, \tau)$ is said to be completely Hausdorff if, for every pair of distinct points $p$, $q$ in $X$, there exists a continuous real function $f$ on $X$, such that $f(p) = 0$, $f(q) = 1$, and $0 \le f(x) \le 1$ for all $x \in X$. That is, every pair of distinct points (*i.e.*, disjoint one-pointic subsets) in $(X, \tau)$ can be separated by a continuous real function on $X$. The following results follow readily.

**Theorem 22 :** *Every Tychonoff space is completely Hausdorff.*

**Proof :** Let $\{p, q\}$ be a pair if distinct points in a Tychonoff space $(X, \tau)$. Since $(X, \tau)$ is a $T_1$-space, the one-pointic set $\{p\}$ is closed; and since $(X, \tau)$ is completely regular, $\{p\}$ and $q$ can be separated by a continuous real function on $X$. Hence $(X, \tau)$ is a completely Hausdorff space.//

It may be noted that *a completely regular space (or even a completely normal space) need not be completely Hausdorff.* This is seen from Example 15 in Miscellaneous Exercises II, page 182.

On the other hand, it can be seen from Example 16 in Miscellaneous Exercises II that *a completely Hausdorff space need not be regular or normal (and hence also not completely regular or completely normal).*

An example of *a semi-regular space, which is not completely Hausdorff,* is given in Example 14 in Miscellaneous Exercises II; and an example of *a completely Hausdorff space, which is not semi-regular,* is given in Example 16 in Miscellaneous Exercises II.

**Theorem 23 :** *Every completely Hausdorff space is necessarily (i) a Urysohn space, and (ii) a Hausdorff space.*

**Proof :** (*i*) Let $(X, \tau)$ be a completely Hausdorff space. Let $\{p, q\}$ be a pair of distinct points in $X$, and $f$ a continuous real function on $X$, which separates $p$ and $q$, such that $f(p) = 0$ and $f(q) = 1$. Let $I$ be the closed unit interval $[0, 1]$, and $\sigma_I$ be the topology on $I$, relative to the usual topology of the real number space. Then $[0, 1/4]$ and $[3/4, 1]$ are two disjoint closed sets in $(I, \sigma_I)$; and since $f: X \to I$ is continuous, $U = f^{-1}[0, 1/4]$ and $V = f^{-1}[3/4, 1]$ are two disjoints closed *nbds* of the points $p$, $q$ in $(X, \tau)$. $(X, \tau)$ is therefore a Urysohn space.

(*ii*) Since every Urysohn space is Hausdorff (by Theorem 36 (*ii*), page 158), the result (*ii*) follows immediately from (*i*).//

The converses of Theorem 23 (*i*) and (*ii*) are not true. In fact, an example of *a Hausdorff space, which is not completely Hausdorff,* is given in Example 14 in Miscellaneous Exercises II. An example of a Urysohn space, which is not completely Hausdorff, is given in Example 20 in Miscellaneous Exercises II.

## MISCELLANEOUS EXERCISES II

**1.** Let $(X, \tau)$ be a topological space, and a binary relation $\rho$ be defined on $X$, such that $p\rho q$ holds, for $p, q \in X$, iff $p \in \overline{\{q\}}$. Show that :

(*i*) $\rho$ is reflexive and transitive, and

(*ii*) $\rho$ is antisymmetric if, and only if, $(X, \tau)$ is a $T_0$-space.

**2.** Let $(X, \tau)$ be a $T_1$-space. Prove that the intersection of every collection of open sets is an open set in $(X, \tau)$ if, and only if, the topology $\tau$ is discrete.

**3.** (*a*) Prove that the property $p \in \overline{\{q\}} \Rightarrow q \in \overline{\{p\}}$ holds in a topological space $(X, \tau)$ if, and only if, every open set contains the closure of every one-pointic subset of it.

(*b*) Let $(X, \tau)$ be a topological space, in which $p \in \overline{\{q\}} \Rightarrow q \in \overline{\{p\}}$. Prove that :

(*i*) $\overline{\{x\}} = \cap \{G : G \in \mathbf{N}_x\}$, where $\mathbf{N}_x$ denotes the *nbd* filter of $x$, for every $x \in X$.

(*ii*) If, in particular, the set $X$ is finite, then $(X, \tau)$ is a regular space.

**4.** Show that an infinite Hausdorff space contains

(*i*) an infinite isolated subset;

(*ii*) a countably infinite discrete subspace;

(*iii*) an infinite sequence of non-empty, disjoint open subsets.

**5.** Let $\tau_1$ and $\tau_2$ be two Hausdorff topologies on a set $X$. Prove that:

(*i*) If $(X, \tau_1 \cap \tau_2)$ is Hausdorff, then the diagonal is a closed subset of $X \times X$ (with product topology); but the converse is not true.

(*ii*) If disjoint $\tau_1$-open sets can be separated by disjoint $\tau_2$-open sets, and vice-versa, then $(X, \tau_1 \cap \tau_2)$ is Hausdorff; but the converse is, in general, not true.

**6.** Let $X$ be an infinite set. Let $a \in X$, and $\tau_\alpha$ consist of

(*i*) all those subsets of $X$ which do not contain the point $a$, and

(*ii*) all those subsets, whose complements are finite subsets (vide Example 11 (*a*), page 103). Show that

(*a*) $\tau_\alpha$ is a Hausdorff topology.

(*b*) For two distinct points $p$, $q$, $\tau_p \cap \tau_q$ is a $T_1$ topology, but not a Hausdorff topology.

**7.** Let **R** denote the set of all real numbers, and let a subset $F$ of $X$ be called closed, if $F$ is bounded and closed in the usual sense. Show that the family of closed subsets, thus determined, satisfies the conditions $C(1) - C(4)$ of Theorem 19, page 61, and that the resulting topology on $X$ is $T_1$, but not Hausdorff.

**8.** Show that a topological space $(X, \tau)$ is a Hausdorff space if, and only if, for very subset $Y \subset X$ and points $y_1, y_2 \in Y$, there is a decomposition $Y = Y_1 \cup Y_2$ with $Y_1 \cap Y_2 = \Phi$, such that $y_2 \notin Y_1'$ and $y_1 \notin Y_2'$.

**9.** (*a*) If every point of a topological space $(X, \tau)$ has a closed *nbd*, which is a $T_2$-space, as a subspace of $(X, \tau)$, then $(X, \tau)$ is a $T_2$-space.

(*b*) Let $(X, \tau)$ be a Hausdorff space, and for every point $x \in X$, there exists a *nbd* $V$, such that $V$ is regular. Show that $(X, \tau)$ is a regular space.

**10.** Prove that the least upper bound of a family of regular topologies on a set $X$ is a regular topology on $X$.

**11.** Let $(X, \tau)$ be an infinite space with cofinite topology, and $(Y, \tau')$ a $T_1$-space. Prove that if $f: X \to Y$ is a continuous surjection, then either $f$ is constant, or $X$ and $Y$ are homeomorphic.

**12.** Show that if $(X, \tau)$ be a $T_3$-space, and $(Y, \tau')$ is obtained from $(X, \tau)$ by identifying a single closed set $F$ in $(X, \tau)$ with a point, then $(Y, \tau')$ is a $T_2$-space.

**13.** Show that a semi-regular $T_1$-space may not be Hausdorff.

**14.** Let $(\mathbf{Z}, \tau)$ denote the topological space, defined in Example 7 (*g*), page 103. Show that $(\mathbf{Z}, \tau)$ is Hausdorff and semi-regular, but neither completely Hausdorff, nor Urysohn, nor regular $T_1$ (therefore not $T_3$).

**15.** Let $(X, \tau)$ denote the topological product of (1) the real number space with usual topology, and (2) the space $\{0, 1\}$ with indiscrete topology. Show that $(X, \tau)$ is completely regular and completely normal (hence also regular and normal), but not $T_0$ and perfectly normal) hence also not $T_1$, $T_2$, $T_3$, $T_4$, $T_5$, Tychonoff, completely Hausdorff and Urysohn).

**16.** *The half-disc topology* (vide Example (*E*), page 125) is completely Hausdorff (hence also Urysohn and Hausdorff), but not semi-regular and normal (hence also not regular, $T_3$, completely regular and Tychonoff, and also not $T_4$, $T_5$, completely normal and perfectly normal).

**17.** Niemytzki's tangent disc topology (vide Example (*F*), page 126, and Example 2, page 172) is $T_2$, and completely regular (hence also regular, $T_3$, semi-regular, Tychonoff, completely Hausdorff and Urysohn), but not normal (hence also not $T_4$, $T_5$, completely normal and perfectly normal).

**18.** Let $X = \{(m, n) : m, n \text{ are integers}\}$, and let a subset $F$ be called closed if, either $(0, 0) \in F$, or $F$ contains finitely many pairs $(m, n)$ for all but a finite number of values of $m$. Show that the closed sets, thus defined, determine a topology $\tau$ on $X$, and that $(X, \tau)$ is a $T_5$-space, but not perfectly normal. [The one-pointic set $\{(0, 0)\}$ is closed, but not a $G_\delta$-set in $(X, \tau)$.]

**19.** Show that if $X \times Y$ is a $T_5$-space, then either every countable subset of $Y$ is closed, or $X$ is perfectly normal. [Vide : M. Katetov—Complete normality of Cartesian products. Fundamenta Mathematica, Volume 35 (1948), pp. 271-274].

**20.** Let $X = S \cup \{(0, 0)\} \cup \{(1/2\, r,\ r\sqrt{2}) : r \in \mathbf{Q},\ 0 < r\sqrt{2} < 1\}$, where $S = \{(x, y) : x, y \in \mathbf{Q}, 0 < x < 1, 0 < y < 1\}$, and $\mathbf{Q}$ is the set of all rational numbers. We define a base for a topology $\tau$ on $X$, by choosing local base for each point as follows :

(*i*) For each point of $S$ we choose a local base induced by the Euclidean topology on $S$.

(*ii*) For the other points $(x, y)$, we choose a local base

$$\{U_n(x, y),\ n = 1, 2, \ldots\} \text{ given by}$$

$$U_n(0, 0) = \{(0, 0)\} \cup \{(x, y) : 0 < x < 1/4,\ 0 < y < 1/n\},$$

$$U_n(1, 0) = \{(1, 0)\} \cup \{(x, y) : 3/4 < x < 1,\ 0 < y < 1/n\},$$

$$U_n(1/2\,r,\ r\sqrt{2}) = \{(x,\ y) : 1/4 < x < 3/4,\ |y - r\sqrt{2}| < 1/n\}.$$

Show that the space $(X,\ \tau)$ is Urysohn but not completely Hausdorff.

**21.** Let $(X,\ \tau)$ be a topological space, and $C(X)$ the set of all bounded continuous real functions on $X$.

(*i*) Let $U(x,\ f,\ \varepsilon) = \{y : |f(x) - f(y)| < \varepsilon\}$, where

$$x \in X,\ f \in C(X) \text{ and } \varepsilon > 0.$$

Show that if $(X,\ \tau)$ be a completely regular $T_2$-space, then

$$\{U(x, f, \varepsilon) : x \in X,\ f \in C(X),\ \varepsilon > 0\}$$

is a subbase of $\tau$.

(*ii*) $(X,\ \tau)$ is completely regular if, and only if, $\tau$ is identical with the weak topology on $X$, induced by the family $C(X)$.

**22.** Let $(X,\ \tau)$ be a Hausdorff space, and $f$ a real function on $X$. Prove that $(X,\ \tau)$ is completely regular if, and only if, each lower semi-continuous function $f$ is the supremum of a suitable family

$$\{f_\alpha : \alpha \in A\}$$

(where $A$ is an indexing set), of continuous real functions on $X$.

**23.** Let $\Omega$ be the first uncountable ordinal, and $\omega$ the first infinite ordinal (*i.e.*, the initial ordinal number of the cardinal number $d$) (vide page 40). Let $[0,\ \Omega]$ and $[0,\ \omega]$ be the corresponding closed ordinal spaces (vide Example $C(b)$, page 123-124); then

$$T = [0,\ \Omega] \times [0,\ \omega] - (\Omega,\ \omega)$$

is known as the Tychonoff plank.

Let $\mathbf{Z}$ denote the set of all integers, and for each $n \in \mathbf{Z}$, let

$$T_n = T \times \{n\},$$

and the elements of $T_n$ be denoted by $(\alpha,\ x;\ n)$. In the topological sum $\sum_{-\infty}^{\infty} T_n$, we make the identifications :

$(\Omega,\ x;\ 2k+1) \sim (\Omega,\ x;\ 2k+2)$ and $(\alpha,\ \omega : 2k+1) \sim (\alpha,\ \omega;\ 2k)$, for each $k \in \mathbf{Z}$, $\alpha \in [0,\ \Omega]$ and $x \in [0,\ \omega]$; and let the resulting space (which is often called the "spiral stair-case") be denoted by $S$. Next, we consider the space $Y = S \cup \{a\} \cup \{b\}$, obtained by adding two elements $a$, $b \notin S$, and choosing the *nbd* filters of $a$ and $b$ as all sets of the form

$\{a\} \cup [S - \cup \{T_n \subset S : n \leq N\}]$ and $\{b\} \cup [S - \cup \{T_n \subset S : n \geq N\}$ respectively. Show that :

(*i*) The Tychonoff plank $T$ is a Tychonoff space, which is not normal.

(*ii*) Any continuous real function on $Y$ has the same value at $a$ and $b$.

(*iii*) The space $Y$ is regular and $T_1$, but not completely regular, and not even completely Hausdorff.

*Chapter 5*

# COUNTABILITY

Topological spaces, in which some prescribed countability properties hold, will be considered in this chapter.

## Art. 1 COUNTABILITY PROPERTIES

Two axioms of countability were formulated by F. Hausdorff in 1914. The axioms are as follows :

A topological space $(X, \tau)$ is said to satisfy the *first axiom of countability*, if there exists a countable open base about every point in $X$ (see page 45). $(X, \tau)$ is then said to be a *first countable* or *first axiom* (or *locally separable*) space.

A topological space $(X, \tau)$ is said to satisfy the *second axiom* of *countability*, if there exists a countable open base of the topology $\tau$. $(X, \tau)$ is then said to be a *second countable* or *second axiom* (or *strongly separable*, or *perfectly separable*, or *completely separable*) space.

The concept of a separable space was introduced by M. Frechet in 1906.

A topological space $(X, \tau)$ is said to be *separable*, if there exists a countable dense subset in $(X, \tau)$; that is, there exists a countable subset $D$, such that $\overline{D} = X$.

Although the property had been used earlier, the term 'Lindeloff space' was introduced by Kuratowski and Sierpinski in 1921.

A topological space $(X, \tau)$ is called a *Lindeloff space* if, for every open covering of $X$, there always exists a countable sub-covering (see page 142).

**Theorem 1 :** *Every second countable space is* (*i*) *first countable,* (*ii*) *separable, and* (*iii*) *Lindeloff.*

**Proof :** Let $(X, \tau)$ be a second countable space, and let $\mathbf{W} = \{W_i : i = 1, 2, \ldots\}$ be a countable open base of the topology $\tau$.

(*i*) Let $p$ be any point in $X$. Then the collection $\mathbf{B}_p$ of all those $W_i \in \mathbf{W}$, which contain the point $p$, forms a countable open base about the point $p$, since for any open set $G$ containing $p$, there exists a $W_j$, such that $p \in W_j \subset G$ (and this $W_j \in \mathbf{B}_p$). Thus there exists a countable open base about every point $p$ in $X$. Hence the space $(X, \tau)$ is first countable.

(*ii*) From each $W_i$ we choose a point $p_i$, for $i = 1, 2, ...,$ such that $D = \{p_i : i = 1, 2, \ldots\}$ is an infinite sequence. Then $D$ is a countable subset of $X$. We shall show that $\overline{D} = X$. Let $x \in X - D$, and $G$ be any open set containing $x$. Then there exists a set $W_j$, such that $x \in W_j \subset G$; and then $x \neq p_j$, since $p_j \in D$ and $x \notin D$. Thus $G$ intersects $D$ in the point $p_j \neq x$; hence $x \in D'$. That is, any point of $X$ either belongs to $D$, or else is a limiting point of $D$; consequently, $X = \overline{D}$. The space $(X, \tau)$ is therefore separable.

(*iii*) Let $\mathbf{V} = \{V_\alpha : \alpha \in A\}$, where $A$ is an indexing set, be an open covering of $X$. Let $x \in X$, and $V$ be member of $\mathbf{V}$, containing the point $x$. Then there exists a set $W \in \mathbf{W}$, such that $x \in W \subset V$. As $x$ and $V$, $x \in V$, run over $X$ and $\mathbf{V}$ respectively, the corresponding sets $W$ form a countable sub-collection $\mathbf{U}$ of $\mathbf{W}$; and let $\mathbf{U} = \{U_i : 1, 2, \ldots\}$. Then each $U_i$ is contained in some $V$ belonging to $\mathbf{V}$; and let $V_i$ denote a set belonging to $\mathbf{V}$, such that $U_i \subset V_i$, for $i = 1, 2, \ldots$ The family of sets

$$\{V_i : i = 1,\ 2, \ldots\},$$

thus distinguished, forms a countable sub-covering of $\mathbf{V}$, since every point $x \in X$ must belong to some $U_i$, and hence also to the corresponding $V_i$. $(X, \tau)$ is therefore a Lindeloff space.//

Theorem 1 (*iii*) is known as the *Lindeloff theorem.*

The converses of (*i*), (*ii*) and (*iii*) in Theorem 1 are not true (vide Theorem 3). Also, excepting the results in Theorem 1 (*i*), (*ii*), (*iii*), there do not exist any other implications among the four countability properties, as can be seen from the following Example 1, Example 2 and Example 4.

**Theorem 2 :** *The real number space **R**, with the usual topology* $\tau$, *is second countable (and, hence, also first countable, separable, and Lindeloff).*

**Proof :** The null set $\Phi$ and the family of open intervals $\{(a, b) : a < b$, and $a$, $b$ are rational numbers$\}$ form a countable open base of the topology $\tau$ on or $\mathbf{R}$ (vide Theorem 8, page 49). Hence $(\mathbf{R}, \tau)$ is a second countable space.//

**Theorem 3:** *The Sorgenfrey line (that is, the real number space **R**, with the lower limit topology σ) is first countable, separable and Lindeloff, but not second countable.*

**Proof:** Let **Q** denote the set of all rational numbers.

(*i*) Let $p$ be any point in **R**, and let $\mathbf{B}_p = \{[p, q) : q \in \mathbf{Q}\}$. Then $\mathbf{B}_p$ is a countable collection; and it also forms an open base about the point $p$. In fact, if $U$ is any open set in (**R**, σ), containing $p$, then there exists a right-half open interval $[a, b)$, where $a$, $b$ are suitable real numbers, such that $p \in [a, b) \subset U$. Then $a \le p$ and $p < b$. Let $q$ be any rational number, satisfying $p < q < b$. Then $[p, q) \in \mathbf{B}_p$, and $p \in [p, q) \subset U$. Thus $\mathbf{B}_p$ forms a countable open base about the point $p$. (**R**, σ) is therefore a first countable space.

(*ii*) **Q** is a countable subset of **R**. We shall show that **Q** is dense in (**R**, σ). Let $p$ be any real number, and let $[a, b)$ be a right-half open interval, containing $p$. Then $[a, b)$ contains a rational number (other than $p$). Thus $p$ is a limiting point of **Q**, so that $\mathbf{R} \subset \mathbf{Q}'$, and therefore $\mathbf{R} = \overline{\mathbf{Q}}$. Consequently, the space (**R**, σ) is separable.

(*iii*) It follows from a more general result, to be proved in Theorem 7, that the space (**R**, σ) s Lindeloff.

(*iv*) We shall now show that the space (**R**, σ) is not second countable. Let us assume, to the contrary, that the space is second countable, and let $\mathbf{B} = \{[p_i, q_i) : i = 1, 2, \ldots\}$ form a countable open base of the topology σ. Let $p$ be any real number, different from the numbers $p_1, p_2, \ldots$, and $q$ be a real number, greater than $p$. Then $[p, q)$ is an open set in (**R**, σ), containing the point $p$. But there does not exist any $[p_j, q_j) \in \mathbf{B}$, satisfying $p \in [p_j, q_j) \subset [p, q)$, because in that case $p \le p_j \le p$, *i.e.*, $p = p_j$, which contradicts the defining property of $p$. Consequently, no countable family **B** can form an open base of σ. Hence the space (**R**, σ) is not second countable.//

**Example 1.** ***The real number space, with the cofinite topology, is separable and Lindeloff, but not first countable.***

**Solution:** Let (**R**, $\sigma_c$) denote the real number space, with the cofinite topology. The space is separable, since the rational numbers form a countable dense subset in it (verify). Next, let **C** form an open covering of **R**, and let $G \in \mathbf{C}$. As the complement of $G$ in **R** is a finite set, we can choose a finite number of members of **C**, which form a covering of $X - G$ (that is, whose union contains $X - G$). Then $G$ together with this finite collection of members of **C** form a finite covering of $X$; and this is a countable sub-covering of the open covering **C**. Hence ($\mathbf{R}$, $\sigma_c$) is a Lindeloff space.

We shall now show that the space ($\mathbf{R}$, $\sigma_c$) is not first countable. Let us assume, to the contrary, that ($\mathbf{R}$, $\sigma_c$) is first countable. Let $p \in \mathbf{R}$, and let $\mathbf{B}_p = \{U_i : i = 1, 2, \ldots\}$ form a countable open base about $p$. Then $\mathbf{R} - U_i$ is a finite set, for $i = 1, 2, \ldots$; and $\cup \{(\mathbf{R} - U_i) : i = 1, 2, \ldots\}$ is, therefore, a countable set. Let $P$ be a finite subset of $\mathbf{R}$, not containing $p$; then $p \in \mathbf{R} - P \in \sigma_c$. Then there exists an $U_j \in \mathbf{B}_p$, satisfying $p \in U_j \subset \mathbf{R} - P$. Thus every finite subset $P$ of $\mathbf{R}$ which does not contain the point $p$, is contained in a set $X - U_j$, for some $j$. The union of all such finite subsets $P$ is the set $\mathbf{R} - \{p\}$, which is uncountable; and this union is contained in the union $\cup \{(\mathbf{R} - U_i) : i = 1, 2, \ldots\}$, which is countable. Thus we arrive at a contradiction. Hence the space ($\mathbf{R}$, $\sigma_c$) cannot be first countable.

**Example 2.** ***Let R be the set of real numbers, and* $\nu$ *consist of (i) all those subsets of R, which do not contain 0, and (ii) the 4 subsets R – {1, 2}, R – {1}, R – {2}, and R. Then (R,* $\nu$*) is a first countable, Lindeloff space, that is not separable.***

**Solution :** It is easy to verify that ($\mathbf{R}$, $\nu$) is a topological space.

Let $x$ be any point in $\mathbf{R}$, and $U(x) = \{x\}$ if $x \neq 0$, and $U(x) = \mathbf{R} - \{1, 2\}$ if $x = 0$. Then $\{U(x)\}$ forms a countable open base about the point $x$ in ($\mathbf{R}$, $\nu$). Hence ($\mathbf{R}$, $\nu$) is first countable.

Any open covering **V** of $\mathbf{R}$ must include at least one of the sets in (*ii*) (in order that 0 may be covered). Let $G$ be such a set for the open covering **V** of $\mathbf{R}$, then $X - G$ consists of at most two points 1 and 2. Let $H_1$ and $H_2$ be two members of **V**, containing the points 1 and 2 respectively. Then $\{G, H_1, H_2\}$ forms a finite (and hence countable) subcovering of **V** for $\mathbf{R}$. Hence, ($\mathbf{R}$, $\nu$) is a Lindeloff space.

Since every one-pointic subset of $\mathbf{R} - \{0\}$ is open, any subset of $\mathbf{R}$, that is dense in ($\mathbf{R}$, $\nu$), must contain the uncountable set $\mathbf{R} - \{0\}$. Hence the space ($\mathbf{R}$, $\nu$) cannot be separable.

**Example 3.** ***[A topological space (X,* $\tau$*), where X is a countable set, is called a countable space].***

***(i) Every countable space is separable and Lindeloff.***

***(ii) A countable space need not be first countable.***

***(iii) A first countable, countable space is second countable.***

**Solution :** Let $(X, \tau)$ be a countable space.

(*i*) Since $X$ is a countable dense subset of $(X, \tau)$, it follows that $(X, \tau)$ is separable.

Next, let **B** be an open covering of $X$ in $(X, \tau)$. For each point $p \in X$, we choose a member $U_p \in \mathbf{B}$, containing $p$. Then the family of subsets $\{U_p : p \in X\}$ forms a covering of $X$, which is a countable sub-covering of **B**. Hence the space $(X, \tau)$ is Lindeloff.

(*ii*) See Example (**G**) (given below).

(*iii*) Let the countable space $(X, \tau)$ be first countable. Let $p \in X$, and $\mathbf{B}_p$ be a countable open base about the point $p$ in $(X, \tau)$. Then

$$\{G \in \mathbf{B}_p : p \in X\}$$

is also a countable collection of subsets, which forms an open base of the topology $\tau$ (verify). $(X, \tau)$ is therefore second countable.

**Example (*G*) :** ***The Appert's space is an example of***

**(*a*)** ***a countable $T_2$-space that is not first countable,***

**(*b*)** ***a separable, Lindeloff space that is not first countable.***

**Solution :** Let $X$ denote the set of all positive integers. For a subset $E$ of $X$, let $N(n, E)$ denote the number of integers in $E$ which are less than a given integer $n$. Let $\tau$ be the family of those subsets $G$ of $X$, for which either (*i*) $1 \notin G$, or (*ii*) $1 \in G$ and $\lim N(n, G)/n = 1$. Then $\tau$ forms a topology on $X$ (verify). The resulting topological space $(X, \tau)$ is known as the *Appert's space.*

(*a*) $(X, \tau)$ is evidently a countable space. It is also a $T_2$-space. In fact, if $p$, $q$ be two distinct points in $X$, then

(*i*) if $p \neq 1$ and $q \neq 1$, then $p$ and $q$ are strongly separated by the pair of disjoint open sets $\{p\}$ and $\{q\}$; on the other hand,

(*ii*) if $p \neq 1$ and $q = 1$, then $p$ and $q$ are strongly separated by the pair of disjoint open sets $\{p\}$ and $X - \{p\}$.

To prove that $(X, \tau)$ is not first countable, let us assume to the contrary; and let $\{V_n : n = 1, 2, \ldots\}$ be a be a countable open base about the point 1. Since each $V_n$ must be an infinite subset, we may choose a point $x_n \in V_n$, such that $x_n > 10^n$; and let $G = X - \cup \{x_n : n = 1, 2, \ldots\}$. Since $N(n, G) \geq n - \log_{10} n$, we have

$$\lim N(n, G)/n \geq \lim (n - \log_{10} n) = 1.$$

Hence $G$ is an open set containing 1. Since $\{V_n : n = 1, 2, ...\}$ forms an open base about 1, there exists a $V_m$, such that $1 \in V_m \subset G$. This is not possible, since $x_m \in V_m$, but $x_m \notin G$. Hence $(X, \tau)$ cannot be first countable.

(*b*) This result now follows immediately from (*a*) and Example 3 (*i*).

## EXERCISES

**1.** Let $(X, \tau)$ be a discrete space. Show that :

(*i*) If $X$ is uncountable, then $(X, \tau)$ is first countable, but not second countable; also, it is neither separable nor Lindeloff.

(*ii*) $(X, \tau)$ is separable if, and only if, it is countable.

(*iii*) $(X, \tau)$ is Lindeloff if, and only if, it is countable.

**2.** Show that the cofinite topology on

(*i*) a countable set is $T_1$ and second countable, but not $T_2$;

(*ii*) an uncountable set is separable and Lindeloff, but not first countable.

**3.** Show that the topology of countable complements on an (uncountable) set is Lindeloff, but neither separable nor first countable.

## 1.1 Subspaces

**Theorem 4 :** (*i*) *Any topological space can be embedded in a separable space.*

(*ii*) *Any $T_1$-space can be embedded in a Lindeloff space.*

**Proof :** (*i*) Let $(X, \tau)$ be an arbitrary topological space, and $z \notin X$. Let $Y = X \cup \{z\}$, and let $\sigma$ consist of $\Phi$ and the subsets $G \cup \{z\}$, for all $G \in \tau$. Then $(Y, \sigma)$ is a topological space (verify), and $(X, \tau)$ is a subspace of $(Y, \sigma)$ (that is, $\tau = \sigma_X$). Also $\{z\}$ is a (countable) dense subset of $(Y, \sigma)$; hence the space $(Y, \sigma)$ is separable.

(*ii*) Let $(X, \tau)$ be a $T_1$-space and $z \notin X$. Let $Y = X \cup \{z\}$, and let a topology $\tau'$ be chosen on $Y$ in terms of *nbd* filters, defined as follows. Let the *nbds* of a point $x \in Y$ be the same as its *nbds* in $(X, \tau)$ if $x \in X$, and the *nbds* of $z$ be the sets $\{z\} \cup E$, where $X - E$ are closed Lindeloff subsets of $X$. It can be readily verified that $(Y, \tau')$ is then a Lindeloff space, having $(X, \tau)$ as a subspace.//

**Theorem 5 :** (*i*) *Every subspace of a first countable space is first countable.*

(ii) Every subspace of a second countable space is second countable.

(iii) A subspace of a separable space need not be separable. But every open subspace of a separable space is separable.

(iv) A subspace of a Lindeloff space need not be Lindeloff. But every closed subspace of a Lindeloff space is Lindeloff.

**Proof :** (i) Let $Y$ be a subset of a first countable space $(X, \tau)$. Let $x$ be any point in $Y$. As $(X, \tau)$ is first countable, there exists a countable open base $W_1(x), W_2(x), ...,$ about the point $x$ in $(X, \tau)$. Correspondingly, $V_1(x), V_2(x), ...$ form a countable base about $x$ in $(Y, \tau_Y)$, where

$$V_i(x) = Y \cap W_i(x),$$

for $i = 1, 2, ...$ (verify). Hence the subspace $(Y, \tau_Y)$ is also first countable.

(ii) Let $(X, \tau)$ be a second countable space, and let $W_1, W_2, ...$ form a countable open base of the topology $\tau$. Let $Y$ be any subset of $X$. Then it follows, from Theorem 58 (i), page 89, that the sets $Y \cap W_1$, $Y \cap W_2, ...$ form a countable open base of the subspace topology $\tau_Y$. Hence the subspace $(Y, \tau_Y)$ is also second countable.

(iii) The subspace $(X, \tau)$ in Theorem 4 (i) may be taken to be non-separable (viz. let $\tau$ be the discrete topology on an uncountable set $X$). It then follows from Theorem 4 (i) that a subspace of a separable space need not be separable.

Next, let $(X, \tau)$ be a separable space, and $D$ a countable subset, dense in $(X, \tau)$; that is, $\overline{D} = X$. Let $Y$ be an open subset of $(X, \tau)$, and $E = Y \cap D$. Then $E$ is a countable subset of $Y$. We shall show that $\overline{E} = Y$. Let $x \in Y - E$. As $x \in X = \overline{D}$, and $x \notin D$ (since $x \in Y - Y \cap D$), it follows that $x \in D'$. Let $G$ be any open set, containing $x$, then $H = Y \cap G$ is also an open set in $(X, \tau)$, containing $x$ (since $Y$ is an open set in $(X, \tau)$). As $x \in D'$, $H$ must intersect $D$ in some point $y (\neq x)$, so that $y \in H \cap D \subset Y \cap D = E$. Hence $x \in E'$, so that $Y \subset \overline{E}$. Hence $Y = Y \cap \overline{E}$; and since the closure of $E$ in $(Y, \tau_Y)$ is $Y \cap \overline{E}$ (vide Theorem 58 (v), page 87), it follows that the subspace $(Y, \tau_Y)$ is separable.

(iv) In Theorem 4 (ii) we may choose $(X, \tau)$ to be a non-Lindeloff space. Then it follows from Theorem 4 (ii) that a subspace of a Lindeloff space need not be Lindeloff. [*Alternative* : As a concrete example, let us consider the Lindeloff space $(\mathbf{R}, v)$, described in Example 2, page 188. Let $S = \mathbf{R} - \{0\}$. Then every one-pointic set is open in $(S, v_S)$. The family of all one-pointic sets forms an uncountable open covering of $S$, for which there does not exist any countable subcovering. Hence the subspace $(S, \nu_S)$ is not Lindeloff].

Let $Y$ be a closed subset of a Lindeloff space $(X, \tau)$. Let $\mathbf{H} = \{H_\alpha : \alpha \in A\}$, $A$ being an indexing set, be an open covering of $Y$ in $(Y, \tau_Y)$. Let $H_\alpha = Y \cap G_\alpha$, $G_\alpha \in \tau$, $\alpha \in A$. Then $Y = \cup \{H_\alpha : \alpha \in A\} \subset \cup \{G_\alpha : \alpha \in A\}$. Hence the subset $X - Y$, together with the family of subsets $\{G_\alpha : \alpha \in A\}$, form an open covering of $X$ in $(X, \tau)$; and since $(X, \tau)$ is a Lindeloff space, there exists a countable sub-covering of $X$, formed by $X - Y$ and a countable sub-family, say $\{G_{\alpha i} : i = 1, 2, \ldots\}$, of the family $\{G_\alpha : \alpha \in A\}$. Then $Y \subset \cup \{G_{\alpha i} : i = 1, 2, \ldots\}$, and therefore $Y = \cup \{H_{\alpha i} : i = 1, 2, \ldots\}$, where $H_{\alpha i} = Y \cap G_{\alpha i}$, for $i = 1, 2, \ldots$ Thus $\{H_{\alpha i} : i = 1, 2, \ldots\}$ forms a countable open covering of $Y$ in $(Y, \tau_Y)$; and this is a sub-covering of the covering $\mathbf{H}$. Hence the subspace $(Y, \tau_Y)$ is also a Lindeloff space.//

It follows from Theorem 5 that whereas first countable spaces and second countable spaces are hereditarily first countable and second countable respectively, the same is not true for separable and Lindeloff spaces. But:

**Theorem 6 :** *Every second countable space is (i) hereditarily separable and (ii) hereditarily Lindeloff; but not conversely.*

**Proof :** Let $(X, \tau)$ be a second countable space, and $(Y, \tau_Y)$ a subspace of it. Then $(Y, \tau_Y)$ is also second countable, by Theorem 5 (*ii*); hence it is also separable and Lindeloff, by Theorem 1. Thus every subspace of $(X, \tau)$ is separable and Lindeloff. In other words, $(X, \tau)$ is (*i*) hereditarily separable and (*ii*) hereditarily Lindeloff.

The Appert's space (vide Example (*G*), page 189) is a countable space, which is not first countable. Since every countable space is separable and Lindeloff (vide Example 3 (*i*), page 188), and since every subspace of a countable space is also a countable space, it follows that the Appert's space is both hereditarily separable and hereditarily Lindeloff. But, as it is not first countable, it is certainly not second countable.//

A typical example of a hereditarily Lindeloff and hereditarily separable space, that is not second countable, is the following :

**Theorem 7 :** *The Sorgenfrey line is hereditarily Lindeloff and hereditarily separable, but not second countable.*

**Proof :** It has been proved in Theorem 3 that the Sorgenfrey line $(\mathbf{R}, \sigma)$ is not second countable. We shall now show that (*a*) it is hereditarily Lindeloff, and (*b*) hereditarily separable.

(*a*) Let $Y$ be a subset of $\mathbf{R}$, and $\mathbf{H} = \{H_\lambda : \lambda \in \Lambda\}$, where $\Lambda$ is an indexing set, be an open covering of $Y$ in $(Y, \sigma_Y)$. Let $H_\lambda = Y \cap G_\lambda$, $G_\lambda \in \sigma$, $\lambda \in \Lambda$. Then $\mathbf{G} = \{G_\lambda : \lambda \in \Lambda\}$ forms an open covering of the subset $Y$ in $(\mathbf{R}, \sigma)$; that is, $Y \subset \cup \{G_\lambda : \lambda \in \Lambda\}$. Then, for any point $x \in$ Y, there exists a half-open interval $J_x = [x, x + \delta_x)$ such that $x \in J_x \subset G_v$, for some $v \in \Lambda$. The aggregate $\{J_x : x \in Y\}$ is then an open covering of $Y$. We shall show that there exists a countable sub-family $\{J_{xi}\}$, such that $\cup \{J_{xi}\} = \cup \{J_x\}$.

Let $n$ be a fixed positive integer, and let $\{J_x^n\}$ be the sub-family, consisting of all those intervals $J_x^n \in \{J_x\}$, for which $\delta_x > 1/n$. Let $y \in J^n = \cup \{J_x^n\}$, and be such that it is not an interior point of any $J_x^n$. Then there exists an interval of length $\geq 1/n$, situated at the left of $y$ and containing no point with $\delta_x > 1/n$. Hence, the set $S$, formed by all such points $y$, is countable and can be covered by the countable family $\{J_y^n : y \in S\}$. If we omit the left end points of all the remaining intervals $J_x^n$, we obtain a family of open intervals $I_x^n$. Since the real number space, with the usual topology, is second countable (Theorem 2), and hence hereditarily Lindeloff (Theorem 5), the (open) set $I^n = \cup \{I_x^n\}$ can be covered by a countable sub-family of $\{I_x^n\}$. Correspondingly, $J^n$ has also a covering by a countable sub-family of $\{J_x^n\}$. But $\cup \{J_x\} = \cup \{J^n\}$, and so $Y \subset \cup \{J_x\}$ is also covered by a countable sub-family $\{J_{xi}\}$ of $\{J_x\}$, and hence also by a countable sub-family $\{G_i\}$ of $\mathbf{G}$. Then $\{H_i\}$ forms a countable subcovering of $\mathbf{H}$ for $(Y, \tau_Y)$. Thus $(Y, \tau_Y)$ is a Lindeloff space. Hence $(\mathbf{R}, \sigma)$ is a hereditarily Lindeloff space.

(*b*) Let $Y$ be a subset of $\mathbf{R}$. Let $Y_1$ be the set of all those points in $Y$, which are not accumulation points of $Y$ from the right; then the set $Y_1$ is at most countable (vide Exercise 1, below). Let $Y - Y_1 = Y_2$; then each point of $Y_2$ is an accumulation point of $Y$ from the right.

Let $Y_0$ be a countable dense subset of $Y$ in the usual Euclidean subspace topology of $Y$. Then $Y_1 \cup Y_0 = B$ is a countable dense subset of $Y$. Let $z \in Y - Y_1 = Y_2$, and consider the half open interval $[z, z + \varepsilon)$, for any $\varepsilon > 0$. Since $z$ is an accumulation point of $Y$ from the right, and $(Y, \sigma_Y)$ is a $T_1$-space, there exist infinitely many points in $[z, z + \varepsilon) \cap Y$. Since $(z, z + \varepsilon) \cap Y$ must contain points of $B$, $[z, z+\varepsilon) \cap Y$ must also contain points of $B$, for arbitrary $\varepsilon > 0$; consequently, $z \in \overline{B}$, where the closure is taken in respect of the topology $\sigma_Y$ on $Y$. Hence the subspace $(Y, \sigma_Y)$ is separable. Thus $(\mathbf{R}, \sigma)$ is a hereditarily separable space.//

## EXERCISES

1. If $A$ is a subset of the Sorgenfrey line $(\mathbf{R}, \sigma)$, and $B$ be the subset consisting of all those points of $A$, which are not accumulation points of $A$ from the right, then the set $B$ is at most countable.

   [**Solution :** If possible, let the set $B$ be uncountable. For any $b \in B$, there exists a half-open interval $[b, v_b)$, such that $[b, v_b) \cap A = \{b\}$. Now, for each $b \in B$, $\{b\} \in (\sigma_A)_B = \sigma_B$. Consider the open covering $\{\{b\} : b \in B\}$ of $B$; then this open covering can have no finite sub-covering, which is impossible, since $(B, \sigma_B)$ is a Lindeloff space, by Theorem 7. Thus the set $B$ is at most countable].

2. A topological space is hereditarily Lindeloff if, and only if, every open subspace of it is Lindeloff.

3. An uncountable Fort space (vide Example 11 (*a*), page 105) is Lindeloff, which is not hereditarily Lindeloff. It is neither first countable nor separable.

4. The closed ordinal space $[0, \Gamma]$ (vide Example $C$ (*b*), page 124) is a Lindeloff space, which is not hereditarily Lindeloff.

## 1.2 Continuous Functions and Countability Properties

**Theorem 8 :** (*a*) *A continuous image of a separable space is separable.*

(*b*) *A continuous image of a Lindeloff space is Lindeloff.*

**Proof :** Let $f$ be a continuous mapping of a topological space $(X, \tau)$ onto a topological space $(Y, \tau')$.

(*a*) Let $(X, \tau)$ be, in particular, a separable space, and let $D$ be a countable dense subset of $(X, \tau)$. Then $f(D)$ is certainly a countable subset of $f(X) = Y$. Also, by Theorem 51, page 84, $f(D)$ is a dense subset of $(Y, \tau')$. Hence the space $(Y, \tau')$ is also separable.

(*b*) Let $(X, \tau)$ be a Lindeloff space. Let $\mathbf{V}$ be an open covering of $Y$ in $(Y, \tau')$. Then $\mathbf{W} = \{f^{-1}(U) : U \in \mathbf{V}\}$ is an open covering of $X$ in $(X, \tau)$, as $f$ is continuous. Since $(X, \tau)$ is a Lindeloff space, there exists for $X$, a countable sub-covering $\mathbf{G}$ of $\mathbf{W}$. Then $\{f(V) : V \in \mathbf{G}\}$ forms a countable open covering of $f(X) = Y$; and this is a (countable) sub-covering of **V**. Hence $(Y, \tau')$ is also a Lindeloff space.//

**Theorem 9 :** (*a*) *A continuous image of a first countable (second countable) space is not necessarily first countable (respectively second countable).*

(*b*) *A continuous open image of a first countable (second countable) space is a first countable (respectively second countable) space.*

**Proof :** (*a*) Let $(X, \tau) = (\mathbf{R}, \tau)$, the real number space with its natural topology. Then $(X, \tau)$ is a second countable and, hence, also a first countable space, by Theorem 2. Let $(Y, \tau') = (\mathbf{R}, \sigma)$, the real number space with the co-finite topology. Then $(Y, \tau')$ is not first countable, and hence also not second countable (vide Example 1, page 187), but $f(x) = x$ is a continuous bijection of $(X, \tau)$ onto $(Y, \tau')$.

(*b*) Let $f$ be a continuous open mapping of a topological space $(X, \tau)$ onto a topological space $(Y, \tau')$.

Let $(X, \tau)$ be first countable. Let $p \in Y$, and $x \in f^{-1}(y)$, so that $f(x) = y$. There exists a countable open base, $\{U_i : i = 1, 2, \ldots\}$, about $x$ in $(X, \tau)$. Then, as $f$ is an open mapping, $\{f(U_i) : i = 1, 2, \ldots\}$ is a countable family of open sets, containing $f(x) = y$ in $(Y, \tau')$. To show that this family forms a countable open base about $y$ in $(Y, \tau')$, let $V$ be any open set containing $y$ in $(Y, \tau')$. As $f$ is continuous, $f^{-1}(V)$ is an open set containing $x$ in $(X, \tau)$; and since $\{U_i : i = 1, 2, \ldots\}$ forms an open base about $x$ in $(X, \tau)$, there exists an $U_j$, such that $x \in U_j \subset f^{-1}(V)$. Hence $y = f(x) \subset f(U_j) \subset V$. This proves that there exists a countable open base about the point $y$; and $(Y, \tau')$ is therefore a first countable space.

Next, let $(X, \tau)$ be a second countable space, and $\{W_i : i = 1, 2, \ldots\}$ be a countable open base of $\tau$. Since $f$ is an open mapping, $\{f(W_i) : i = 1, 2, \ldots\}$ is a countable family of open sets in $(Y, \tau')$; we shall show that this (countable) family forms an open base of $\tau'$. Let $V \in \tau'$, then $f^{-1}(V) \in \tau$, since $f$ is continuous; and it can be expressed as a union, $f^{-1}(V) = \cup \{W_k : k \in B \subset \mathbf{N}\}$, where **N** denotes the set of all natural numbers, and $B$ is a subset of **N**. As $f$ is surjective, $V = f(f^{-1}(V)) = \cup \{f(W_k) : k \in B \subset \mathbf{N}\}$. Thus $\{f(W_i) : i = 1, 2, \ldots\}$ forms a countable open base of $\tau'$; $(Y, \tau')$ is therefore a second countable space.//

It now follows directly from Theorems 8 and 9 (*b*) that :

**Theorem 10 :** *The property that a topological space is (i) first countable, (ii) second countable, (iii) separable, or (iv) a Lindeloff space is a topological property.//*

## EXERCISES

**1.** Show that separability and Lindeloff property are preserved under a weakening of the topology; but this is not true for first and second countability.

## 1.3 Topological Products

**Theorem 11 :** *Let $(X, \tau)$ be the topological product of a family of topological spaces $(X_a, \tau_a) : \alpha \in A$, where $A$ is an indexing set. Then :*

(*a*) *$(X, \tau)$ is first countable if, and only if, each space $(X_a, \tau_a)$ is first countable, and all but a countable number of the spaces $(X_a, \tau_a)$ are indiscrete.*

(*b*) *$(X, \tau)$ is second countable if, and only if, each space $(X_a, \tau_a)$ is second countable, and all but a countable number of the spaces $(X_a, \tau_a)$ are indiscrete.*

(*c*) *$(X, \tau)$ is separable if, and only if, each $(X_\alpha, \tau_\alpha)$ is a separable Hausdorff space with at last two elements, and card $(A) \le c$.*

**Proof :** (*a*) Let $(X, \tau)$ be first countable. Since each coordinate space $(X_a, \tau_a)$ is homeomorphic with a certain subspace of the product space $(X, \tau)$, and every subspace of a first countable space is first countable, and first countability is a topological property, it follows that each space $(X_a, \tau_a)$ is first countable. Let $x$ be a point in $(X, \tau)$, such that its $\alpha$th co-ordinate $x_\alpha$ is chosen in the following manner :

(*i*) if space $(X_\alpha, \tau_\alpha)$ is not indiscrete, then $x_a$ is a point in it, such that $x_\alpha \in G_\alpha$, where $G_\alpha \in \tau_\alpha$ and $G_\alpha \neq X$, and

(*ii*) if a space $(X_\alpha, \tau_\alpha)$ is indiscrete, then any point in it is chosen as $x_\alpha$.

Let $\{W_n : n = 1, 2, \ldots\}$ be a countable open base about the point $x$ in $(X, \tau)$. For each integer $n$, $W_n$ must contain a member of the base, which generates the product topology $\tau$, say II $\{V_\alpha : \alpha \in A\}$, where $V_\alpha \in \tau_\alpha$ for all $\alpha \in A$, and $V_\alpha = X_\alpha$ for all but a finite number of values of $\alpha$, say $\{\alpha_i^n : i = 1, 2, \ldots\}$. The collection of all the excepted values of $\alpha$, *i.e.*, $\{\alpha_i^n : i = 1, 2, \ldots$ and $n = 1, 2, \ldots\}$, is countable. For any other value of

$\alpha$ we may choose $x_\alpha \in G_\alpha \neq X_\alpha$, if $X_\alpha$ is not indiscrete, and then $P_a^{-1}(G_\alpha)$ would be an open set in $(X, \tau)$, containing $x$ but not containing any member of the open base $\{W_n : n = 1, 2, \ldots\}$ about the point $x$, which is impossible. Hence, for every value of $\alpha$ other than the countable collection of excepted values $\{\alpha_i^n : i = 1, 2, \ldots$ and $n = 1, 2, \ldots\}$, the spaces $(X_\alpha, \tau_a)$ must be indiscrete.

Conversely, let $(X_a, \tau_\alpha)$ be a first countable space, for each $\alpha \in A$, and indiscrete for $\alpha \notin \{\alpha_i : i = 1, 2, \ldots\}$. Let $x = \{x_\alpha : \alpha \in A\}$ be an arbitrary point in $(X, \tau)$, and let $\{V_{\alpha, i} : i = 1, 2, \ldots\}$ be a countable open base about the point $x_\alpha$ in $(X_\alpha, \tau_\alpha)$. Then $V_{\alpha, j} = X_\alpha$, for all values of $j$, if $\alpha \neq \alpha_i$, $i = 1, 2, \ldots$ The family $\{\pi_{\alpha i}^{-1}(V_{\alpha\, i, j}) : i = 1, 2, \ldots$ and $j = 1, 2, \ldots\}$ is a countable collection of open sets in the product space $(X, \tau)$. The set formed by all finite intersections of members of this collection is certainly countable, and it forms an open base about the point $x$ in $(X, \tau)$. The product space $(X, \tau)$ is therefore first countable.

*(b)* Let $(X, \tau)$ be second countable. Since each coordinate space $(X_\alpha, \tau_\alpha)$ is homeomorphic with a certain subspace of the product space $(X, \tau)$, and every subspace of a second countable space is second countable, and second countability is a topological property, it follows that $(X_\alpha, \tau_\alpha)$ is second countable, for each $\alpha \in A$. Again, since $(X, \tau)$ is necessarily first countable, it follows from *(a)* that all but a countable number of the spaces $(X_\alpha, \tau_\alpha)$ are indiscrete.

Conversely, let the space $(X_\alpha, \tau_\alpha)$ be second countable, and let $\{W_{\alpha n} : n = 1, 2, \ldots\}$ be a countable open base of $\tau_\alpha$, for each $\alpha \in A$; and $(X_\alpha, \tau_\alpha)$ be indiscrete for all $\alpha \notin \{\alpha_i : i = 1, 2, \ldots\}$. Then the sets of the form $W_{\alpha 1, n1} \times \ldots \times W_{\alpha k, nk} \times \Pi\{X_\alpha : \alpha \neq \alpha_{1, \ldots,} \alpha_k\}$ form a countable open base of the product topology $\tau$. $(X, \tau)$ is therefore a second countable space.

*(c)* Let $(X, \tau)$ be a separable space. Since projection mappings are continuous, it follows, by Theorem 8 *(a)*, that $(X_\alpha, \tau_\alpha)$ is also separable, for each $\alpha \in A$. We shall show that card $(A) \leq c$. Since each $(X_\alpha, \tau_\alpha)$ is a Hausdorff space, and $X_\alpha$ has at least two points, for each $\alpha \in A$, there exists a pair of disjoint non-empty open sets $G_\alpha$, $H_\alpha$ in $(X_a, \tau_\alpha)$. Let $D$ be a countable dense subset in $(X, \tau)$; and, for each $\alpha \in A$, let $D_\alpha = D \cap \pi_\alpha^{-1}(G_\alpha)$. Then $D$ is non-empty for each

$\alpha$; and for distinct $\alpha$ and $\beta$, $D_\alpha \neq D_\beta$, since any point in $\pi_\beta^{-1}(G_\alpha) \cap \pi_\beta^{-1}(H_\beta)$, which belongs to $D$, will belong to $D_\alpha$ but not to $D_\beta$. Thus the mapping $f : A \to \mathbf{P}(D)$, defined by $f(\alpha) = D_\alpha$, is injective, and therefore card $(A) \leq$ card $(\mathbf{P}(D)) = 2^d = \mathbf{c}$.

Conversely, let $(X_\alpha, \tau_\alpha)$ be a separable space, and $D_\alpha = \{g_{\alpha}i : i = 1, 2, \ldots\}$ be a countable dense subset in $(X_\alpha, \tau_\alpha)$, for each $\alpha \in A$, where the indexing set $A$ has cardinal number $\leq \mathbf{c}$. $A$ can be considered as a subset of the closed unit interval $I$. For each sequence $J_1, \ldots, J_k$ of pair-wise disjoint closed intervals with rational endpoints, and each sequence of positive integers $n_1, \ldots n_k$, we define a point $p(J_1, \ldots, J_k; n_1, \ldots, n_k)$ as follows :

$$p_\alpha = g_{\alpha ni} \text{ if } \alpha \in J_i, \text{ and } p_\alpha = g_{\alpha 1} \text{ otherwise.}$$

The points $p$, defined as above, form a countable set $D$. Also, $D$ is a dense subset in $(X, \tau)$. In fact, the sets of the form $V = \pi_{\alpha 1}^{-1}(U_{\alpha 1}) \cap \ldots \cap \pi_{\alpha m}^{-1}(U_{\alpha m})$, where $U_{\alpha i} \in \tau_{\alpha i}$, $i = 1, 2, \ldots, m$, constitute an open base of the product topology $\tau$. $U_{\alpha i}$ contains a point $g_{\alpha i, ni}$ of $D_{\alpha i}$, for each $i$, and there are disjoint closed intervals $J_1, \ldots, J_m$, with rational end-points, containing the points $\alpha_1, \ldots, \alpha_m$ respectively. The point $p(J_1, \ldots, J_m; n_1, \ldots, n_m)$ belongs to $V$, since $p_{\alpha i} = g_{\alpha i, ni}$, for $i = 1, \ldots, m$. Hence $D$ is a dense subset in $(X, \tau)$. $(X, \tau)$ is therefore a separable space.//

**Theorem 12 :** *Let $(X, \tau)$ be the topological product of a family of topological spaces $\{X_\alpha, \tau_\alpha) : \alpha \in A\}$, where $A$ is an indexing set. If $(X, \tau)$ is a Lindeloff space, then each coördinate space $(X_\alpha, \tau_\alpha)$ is a Lindeloff space.*

*The converse is not true; in fact, the topological product of even two Lindeloff spaces may not be a Lindeloff space.*

**Proof :** Let $(X, \tau)$ be a Lindeloff space. Since each coordinate space $(X_\alpha, \tau_\alpha)$ is homeomorphic with a closed subspace of the product space $(X, \tau)$, and every closed subspace of a Lindeloff space is also a Lindeloff space, and the property that a space is a Lindeloff space is a topological property, it follows that $(X_\alpha, \tau_\alpha)$ is a Lindeloff space, for each $\alpha \in A$.

It has been proved in Theorem 3 that the Sorgenfrey line $(\mathbf{R}, \tau_1)$ is a Lindeloff space. We shall show that the product $(\mathbf{R}, \tau_1) \times (\mathbf{R}, \tau_1)$ is not a Lindeloff space. Let $S$ be the subset of $\mathbf{R} \times \mathbf{R}$, given by $S = \{(x, -x) : x \in \mathbf{R}\}$. Then $S$ is a closed subset of $(\mathbf{R}, \tau_1) \times (\mathbf{R}, \tau_1)$ (vide proof of Theorem 28, page 149). The sets $\{(p, q) : p \geq x \text{ and } q \geq -x\}$, as $x$ runs over $\mathbf{R}$, form an open covering of $S$, for which there does not exist any countable subcovering, since each member of the above covering contains exactly one point of $S$, and $S$ is an uncountable set. Thus the closed subspace $(S, \sigma)$, where $\sigma$ is the subspace topology on $S$, is not a Lindeloff space. Hence, by Theorem 5 (*iv*), $(\mathbf{R}, \tau_1) \times (\mathbf{R}, \tau_1)$ is not a Lindeloff space.//

Since the real number space and the closed unit interval $\mathbf{I} = [0, 1]$ are second countable (vide Theorems 2 and 5 (*ii*)), it follows, by Theorem 11 (*b*), that :

**Corollary :** (*i*) *The Euclidean n-space* $E^n$, *and* (*ii*) *the unit cube* $I^d$, *are second countable, and hence also first countable, separable and Lindeloff.*

**Example 4.** ***The Sorgenfrey line*** $(R, \tau_l)$ ***is first countable and separable (and also Lindeloff), by Theorem 3. Hence, by Theorem 11 (a) and (c), the product space*** $(R, \tau_l) \times (R, \tau_l)$ ***is first countable and separable; but it is not Lindeloff (by Theorem 12).***

## Art 2. PARTICULAR PROPERTIES

### 1.1 Regular Lindeloff Spaces

**Theorem 13 :** *A regular Lindeloff space is normal.*

**Proof :** Let $(X, \tau)$ be a regular Lindeloff space, and let $P$, $Q$ be a pair of disjoint closed sets in $(X, \tau)$. Then $P \subset X - Q$ and $Q \subset X - P$. Now, for any point $p$ in $P$, there exists an open set $U(p)$, such that $p \in U(p)$ and $\overline{U(p)} \subset X - Q$ (by the regularity criterion of Tychonoff), so that $Q \subset X - \overline{U(p)} = \text{Ext}\, U(p)$. As $p$ runs over $P$, the corresponding open sets $\{U(p)\}$ form an open covering of $P$. Since $(X, \tau)$ is a Lindeloff space and $P$ is a closed subset, $P$ is also a Lindeloff space (by Theorem 5 (*iv*)). Hence there exists an open covering of $P$ by a countable subfamily of the sets $\{U(p)\}$–say by the sets $U_1, U_2, \ldots$ Thus

$$P \subset U_1 \cup U_2 \cup \ldots, \text{ where } Q \subset \text{Ext}\,(U_i), \text{ for } i = 1, 2, \ldots \qquad \ldots(1)$$

Similarly, there exists a countable family of open sets $V_1, V_2, \ldots$, such that

$$Q \subset V_1 \cup V_2 \cup \ldots, \text{ where } P \subset \text{Ext}\,(V_i), \text{ for } i = 1, 2, \ldots \qquad \ldots(2)$$

Let us now define two sequences of open sets $\{G_n : n = 1, 2, \ldots\}$ and $\{H_n : n = 1, 2, \ldots\}$ by

$$G_n = U_n \cap \text{Ext}\,(H_1) \cap \ldots \cap \text{Ext}\,(H_{n-1}),$$
$$H_n = V_n \cap \text{Ext}\,(G_1) \cap \ldots \cap \text{Ext}\,(G_n),$$

Next, let $G = G_1 \cup G_2 \cup \ldots$, and $H = H_1 \cup H_1 \cup \ldots$ Then $G$ and $H$ are two open sets; and they are disjoint. In fact, for $i \le k$,

$$G_i \cap H_k = G_i \cap [V_k \cap \text{Ext}\,(G_1) \cap \ldots \cap \text{Ext}\,(G_i) \cap \ldots \cap \text{Ext}\,(G_k)]$$
$$= \Phi, \text{ since } G_i \cap \text{Ext}\,(G_i) = \Phi;$$

and for $i > k$,

$$G_i \cap H_k = H_k \cap [U_i \cap \text{Ext}\,(H_1) \cap \ldots \cap \text{Ext}\,(H_k) \cap \ldots \cap \text{Ext}\,(H_{i-1})]$$
$$= \Phi, \text{ since } H_k \cap \text{Ext}\,(H_k) = \Phi.$$

Again, $H_i \subset V_i$ implies $\text{Ext}\,(V_i) \subset \text{Ext}\,(H_i)$, for $i = 1, 2, \ldots$ Hence, if $p$ is any point in $P$, then

$$p \in P \subset \text{Ext}\,(V_i) \text{ [by (2)]} \subset \text{Ext}\,(H_i), \text{ for } i = 1, 2, \ldots$$

Also, the point $p$ must be contained in some $U_m$ (since the sets $U_1$, $U_2$, ... form a covering of $P$), and correspondingly, $p$ is contained in $G_m$, and therefore in $G$. Thus $P \subset G$; similarly, $Q \subset H$.

Thus the pair of disjoint closed sets $\{P, Q\}$ is separated strongly by the two disjoint open sets $G$ and $H$ in $(X, \tau)$. Hence $(X, \tau)$ is a normal space.//

**Corollary 1 :** *Every hereditarily Lindeloff, regular space is completely normal.*

**Proof :** Let $(X, \tau)$ be a hereditarily Lindeloff, regular space. Then every subspace of $(X, \tau)$ is a regular Lindeloff space, and hence a normal space. Consequently, $(X, \tau)$ is completely normal.//

**Corollary 2 :** *(Tychonoff theorem) : Every regular second countable space is completely normal.*

**Proof :** The proposition is an immediate consequence of Corollary 1, by virtue of the property that every second countable space is heriditarily Lindeloff (vide Theorem 6 (*ii*)).//

One obtains stronger results in the following two Theorems :

**Theorem 14 :** *In a regular second countable space, every closed set is a $G_\delta$-set.*

**Proof :** Let $F$ be a closed set in a regular second countable space $(X, \tau)$. If $F = X$, then $F$ is obviously a $G_\delta$-set. Next, let $F \neq X$. Then $X - F = N_x$ is an open neighbourhood for every point $x \in X - F$. By the regularity criterion, there exists, correspondingly, an open neighbourhood $N_x^*$ of the point $x$, such that $\overline{N_x^*} \subset X - F$ (for each point $x \in X - F$). Again, since $(X, \tau)$ is second countable, every $N_x^*$ can be expressed as the union of a sub-family $\{W_{ki},\ i = 1, 2, \ldots\}$ of a countable family $\{W_i\}$ which forms an open base of the topology $\tau$; say $N_x^* = W_{k1} \cup W_{k2} \cup \ldots$ As $x \in N_x^*$, $x$ must belong to some $W_{ki}$, say $x \in W_{k(x)} \subset N_x^* \subset X - F$, so that $F \subset X - W_{k(x)}$. Consequently, $F = \cap \{X - W_{k(x)} : x \in X - F\}$. The set $k(x)$ is countable; hence $F$ is a $G_\delta$-set.//

Since every second countable space is Lindeloff (Theorem 1), it now follows by Theorems 13 and 14 that every regular second countable space is a normal space, in which every closed set is a $G_\delta$-set. Hence, by Theorem 19, page 177, we get :

**Theorem 15 :** *Every regular second countable space is perfectly normal.*

## EXERCISES

1. An $F_\sigma$ subset of a regular Lindeloff space is Lindeloff.
2. Every open subset of a hereditarily Lindeloff, regular space is an $F_\sigma$-set.

## 2.2 Convergence in a First Countable Space

**Lemma :** *Let $x$ be a point in a first countable space $(X, \tau)$, and let $\{W_i : i = 1, 2, \ldots\}$ form a countable open base about the point $x$, then there exists an infinite sub-sequence $\{V_i : i = 1, 2, \ldots\}$ of the sequence $\{W_i\}$, such that*

(*i*) *for any open set $U$, containing $x$, there exists a suffix $m$ such that $V_i \subset U$ for all $i \geq m$; and*

(*ii*) *if $(X, \tau)$ be, in particular, a $T_1$-space, then*

$$\cap \{V_i : i = 1, 2, \ldots\} = \{x\}.$$

**Proof :** $W_1 \cap W_2 \cap \ldots \cap W_k$ is an open set, containing the point $x$. As the sets $\{W_i\}$ forms an open base about $x$, there exists one among the sets $\{W_i\}$, which we shall denote by $V_k$, such that $x \in V_k \subset W_1 \cap W_2 \cap \ldots \cap W_k$, for $k = 1, 2, \ldots$ The sequence $\{V_i : i = 1, 2, \ldots\}$, thus obtained, has the required properties. In fact,

(*i*) If $U$ is any open set, containing $x$, then there exists a set $W_m$ say, belonging to the family $\{W_i\}$, such that $x \in W_m \subset U$. Also, since $V_i \subset W_m$, for all $i \geq m$, it follows that $V_i \subset U$, for all $i \geq m$.

(*ii*) Next, let $(X, \tau)$ be a $T_1$-space, and let $\cap \{V_i\} = M$. As $x$ is contained in each $V_i$, it follows that $x \in M$. Let $y$ be any point in $X$, different from $x$. By the $T_1$ separation axiom, there exists an open set $U$, such that $x \in U$ and $y \notin U$. There exists a suffix $m$, such that $V_i \subset U$ for all $i \geq m$. Consequently, $y \notin V_i$ for all $i \geq m$; hence $y \notin M$. Thus $M$ consists of the point $x$ only.//

Sequential convergence is not adequate in many cases in a general topological space (as pointed out in pages 78-79). But it is found to possess some satisfactory properties in a first countable space (and hence also in a metric space) as given in the following three theorems.

**Theorem 16 :** *A first countable space $(X, \tau)$ is Hausdorff if, and only if, every convergent sequence has a unique limit in it.*

**Proof :** If $(X, \tau)$ be a Hausdorff space, then every convergent sequence has a unique limit in the space (vide Theorem 13, page 135).

On the other hand, if $(X, \tau)$ is a first countable space, which is not Hausdorff, then there exist two distinct points $p$ and $q$, which cannot be strongly separated in $(X, \tau)$. As $(X, \tau)$ is first countable, there exist, by the above Lemma, a sequence of open sets $\{V_i(p) : i = 1, 2, \ldots\}$, belonging to an open base about the point $p$, and a sequence of open sets $\{V_i(q) : i = 1, 2, \ldots\}$, belonging to an open base about the point $q$, possessing the properties stated in the Lemma. Then $V_i(p) \cap V_i(q) \neq \Phi$ (as $p$ and $q$ cannot be strongly separated) and let $x_i \in V_i(p) \cap V_i(q)$, for $i = 1, 2, \ldots$ Let $U$ and $V$ be two arbitrary open sets, containing the points $p$ and $q$ respectively. Then, by the Lemma, there exists a positive integer $m$, such that $V_i(p) \subset U$ and $V_i(q) \subset V$, for all $i \geq m$. Hence $\lim x_n = p$ and $\lim x_n = q$; thus there exists a convergent sequence, whose limit is not unique in $(X, \tau)$. This proves the converse part.//

**Theorem 17 :** *A point $x$ in a first countable space $(X, \tau)$ is an accumulation point of a subset $E$ if, and only if, there exists an infinite sequence of points $x_1, x_2, \ldots, (x_i \neq x)$ in $E$, such that $\lim x_n = x$.*

**Proof :** Let the condition hold for an infinite sequence of points $x_1, x_2, \ldots$ in $E$. Then, for any open set $U$, containing $x$, there exists a suffix $m$, such that $x_i \in U$, for all $i \geq m$. Since the points $x_1, x_2, \ldots$ belong to $E$, and $x_i \neq x$, it follows that $x$ is an accumulation point of the subset $E$.

Conversely, let $x$ be an accumulation point of the subset $E$. Then, by the Lemma, there exists a sequence of open sets $\{V_i : i = 1, 2, \ldots\}$, belonging to an open base about the point $x$, such that for any open set $U$ containing $x$, there exists a suffix $m$, such that $V_i \subset U$ for all $i \geq m$. Now, since

$V_n$ is an open set containing $x$, and $x$ is an accumulation point of the subset $E$, there exists a point $x_n$ in $E$ different from $x$, such that $x_n \in V_n$. For $n = 1, 2, \ldots$, we thus obtain a sequence of points $x_1, x_2, \ldots$, belonging to $E$. Then $x$ is a limit of this sequence of points. In fact, if $U$ is any open set, containing $x$, then there exists a suffix $m$, such that $V_i \subset U$ for all $i \geq m$, and therefore $x_i \in U$ for all $i \geq m$. Hence $\lim x_n = x$.//

**Note :** If $(X, \tau)$ is a $T_1$-space, the points $x_1, x_2, \ldots$ may be selected in such a manner that these are distinct. To do so, we may choose $x_i$ to be a point in $E$, different from $x$, and contained in the open set $V_i - \{x_1, x_2, \ldots, x_{i-1}\}$ which contains the point $x$.

**Theorem 18 :** *(Heine's continuity criterion) : A function f, defined on a subset E of a first countable space* $(X, \tau)$, *with values in a topological space* $(Y, \tau')$, *is continuous at a point x in E if, and only if, for any infinite sequence of points* $x_1, x_2, \ldots$ *in E,* $\lim x_n = x$ *implies* $\lim f(x_n) = f(x)$ *in* $(Y, \tau')$.

**Proof :** If $f$ is continuous at the point $x$, then $\lim x_n = x$ implies $\lim f(x_n) = f(x)$, by Theorem 49, page 80. This proves the necessity of the condition.

To prove the converse part, let us assume that the function $f$ is not continuous at the point $x$ in $E$. Then there exists an open set $H \in \tau'$, containing $f(x)$, such that every open set $G \in \tau$, containing $x$, contains at least one point $y$ in $E$ for which $f(y) \notin H$. By the Lemma, there exists an infinite sequence of open sets, $V_1, V_2, \ldots$, containing $x$ (and belonging to an open base about the point $x$), such that for any open set $G$ containing $x$ in $X$, there exists a suffix $m$, for which $V_i \subset G$ for all $i \geq m$. Every $V_i$ contains (at least) one point $x_i$, such that $f(x_i) \notin H$. Then, for the infinite sequence of points $x_1, x_2, \ldots$, thus determined, $\lim x_n = x$, since any open set $G$ containing $x$, contains all $V_i$ for all $i \geq$ some positive integer $m$, so that $G$ contains all the points $x_i$ for $i \geq m$.

But $\lim f(x_n) \neq f(x)$, since the open set $H$, containing $f(x)$ in $(Y, \tau')$, does not contain any point $f(x_i)$, for $i = 1, 2, \ldots$ Thus we find that if $f$ is not continuous at the point $x$, then there exists an infinite sequence of points $\{x_i : i = 1, 2, \ldots\}$ in $E$, such that whereas $\lim x_n = x$ holds, but $\lim f(x_n) \neq f(x)$. Hence the given condition is sufficient to ensure that $f$ is continuous at the point $x$.//

## 2.3 Cardinality Properties in a Second Countable Space

In this section $(X, \tau)$ shall denote a second countable space, and $\{W_i : i = 1, 2, \ldots\}$ a countable open base of the topology $\tau$.

**Theorem 19 :** *The cardinality of the set of all open sets in a second countable space is at most equal to* **c** *(the power of the continuum).*

**Proof :** Let $U$ be any open set in $(X, \tau)$. Then $U$ is the union of a certain sub-collection of the countable collection $\{W_i : i = 1, 2, \ldots\}$. Hence the cardinality of the set of all open sets in $(X, \tau)$ is not greater than the cardinality of the set of all sub-collections of the countable collection $\{W_i : i = 1, 2, \ldots\}$. Thus the cardinality of $\tau \leq \mathbf{c}$.//

Since every closed set in $(X, \tau)$ is the complement of an open set, it follows that :

**Corollary :** *The cardinality of the set of all closed sets in a second countable space is at most equal to* **c**.

**Theorem 20 :** *(Souslin condition) : Any collection of mutually disjoint open sets in a second countable space is at most countable.*

**Proof :** Let $\Gamma$ denote any collection of mutually disjoint open sets in $(X, \tau)$. Let $U \in \Gamma$, then there exists at least one set $W_m$, belonging to the collection $\{W_i : i = 1, 2, \ldots\}$, such that $W_m \subset U$. Let $n$ be the smallest suffix for which $W_n \subset U$. Since the sets $U$ in $\Gamma$ are mutually disjoint, it follows that, for different sets $U \in \Gamma$, there correspond sets $W_n$ with different suffices $n$. Hence the sets in $\Gamma$ are in a (1, 1)-correspondence with a sub-collection of the countable collection $\{W_i : i = 1, 2, \ldots\}$; consequently, the cardinality of $\Gamma$ is less than or equal to **d**.//

The Souslin condition, though necessary, is not sufficient for a topological space to be second countable, as can be seen from the following example.

**Example 5.** ***Since the Souslin condition is evidently satisfied in a countable space, it follows by Example (G), page 189 that the Souslin condition holds in the Appert's space, which is not second countable.***

The results proved in Theorems 19 and 20 can be considerably sharpened in the case of a Hausdorff space.

**Theorem 21 :** *There exits in a Hausdorff space an infinite sequence of mutually disjoint non-null open sets.*

**Proof :** Let $(X, \tau)$ be a Hausdorff space. If the space $(X, \tau)$ be discrete, then every one-pointic subset is open; and these constitute an infinite set of mutually disjoint non-null open subsets of $X$, and an infinite sequence can be chosen from this infinite set.

Next, let $(X, \tau)$ be not a discrete space; then there exists a point $q$ in $X$ which is not a limiting point of $X$. Let $p_1$ be a point in $X$, different from $q$, then by the separation axiom $T_2$, there exist two open sets $U_1$ and $V_1$, such that $p_1 \in U_1$, $q \in V_1$ and $U_1 \cap V_1 = \Phi$. Again, since $q \in V_1$ and $q$ is a limiting point of $X$, there exists a point $p_2$ in $V_1$, different from $q$. By the separation axiom $T_2$, there exist two open sets $U_2$ and $V_2$, such that $p_2 \in U_2, q \in V_2$ and $U_2 \cap V_2 = \Phi$; and it may be so arranged that $U_2 \subset V_1$ and $V_2 \subset V_1$, for otherwise we could take $U_2 \cap V_1$ and $V_2 \cap V_1$ in place of $U_2$ and $V_2$ respectively. Suppose that we have, by continuing the above process, determined successively the points $p_1, p_3, ..., p_n$ and correspondingly the open sets $U_1, U_2, ..., U_n$ and $V_1, V_2, ..., V_n$, such that, for $k = 1, 2, ..., n$, $p_k \in U_k \subset V_{k-1}$, $q \in V_k \subset V_{k-1}$ and $U_k \cap V_k = \Phi$.

Now, since $q$ is a limiting point of $X$ and $V_n$ an open set containing $q$, there exists a point $p_{n+1}$ in $X$, different from $q$ and lying within $V_n$. Then by the separation axiom $T_2$, there exist two open sets $U_{n+1}$ and $V_{n+1}$ satisfying the conditions

$$p_{n+1} \in U_{n+1},\ q \in V_{n+1} \text{ and } U_{n+1} \cap V_{n+1} = \Phi;$$

and as above, it may be so arranged that $U_{n+1} \subset V_n$ and $V_{n+1} \subset V_n$. The infinite sequence of open sets $U_1, U_2, ...,$ thus defined by induction, are mutually disjoint, since $U_{n+1} \subset V_n$, $U_n \cap V_n = \Phi$, and $V_n \subset V_{n+1}$, for $n = 1, 2, \ldots$//

It now follows readily that :

**Theorem 22 :** *The set of all open sets in a second countable Hausdorff space has the cardinality* **c**.

**Proof :** Let $(X, \tau)$ be a second countable Hausdorff space. Then, by Theorem 21, there exists in $(X, \tau)$ an infinite sequence of mutually disjoint non-null open sets $U_1, U_2, \ldots$ Different sub-sequences of the sequence $\{U_i : i = 1, 2, \ldots\}$ will determine, as their unions, different open sets. But since the set of all subsets of a countable set has the cardinality **c**, it follows that the set of all open sets in $(X, \tau)$ has the cardinality $\geq$ **c** (all the open sets may not be obtainable as unions of the sub-sequences of the sequence $\{U_i,\ i = 1, 2, \ldots\}$). Again, by Theorem 19, the cardinality of the set of all open sets in $(X, \tau) \leq$ **c**. Consequently, the cardinality of the set of all open sets in $(X, \tau)$ is exactly equal to **c**.//

As an immediate consequence of Theorem 22, we have :

**Corollary :** *The set of all closed sets in a second countable Hausdorff space has the cardinality* **c**.

We shall now prove the following Theorem on the cardinality of a second countable $T_1$-space.

**Theorem 23 :** *The set of all points in a second countable $T_1$-space has cardinality at most equal to* **c** *(i.e., has the power of the continuum at most).*

**Proof :** Let $(X, \tau)$ be a second countable $T_1$-space, for which the sets $W_1, W_2, \ldots$ form a countable open base of $\tau$. Then, for any given point $p \in X$, $W_{p1}, W_{p2}, \ldots$ form a countable open base about the point $p$ in $(X, \tau)$, where $W_{pi}$, for $i = 1, 2, \ldots$, is a sub-sequence of the sequence $\{W_n : n = 1, 2, \ldots\}$, consisting of all those $W_n$ which contain the point $p$. Since a second countable space is necessarily first countable, by the Lemma, page 201, corresponding to the point $p$, there exists an infinite sequence of open sets $\{V_i : i = 1, 2, \ldots\}$, which is a subsequence of the sequence $\{W_{pn} : n = 1, 2, \ldots\}$, and therefore also a subsequence of the sequence $\{W_n : n = 1, 2, \ldots,$ such that $\cap \{V_i : i = 1, 2, \ldots\} = \{p\}$.

Thus to each point $p$ in $X$, there corresponds a sub-sequence $\{V_i : i = 1, 2, \ldots\}$ of the sequence $\{W_n : n = 1, 2, \ldots\}$; and to two different points there correspond two such different sub-sequences $\{V_i : i = 1, 2, \ldots\}$. Hence the set of all points in $X$, *i.e.*, the set $X$, has the same cardinality as that of a certain sub-collection of the collection of all sub-sequence of the sequence $\{W_n : n = 1, 2, \ldots\}$. Thus the cardinality of $X$ is less than or equal to **c**. In other words, the set $X$ has the power of the continuum at most.//

### 2.3.1 Points of Condensation in a Second Countable Space

A point of condensation has been defined in page 59. We now prove that:

**Theorem 24 :** *In a second countable space, every uncountable subset contains a point of condensation.*

**Proof :** Let $(X, \tau)$ be a second countable space, and $\{W_i : i = 1, 2, \ldots\}$ be a countable open base of $\tau$. Let $A$ be a subset of $X$, such that $A$ does not contain any point of condensation. For each point $p \in A$, $p$ is not a point of condensation of $A$; hence there exists an open set $G$, containing $p$, such that $G \cap A$ is countable at most. There exists a suffix $n(p)$, such that $p \in W_{n(p)} \subset G$; and then $A \cap W_{n(p)}$ is also countable at most. But we can express $A$ in the form $A = \cup \{p : p \in A\} \subset \cup \{A \cap W_{n(p)} : p \in A\}$, and there can be at most a countable number of different suffices. So, $A$ is at most a countable union of countable sets; that is, $A$ is at most a countable subset of $X$. Consequently, if $A$ is an uncountable subset, then it must possess a point of condensation.//

**Theorem 25 :** *If E is an uncountable subset of a second countable space, then the subset D, consisting of all those points of E which are not points of condensation of E, is at most countable.*

**Proof :** Let $(X, \tau)$ be a second countable space, and $\{W_i : i = 1, 2, ...\}$ be a countable open base of $\tau$. Let $\{V_i : i = 1, 2, \dots\}$ be a sub-sequence of the sequence $\{W_i : i = 1, 2, \dots\}$, consisting of all those sets $W_j$ for which $W_j \cap E$ is at most countable. Then $\{V_i \cap E : i = 1, 2, \dots\}$ is at most countable. We shall show that $D = E \cap (V_1 \cup V_2 \cup \dots)$.

Let $x \in D$, then $x$ is not a point of condensation of $E$; hence there exists a *nbd* $U$ of $x$, such that $U \cap E$ is at most countable. Also, there exists a set $W_j$, belonging to the sequence $\{W_i : i = 1, 2, \dots\}$, satisfying $x \in W_j \subset U$. Then $W_j \cap E$ is at most countable, and so $W_j$ must be one of the sets $V_i$, and therefore $x \in V_i$. Again, since $x \in E$, it follows that

$$x \in E \cap (V_1 \cup V_1 \cup \dots).$$

Next, let $y \in E \cap (V_1 \cup V_2 \cup \dots)$, then $y$ belongs to some $V_i$. Now, as $V_i$ is a *nbd* of $y$, and $V_i \cap E$ is at most countable, $y$ cannot be a point of condensation of $E$. Also, as $y \in E$, it follows that $y \in D$. Thus $D = E \cap (V_1 \cup V_2 \cup \dots)$.

Now $D = E \cap (V_1 \cup V_2 \cup \dots) = (E \cap V_1) \cup (E \cap V_2) \cup ...$; and since each $E \cup V_i$ is at most countable, it follows that $D$ is also at most countable.//

We can now deduce a number of important results as Corollaries of the above Theorem.

**Corollary 1 :** *If E is an uncountable subset of a second countable space; and if B is the subset of E, consisting of the points of condensation of E which lie inside E, then B is uncountable and dense-in-itself.*

**Proof :** With the same notations as in the above Theorem, we have $B = E - D$. Since $E$ is uncountable and $D$ is at most countable, it follows that $B$ is necessarily uncountable.

Next, let $p$ be any point in $B$, and $U$ be any *nbd* of $p$. Since $p$ is a point of condensation of $E$, the set $E \cap U$ is uncountable. But $B \cap U \supset E \cap U - D \cap U$, and $D \cap U$ is at most countable (since $D$ is at most countable, by Theorem 24). Hence $B \cap U$ is uncountable, consequently, $p$ is a point of condensation of $E$. Thus every point of $B$ is a point of condensation (and there also an accumulation point) of $B$. Hence the set $B$ is dense-in-itself.//

**Corollary 2 :** *Every scattered set, contained in a second countable space, is at most countable.*

**Proof :** Let $S$ be a scattered set, contained in a second countable space $(X, \tau)$. If $S$ is uncountable then, by Corollary 1, the points of condensation of $S$ would form an uncountable dense-in-itself subset $B$ of $S$. But this is not possible as a scattered set does not contain any dense-in-itself subset.//

**Corollary 3 :** *Every isolated set, contained in a second countable space, is at most countable.*

**Proof :** This is an immediate consequence of Corollary 2, as every isolated set is scattered.//

## EXERCISES

1. **(Cantor Bendixon theorem)** : Show that in a topological space every closed set can be expressed uniquely as the union of a perfect set and a scattered set (which may be empty) disjoint with it.

   [**Hints :** Let $F$ be a closed set in a topological space $(X, \tau)$. Let $N$ be the nucleus of $F$, then $N$ is a perfect set, and $S = F - N$ is scattered. Thus $F = N \cup S$, where $N$ and $S$ are disjoint, $N$ is perfect and $S$ is scattered ($S$ may, in particular, be empty). Then it can be shown that the above representation is unique].

2. Show that every closed set in a second countable space can be expressed uniquely as the union of a perfect set and a set which is at most countable (*Cantor Bendixon*).

   [**Hints :** The result follows immediately from Example 1 and Corollary 2 of Theorem 25].

*Chapter 6*

# COMPACTNESS

Bounded and closed subsets of Euclidean spaces have many interesting properties. Compactness is intended to be a property which would endow a subset with properties similar to those of a bounded and closed subset in a Euclidean space. Boundedness cannot be satisfactorily defined in a general topological space.

In a Euclidean space, bounded and closed sets be characterised in several ways, e.g. by

(*i*) Bolzano-Weierstrass theorem,

(*ii*) Heine-Borel theorem, or

(*iii*) Cantor's theorem on nested closed sets.

These can, however, be utilised to define "compactness" in a general topological space, because these theorems are formulated in terms of topological properties. These theorems, which are equivalent in a Euclidean space, are however not so in an arbitrary topological space. Thus we arrive at different types of compactness in a general topological space.

## Art. 1 Different types of compactness

We shall now introduce four different kinds of compactness.

A topological space $(X, \tau)$ is said to be

(*i*) *compact*, if every open covering of $X$ has a finite sub-covering;

(*ii*) *countably compact*, if every countable open covering of $X$ has a finite sub-covering;

(*iii*) *sequentially compact*, if every infinite sequence in $X$ contains a convergent subsequence;

(*iv*) *Frechet compact* (or *B-W compact, i.e., Bolzano-Weierstrass compact*), if every infinite subset of $X$ has an accumulation point.

The above types of compactness may also be defined for the subsets of $X$ with respect to their subspace topologies.

Lindeloff property may as well be considered as a kind of compactness. The four kinds of compactness and the Lindeloff property are not equivalent; but they are not independent of each other, as will be shown in Art 1.4.

Other types of compactness, viz. meta- and para-compactness are defined in terms refinements (instead of sub-coverings) of coverings. All the types of compactness, meta- and para-compactness are characterised more appropriately in terms of nets and filters, and will be considered in a later chapter.

Of the four types of compactness, mentioned above, sequential compactness and Frechet compactness are considered to be less important. Also, properties of a countably compact space are almost similar to the corresponding properties of a compact space, with the limitation of countability. Accordingly, we shall develop the properties of a compact space in details, and then obtain similar properties for a countably compact space. The properties for sequentially and Frechet compact spaces will mostly be stated (without proofs).

## 1.1 Compactness

Compactness can be characterised in terms of "the finite intersection property" of closed sets.

A collection of subsets $\{F_\nu : \nu \in \Lambda\}$ of a given set $X$ ($\Lambda$ being an indexing set) is said to possess the *finite intersection property*, if every finite subcollection of $\{F_\nu\}$ has non-empty intersection.

**Theorem 1 :** *A topological space $(X, \tau)$ is compact if, and only if, for every collection of closed sets $\{F_\nu : \nu \in \Lambda\}$ in $(X, \tau)$, possessing the finite intersection property, the intersection $\cap \{F_\nu : \nu \in \Lambda\}$ of the entire collection is non-empty.*

**Proof :** Let $(X, \tau)$ be a compact space, and let $\{F_\nu : \nu \in \Lambda\}$, $\Lambda$ being an indexing set, be a family of closed sets in $(X, \tau)$, possessing the finite intersection property. If possible, let $\cap \{F_\nu : \nu \in \Lambda\} = \Phi$. Then, by taking complements, we get $\cup \{X - F_\nu : \nu \in \Lambda\} = X$, so that the collection $\{X - F_\nu : \nu \in \Lambda\}$ forms an open covering of $X$. And, since the space $(X, \tau)$ is compact, there exists a finite subcovering of $X$, say formed by the sets $X - F_1, ..., X - F_n$. Then

$$X = (X - F_1) \cup \dots \cup (X - F_n) = X - (F_1 \cap F_2 \cup \dots \cap F_n)$$

implies $F_1 \cap F_2 \cap \dots \cap F_n = \Phi$, contrary to the assumption that the family $\{F_\nu : \nu \in \Lambda\}$ possesses the finite intersection property. Hence $\cap \{F_\nu : \nu \in \Lambda\}$ is non-empty.

Conversely, let the condition hold in a topological space $(X, \tau)$. Let $\{G_\alpha : \alpha \in \Lambda\}$ be an open covering of $X$, so that $X = \cup \{G_\alpha : \alpha \in \Lambda\}$. Taking complements, we get $\Phi = X - \cup \{G_\alpha : \alpha \in \Lambda\} = \cap \{X - G_\alpha : \alpha \in \Lambda\}$.

$\{X - G_\alpha : \alpha \in \Lambda\}$ forms a family of closed sets in $(X, \tau)$. And, since $\cap \{X - G_\alpha : \alpha \in \Lambda\} = \Phi$, it follows, by the given condition, that the family $\{X - G_\alpha : \alpha \in \Lambda\}$ cannot possess the finite intersection property. Hence a certain finite sub-collection, say $X - G_1, X - G_2, ..., X - G_n$, must have an empty intersection, *i.e.*,

$$\Phi = (X - G_1) \cap (X - G_2) \dots \cap (X - G_n) = X - (G_1 \cup G_2 \cup \dots \cup G_n),$$

and therefore $X = G_1 \cup G_2 \cup \dots \cup G_n$. Thus $\{G_1, G_2, ..., G_n\}$ forms a finite sub-covering of the covering $\{G_\alpha : \alpha \in \Lambda\}$ for $X$. Hence $(X, \tau)$ is a compact space.//

The real number space $(\mathbf{R}, \tau)$, with the usual topology, is not compact, since for the open covering $\{(-n, n) : n = 1, 2, \dots\}$ of $\mathbf{R}$, there does not exist any finite sub-covering.

The closed intervals in $\mathbf{R}$ are however compact :

**Theorem 2 :** *(Heine-Borel theorem) : Every closed interval in the real number space* $(\boldsymbol{R}, \tau)$, *with the usual topology, is compact.*

**Proof :** Let $[a, b]$, with $a < b$, be a closed interval in $\mathbf{R}$, and let $\mathbf{C}$ be an open covering of $[a, b]$ in $(\mathbf{R}, \tau)$. Let $P$ consist of all those elements $x \in [a, b]$, such that there exists a finite sub-covering of $\mathbf{C}$ for $[a, x]$. Then $a \in P$. Also, $b$ is an upper bound of the set $P$; hence $P$ has a least upper bound $c$, say. As $a$ is contained in some member of $\mathbf{C}$, which must contain points in $(a, b]$ also (since an open set is the union of some open intervals, and any open interval, containing $a$ cannot have $a$ as its (right) end-point), it follows that $a < c$. Let $U$ be a member of $\mathbf{C}$, such that $c \in U$. Let $d \in (a, c) \cap U$, then $[a, d)$ has an open covering by a finite sub-collection $\mathbf{C}'$ of $\mathbf{C}$. Then $\mathbf{C}' \cup U$ forms a finite subcovering of $\mathbf{C}$ for $[a, c]$. If $c \neq b$, then there exits a number $e$, such that $e \in U$ and $(c, e) \subset U$; and in that case, $\mathbf{C}' \cup U$ forms a finite subcovering of $\mathbf{C}$ for $[a, e]$, so that $e \in P$, which is impossible as $e > c$ (and $c$ is the *l.u.b.* of $P$). Thus $c = b$, and therefore $\mathbf{C}' \cup U$, forms a finite sub-covering of $\mathbf{C}$ for $[a, b]$. Hence the closed interval $[a, b]$ is compact in $(\mathbf{R}, \tau)$.//

**Example 1.** *An open interval in the real number space* $(\boldsymbol{R}, \tau)$ *may not be compact.* In fact, for the open covering $\{(1/n, n/n + 1) : n = 2, 3, ...\}$ of the open interval $(0, 1)$, there does not exist any finite sub-covering.

The complete characterisation of the compact subspaces of the real number space (with its usual topology) is given below in Corollary 2 of Theorem 3.

**Compactness in a linearly ordered space.**

**Theorem 3 :** *(Haar-Konig theorem) : A linearly ordered set, with its interval topology, is compact if, and only if, it is order-complete (i.e., iff every non-empty subset of it has a least upper bound and a greatest lower bound).*

**Proof :** Let $X$ be a linearly ordered set, which is order-complete, and let $\tau$ denote the interval topology on $X$. $X$ possesses a smallest element $a$ and a largest element $b$. Proceeding, exactly in the same way as in the proof of Theorem 2, we can show that $(X, \tau)$ is compact.

Next, let $X$ be a linearly ordered set, and $\tau$ be its interval topology. Let $X$ be *not* order-complete. We shall show that $(X, \tau)$ is not compact. Let $Y$ be a non-empty subset of $X$, such that $Y$ has no least upper bound in $X$. Then the family of open intervals :

$(-\infty, y)$, for every $y \in Y$, and

$(u, +\infty)$, for every upper bound $u$ of $Y$

forms an open covering of $X$. This covering does not contain any finite sub-covering, since in that case the largest $y$ or the smallest $u$ occuring in that sub-covering would be the least upper bound of the set $Y$. Hence $(X, \tau)$ cannot be compact. (On the other hand, if $Y$ is a non-empty subset, having no greatest lower bound in $X$, then similar reasoning will show that $(X, \tau)$ cannot be compact.//

As immediate consequence of the above Theorem, we obtain :

**Corollary 1 :** *A well-ordered set, with its interval topology, is compact if, and only if, it contains a maximal element.*

**Proof :** The necessity of the condition follows from Theorem 3, while the sufficiency is proved in exactly the same way as in Theorem 3.//

**Corollary 2 :** *A subset X of the real number space, with its usual (interval) topology, is compact if, and only if, X is closed and bounded.*

**Proof :** The result follows directly from Theorem 3 in view of the fact that a subset $X$ of the real number space is order-complete if, and only if, it is closed and bounded.//

**Subspaces of a compact space.**

**Theorem 4 :** (*a*) *A subspace of a compact space need not be compact.*

(*b*) *Any closed subspace of a compact space is compact.*

(*c*) *The union of a finite collection of compact subspaces of a topological space is compact.*

**Proof :** (*a*) It follows, by Theorem 2 and Example 1, that the subspace (0, 1) of the compact space [0, 1] is not compact.

(*b*) Let $(X, \tau)$ be a compact space. Let $A$ be a closed subset in $(X, \tau)$, and let $\{F_\nu : \nu \in \Lambda\}$ be a family of closed sets in $(A, \tau_A)$, possessing the finite intersection property. The sets $F_\nu$ are also closed in $(X, \tau)$, since $A$ is a closed subset in $(X, \tau)$ (vide example 2 (*i*), page 74). Thus $\{F_\nu : \nu \in \Lambda\}$ is a family of closed sets in $(X, \tau)$, possessing the finite intersection property; and as $(X, \tau)$ is compact, it follows by Theorem 1 that $\cap \{F_\nu : \nu \in \Lambda\}$ is non-empty. And, it now follows, by Theorem 2, that the subspace $(A, \tau_A)$ is compact.

(*c*) Let $Y_1, Y_2, ..., Y_n$ be a finite collection of compact subspaces (with their respective subspace topologies) of a topological space $(X, \tau)$. Let $Y = Y_1 \cup Y_2 \cup \dots \cup Y_n$. Let **V** be an open covering of $Y$, then **V** is also an open covering for $Y_i$; and since $Y_i$ is compact, there exists a finite sub-covering $\mathbf{U}_i = \{U_{i,1}, ..., U_i, r_i\}$ of **V** for $Y_i$, for $i = 1, 2, ..., n$. Then the sets belonging to $\mathbf{U}_1, ..., \mathbf{U}_n$ form a finite open covering of $Y$. Hence the space $(Y, \tau_Y)$ is compact.//

**Tychonoff product :** Topological product of compact spaces is known as the Tychonoff product. We shall use the following Theorem for proving Tychonoff product theorem.

**Theorem 5 :** *(Alexander's subbase theorem) : Let* **S** *be a subbase of the topology* $\tau$ *of a topological space* $(X, \tau)$*. If, for every covering of* $X$ *by a collection of members of* **S***, there exists a finite sub-covering, then* $(X, \tau)$ *is a compact space.*

**Proof :** Let **B** be a family of open sets in $(X, \tau)$, such that no finite subcollection of **B** forms a covering of $X$, then **B** is of finite character. Hence, by Tukey's lemma (vide page 17), there exists a maximal family **A** with the above property; and it will be sufficient for the purpose of our proof to show that **A** does not form an open covering of $X$.

Let **C** be the collection of all those members of **A** which belong to the subbase **S**. Then no finite sub-collection of **C** forms a covering of $X$; hence, by the hypothesis, **C** does not form a covering of $X$.

Again, since **S** forms a subbase of $\tau$, for any point $x \in A$, where $A \in \mathbf{B}$, there exists a finite sub-collection of **S**, $S_1, S_2, ..., S_n$ (say), such that

$$x \in S_1 \cap S_2 \cap \dots \cap S_n \subset A. \qquad ...(1)$$

If none of the sets $S_1, S_2, ..., S_n$ belongs to **A**, then it follows from the maximal property of **A** that for the collection $A \cup S_i$, there would exist a finite sub-collection forming a covering of $X$.

$$S_i \cup A_{i,1} \cup A_{i,2} \cup ... \cup A_{i,r_i} = X, \text{for } i = 1, 2, ..., n. \text{ Then}$$

$$(S_1 \cup S_2 \cup ... \cup S_n) \cup (A_{1,1} \cup ... \cup A_{1,r_1}) \cup ... \cup (A_{n,1} \cup ... \cup A_{n,r_n}) = X,$$

from which it follows by (1) that

$$A \cup (A_{1,1} \cup ... \cup A_{1,r_1}) \cup ... \cup (A_{n,1} \cup ... \cup A_{n,r_n}) = X,$$

which is impossible, as no finite sub-collection of **A** would form a covering of $X$. Hence one at least of the sets $S_1, S_2, ..., S_n$, say $S_k$, must belong to **A**. Then $S_k \in$ **C**, so that $x \in \cup\{\mathbf{C}\}$; consequently $\cup\{\mathbf{A}\} = \cup\{\mathbf{C}\}$. But **C** does not form a covering of $X$; hence **A** also does not form a covering of $X$.//

**Theorem 6 :** *(Tychonoff product theorem) : Topological product of any collection of compact spaces is compact.*

**Proof :** Let $(X, \tau)$ be the topological product of a given collection of compact spaces $\{(X_\nu, \tau_\nu) : \nu \in \Lambda\}$. Then $\mathbf{S} = \{p_\nu^{-1}(U_\nu) : U_\nu \in \tau_\nu, \nu \in \Lambda\}$ forms a subbase for the product topology $\tau$, $p_\nu$ being the projection mapping $X \to X_\nu$, $\nu \in \Lambda$. If follows, by Alexander's subbase Theorem, that $X$ will be compact if each sub-family **A** of **S**, such that no finite sub-collection of **A** forms a covering of $X$, fails to form a covering of $X$.

For each index $a$, let $\mathbf{B}_a$ be the aggregate of all those open sets $U_a$, belonging to $\tau_a$, for which $p_a^{-1}(U_a) \in \mathbf{A}$. Then no finite sub-collection of $\mathbf{B}_a$ forms a covering of $X_a$; and so the entire collection cannot form a covering of $X_a$, since $(X_a, \tau_a)$ is compact. Hence there exists a point $x_a$ in $X_a$ which is not contained in any open set $U_a$ belonging to $\mathbf{B}_a$. Then the point $x$ in $X$, whose $a$th co-ordinate is $x_a$, does not belong to any member of **A**; consequently, **A** does not form a covering of $X$.//

**Corollary :** *(i) The n-cube $\boldsymbol{I}^n$ is compact, for any positive integer n, where $\boldsymbol{I}$ is any closed interval in the real number space.*

*(ii) The cube $I^\alpha$, where $\alpha$ is any transfinite cardinal number, is compact.*

**Proof :** Let **I** denote a closed interval $[a, b]$ in the real number space $(\mathbf{R}, \tau)$, $\tau$ being the usual (order) topology on **R**. Then $(\mathbf{I}, \tau_\mathbf{I})$ is a compact space, by Heine-Borel theorem. Let $A$ be a set, whose cardinal number is $\beta$. Then $\mathbf{I}^\beta$ is the topological product of the family of compact spaces $\{(X_\nu, \tau_\nu) : X_\nu = \mathbf{I}, \tau_\nu = \tau_{\mathbf{I}}, \text{ for each } \nu \in A\}$. Hence, by the Tychonoff product theorem, $\mathbf{I}^\beta$ is a compact space. If $A$ is the finite set, $A = \{1, 2, ..., n\}$, then $\beta = n$, and we get the result (*i*). And, for an infinite set $A$, for which $\beta = \alpha$, we get the result (*ii*).//

### 1.1.1 Compactness in a Hausdroff Space

We shall first prove a useful lemma.

**Lemma :** *Let* $(X, \tau)$ *be a topological space.*

(*i*) *If* $B$ *is a compact subset in* $(X, \tau)$, *and* $x$ *be a point in* $X$, *such that* $x$ *can be strongly separated from every point* $y$ *in* $B$, *then* $x$ *and* $B$ *can also be strongly separated in* $(X, \tau)$.

(*ii*) *If* $A$ *and* $B$ *are two compact subsets in* $(X, \tau)$, *such that every point* $x$ *in* $A$ *can be strongly separated from every point* $y$ *in* $B$, *then* $A$ *and* $B$ *can be strongly separated in* $(X, \tau)$.

**Proof :** (*i*) Let $U_y(x)$ and $V_x(y)$ separate strongly the point $x$ from a point $y \in B$. As $y$ runs over $B$, the corresponding sets $V_x(y)$ form an open covering of $B$, for which there exists a finite sub-covering, $V_x(y_1)$, $V_x(y_2), ..., V_x(y_n)$, say, since $B$ is compact. Let $U_{y_1}(x), U_{y_2}(x), ..., U_{y_n}(x)$ be the corresponding open sets containing the point $x$. Let

$$U_B(x) = U_{y_1}(x) \cap \ldots \cap U_{y_n}(x), \text{ and } V_x(B) = V_x(y_1) \cup \ldots \cup V_x(y_n).$$

Then $x \in U_B(x)$ and $B \subset V_x(B)$. Also, since $U_{y_i}(x) \cap V_x(y_i) = \Phi$, and $U_B(x) \subset U_{y_i}(x)$, for $i = 1, 2, ..., n$, it follows that $U_B(x) \cap V_x(B) = \Phi$. Thus $x$ and $B$ are separated strongly by the pair of disjoint open sets $U_B(x)$ and $V_x(B)$ in $(X, \tau)$.

(*ii*) Let $x$ run over $A$, then the corresponding sets $U_B(x)$ form a covering of $A$, for which there exists a finite sub-covering $\{U_B(x_1), U_B(x_2), ..., U_B(x_m)\}$, say, for $A$ (since $A$ is compact). Let $V_{x_1}(B)$, $V_{x_2}(B), ..., V_{x_m}(B)$ be the corresponding open sets containing $B$. Then $U(A) = U_B(x_1) \cup \ldots \cup U_B(x_m)$, and $V(B) = V_{x_1}(B) \cap \ldots \cap V_{x_m}(B)$ are two disjoint open sets (as in (*i*)), which separate $A$ and $B$ strongly in $(X, \tau)$.//

**Theorem 7 :** *A compact Hausdorff space is a* $T_4$*-space.*

**Proof :** Let $P$ and $Q$ be two disjoint closed sets in a compact Hausdorff space $(X, \tau)$. Then, by Theorem 4 (*b*), the subsets $P$ and $Q$ are compact. Again, since $(X, \tau)$ is a Hausdorff space, every point in $P$ can be strongly separated from every point in $Q$ (as $P$, $Q$ are disjoint). Hence, by the result (*ii*) of the above Lemma, it follows that $P$ and $Q$ can be strongly separated in $(X, \tau)$. Consequently, the space $(X, \tau)$ is normal, and therefore also a $T_4$-space (since it is a Hausdorff space).//

**Theorem 8 :** *A compact subset in a Hausdorff space is closed.*

**Proof :** Let $B$ be a compact subset in a Hausdorff space $(X, \tau)$, and let $x$ be a point in $X$, lying outside $B$. Then $x$ can be strongly separated from every point $y$ in $B$, as $(X, \tau)$ is a Hausdorff space. Hence, by the result (*i*) of the above Lemma, $x$ can be strongly separated from $B$, so that there exists an open set $U$, containing $x$, such that $U \cap B = \Phi$. Consequently, $x$ cannot be an accumulation point of $B$. Hence $B$ must be a closed set in $(X, \tau)$.//

Now, combining the results in Theorems 4 (*ii*) and 8, we get :

**Theorem 9 :** *In a compact Hausdorff space, the compact subsets are the same as the closed subsets.*

It may however be noted that a compact subset in a $T_1$-space need not be closed, as can be seen from the following Example.

**Example 2. *Example of a compact subset in a $T_1$-space that is not closed.***

**Solution :** Let $\sigma$ denote the topology of finite complements on the set of real numbers $\mathbf{R}$ (vide Example 3, page 44). Since the finite sets are closed in $(\mathbf{R}, \sigma)$, $(\mathbf{R}, \sigma)$ is a $T_1$-space. Let $A = \mathbf{R} - \{0\}$; that is, $A$ consists of all non-zero real numbers. Then $(A, \sigma_A)$ is a compact subspace of $(\mathbf{R}, \sigma)$ (since, if $\mathbf{C}$ be any open covering of $A$, and $U$ be any member of $\mathbf{C}$, then $A - U$ is a finite set). But $A$ is not closed in $(\mathbf{R}, \sigma)$, since 0 is an accumulation point of $A$.

**Theorem 10 :** *Every subset of a compact Hausdorff space is a Tychonoff space. Conversely, any Tychonoff space is homeomorphic with a subset of a compact Hausdorff space.*

**Proof :** A compact Hausdorff space $(X, \tau)$ is, by Theorem 7, a $T_4$-space, and hence also a Tychonoff space. Also, every subspace of a Tychonoff space is a Tychonoff space. Hence every subspace of $(X, \tau)$ is a Tychonoff space.

Conversely, let $(X, \tau')$ be a Tychonoff space. Then, by Theorem 17, page 174, $(X, \tau')$ is homeomorphic to a subset of a unit cube $I^{\alpha}$. Also,

$I^{\alpha}$ is a compact space, by Corollary (*ii*) of Theorem 6; and $I^{\alpha}$ is a Hausdorff space. Thus $(X, \tau')$ is homeomorphic with a subset of the compact Hausdorff space $I^{\alpha}$.//

## 1.2 Countable Compactness

We shall obtain several necessary and sufficient conditions for a topological space to be countably compact. One such condition is given in terms of the concept of a cluster point of a sequence. A point $p$ is a called a *cluster point of an infinite sequence* $\{x_n : n = 1, 2, \ldots\}$ in a topological space $(X, \tau)$ if, for any given open set $U$, containing $p$, and any positive integer $r$, there always exists a positive integer $m > r$, such that $x_m \in U$.

**Theorem 11 :** *A topological space $(X, \tau)$ is countably compact if, and only if, any one of the following conditions holds :*

(*a*) *Every countable aggregate of closed sets, possessing the finite intersection property, has a non-empty intersection in $(X, \tau)$.*

(*b*) *Every descending chain of non-empty closed sets, $F_1 \supset F_2 \supset \ldots$, has a non-empty intersection in $(X, \tau)$.*

*(Cantor's intersection theorem).*

(*c*) *Every infinite sequence in X has a cluster point in X,*

(*d*) *Every infinite set $S \subset X$ has an $\omega$-accumulation point in X.*

**Proof :** (*a*) The proofs of both the necessary and sufficient parts are the same as those for Theorem 1, the indexing set $\Lambda$ being taken to be a countable set.

(*b*) Let $(X, \tau)$ be countably compact, and let $F_1 \supset F_2 \supset \ldots$ be a descending chain of non-empty closed sets in $(X, \tau)$. Then $\{F_i : i = 1, 2, \ldots\}$ is a countable aggregate of closed sets in $(X, \tau)$, possessing the finite intersection property. Hence $\cap \{F_i : i = 1, 2, \ldots\}$ is non-empty, by the property (*a*).

Conversely, let the condition hold in $(X, \tau)$. Let $\{K_i; i = 1, 2, \ldots\}$ be a countable aggregate of closed sets possessing the finite intersection property in $(X, \tau)$. Let $F_i = K_1 \cap K_2 \cap \ldots \cap K_i$, for $i = 1, 2, \ldots$; then $F_1 \supset F_2 \supset \ldots$ is a descending chain of non-empty closed sets in $(X, \tau)$. Hence, by hypothesis, $\cap \{F_i; i = 1, 2, \ldots\}$ is non-empty. Consequently, $\cap \{K_i : i = 1, 2, \ldots\} = \cap \{F_i : i = 1, 2, \ldots\}$ is also non-empty. Hence the space $(X, \tau)$ is countably compact, by the property (*a*).

(*c*) Suppose $\{x_i : i = 1, 2, \ldots\}$ is an infinite sequence in $(X, \tau)$ with no cluster points. Then, for each point $p \in X$, there exits an open set $U_p$, such that $p \in U_p$ and $U_p \cap \{x_{r+1}, x_{r+2}, \ldots\} = \Phi$. For each positive integer $n$, let $G_n$ be the union of all sets $U_p$ which do not contain any term of $\{x_{n+1}, x_{n+2}, \ldots\}$. Then $\{G_1, G_2, \ldots\}$ forms a countable open covering of $X$, for which there does not exist any finite sub-covering. Hence the space $(X, \tau)$ cannot be countably compact.

Next, suppose that the space $(X, \tau)$ is not countably compact. Then there is a countable open covering $\{G_1, G_2, \ldots\}$ of $X$, for which there does not exist any finite sub-covering. Let $H_1 = G_1$ and $x_1 \in H_1$. For $n > 1$, let $H_n$ be the first one of $G_2, G_3, \ldots$, which is not contained in $H_1 \cup \ldots \cup H_{n-1}$, and let $x_n \in H_n - (H_1 \cup \ldots \cup H_{n-1})$. Let $p \in X$, then there exists a positive integer $r$, such that $p \in H_r$. Then $p$ is not a cluster point of the infinite sequence $\{x_i : i = 1, 2, \ldots\}$, since $H_r \cap \{x_{r+1}, x_{r+2}, \ldots\} = \Phi$. Thus the infinite sequence $\{x_i : i = 1, 2, \ldots\}$ has no cluster point in $X$.

(*d*) Let $X$ contain an infinite subset $S$, which has no $\omega$-accumulation point in $X$. Then there exists a denumerable subset $T$ of $S$, such that $T$ has no $\omega$-accumulation point in $X$. Hence, for every $p \in X$, there exists an open set $G_p$ such that $p \in G_p$ and $G_p \cap T$ is a finite set. For every finite subset $F$ of $T$ (including the case $F = \Phi$), let $G_F = \cup \{G_p : G_p \cap T = F\}$. The aggregate of the sets $\{G_F\}$ forms a countable open covering of $X$. But, since each $G_F$ intersects the infinite set $T$ in only finitely many points, for the countable open covering $\{G_F\}$ of $X$, there does not exist any finite sub-covering. Hence the space $(X, \tau)$ is not countably compact.

Next, suppose that $(X, \tau)$ is not countably compact. Then there exists a countable open covering $\{V_i : i = 1, 2, \ldots\}$ of $X$, for which there is no finite sub-covering. Then, for any positive integer $n$, $X - (V_1 \cup \ldots \cup V_n)$ is non-empty; let $x_n \in X - (V_1 \cup \ldots \cup V_n)$, for $n = 1, 2, \ldots$ Let $p$ be any point in $X$, then $p \in V_m$, for some $m$. The point $p$ cannot be an $\omega$-accumulation point of the infinite subset $S = \{x_1, x_2, \ldots\} \subset X$, since $p \in V_m$ and $V_m \cap S$ is a finite subset.//

**Theorem 12 :** (*a*) *A subspace of a countably compact space need not be countably compact.*

(*b*) *Every closed subspace of a countably compact space is countably compact.*

(*c*) *The union of a finite collection of countably compact subspaces of a topological space is a countably compact subspace.*

**Proof :** (*a*) The closed unit interval [0, 1] is compact, by Heine-Borel theorem; hence it is also countably compact (vide Theorem 14). The subspace (0, 1) of [0, 1] is, however, not countably compact. In fact, it can be seen from Example 1 that (0, 1) is not countably compact.

(*b*) and (*c*) The proofs for the cases (*b*) and (*c*) are the same as that those for Theorem 4 (*b*) and (*c*), the indexing set $\Lambda$ being taken to be a countable set.//

It has been shown by J. Novak (vide : On the Cartesian Product of two Compact Spaces, Fundamental Math, Volume 40, pp. 106-112, 1953) that, unlike Theorem 6, *the topological product of two countably compact spaces need not be countably compact.*

## 1.3 Sequentially Compact and Frechet Compact Spaces

The properties (*b*) in Theorem 4 and 12 hold also for sequential compact, Frechet compact and Lindeloff spaces. Thus

**Theorem 13 :** (*a*) *Every closed subspace of a sequentially compact space is sequentially compact.*

(*b*) *Every closed subspace of a Frechet compact space is Frechet compact.*

(*c*) *Every closed subspace of a Lindeloff space is a Lindeloff space.*

**Proof :** The result (*c*) was proved earlier in Theorem 5 (*iv*), page 191, Results (*a*) and (*b*) are left as exercises.//

By virtue of the mutual dependence among the different types of compactness, as will be established in Art. 1.4, it follows from the following Example that :

(*i*) *a subspace of a sequentially compact space need not be sequentially compact, and*

(*ii*) *a subspace of a Frechet compact space need not be Frechet compact.//*

**Example 3.** Example of a compact and sequentially compact space, a subspace of which is neither Frechet compact nor a Lindeloff space.

**Solution :** Let ($\mathbf{R}$, $v$) be the topological space, described in Example 2, page 188. It has been shown in that example that for any open covering $\mathbf{V}$ of $\mathbf{R}$, there always exists a finite sub-covering; hence the space ($\mathbf{R}$, $v$) is compact. Let $S = \mathbf{R} - \{0\}$; then it has been shown there that the subspace ($S$, $v_S$) is not a Lindeloff space.

As $\nu_S$ is the discrete topology on $S$, $S$ is an infinite set having no accumulation point in $S$. Hence the subspace $(S, \nu_s)$ is not Frechet compact.

The space $(\mathbf{R}, \nu)$ is also sequentially compact. In fact, any infinite sequence $\{x_i : i = 1, 2, ...\}$ in $\mathbf{R}$ is of any one of the following two types:

(*i*) $x_i \neq 1$ and 2 for all $i$, except for finitely many values of $i$; and the sequence $\{x_i : i = 1, 2, \ldots\}$ is itself convergent, converging to the limit 0;

(*ii*) $x_i = 1$ or 2 for infinitely many values of $i$, and then there exists an infinite subsequence of $\{x_i : i = 1, 2, \ldots\}$, which converges to the limit 1 or 2.

## 1.4 Mutual Dependence of Different Types of Compactness

**Theorem 14 :** (*a*) *Every compact space is countably compact, and also a Lindeloff space.*

(*b*) *A countably compact Lindeloff space is compact.*

**Proof :** (*a*) Let $(X, \tau)$ be a compact space. Since for every open covering of $X$, there exists a finite sub-covering, the same is true for every countable open covering. Hence $(X, \tau)$ is countably compact. Also, since a finite sub-covering is necessarily a countable sub-covering, it follows that $(X, \tau)$ is also a Lindeloff space.

(*b*) Let $(X, \tau)$ be a countably compact, Lindeloff space. Let $\mathbf{U}$ be any open covering of $X$. As $(X, \tau)$ is a Lindeloff space, there exists a countable sub-covering $\mathbf{V}$ of $\mathbf{U}$ for $X$. Again, since $(X, \tau)$ is countably compact, for the countable open covering $\mathbf{V}$ of $X$, there exists a finite sub-covering $\mathbf{W}$. Then $\mathbf{W}$ is a finite sub-covering of $\mathbf{U}$ for $X$. Hence the space $(X, \tau)$ is compact.//

**Theorem 15 :** (*a*) *A countably compact space is Frechet compact.*

(*b*) *Any Frechet compact $T_1$-space is countably compact.*

**Proof :** (*a*) Let $(X, \tau)$ be a countably compact space. Then every infinite subset $S$ of $X$ has an ω-accumulation point in $X$ (by Theorem 11 (*d*)); thus $S$ has an accumulation point in $X$. Hence $(X, \tau)$ is a Frechet compact space.

(*b*) Let $(X, \tau)$ be a Frechet compact, $T_1$-space, and $S$ be an infinite subset of $X$. As $(X, \tau)$ is Frechet compact, $S$ has an accumulation point $x$ (say) in $X$; and since $(X, \tau)$ is a $T_1$-space, the accumulation point $x$ is an ω-accumulation point (by Theorem 10 (*vi*), page 132). Hence $(X, \tau)$ is countably compact, by Theorem 11 (*d*).//

**Theorem 16 :** (*a*) *A sequentially compact space is countably compact.*

(*b*) *Any countably compact, first countable space is sequentially compact.*

**Proof :** (*a*) Let $(X, \tau)$ be a sequentially compact space, and let $\{x_i : i = 1, 2, \ldots\}$ be any infinite sequence in $X$. Then the sequence $\{x_i : i = 1, 2, \ldots\}$ contains a convergent subsequence. The limit of the convergent subsequence is a cluster point of the sequence $\{x_i : i = 1, 2, \ldots\}$. Hence $(X, \tau)$ is countably compact, by Theorem 11 (*c*).

(*b*) Let $(X, \tau)$ be a countably compact, first countable space. Let $\{x_i : i = 1, 2, \ldots\}$ be an infinite sequence in $X$. Since $(X, \tau)$ is countably compact, it is also Frechet compact (by Theorem 15 (*a*)); hence the infinite sequence $\{x_i : i = 1, 2, \ldots\}$ has an accumulation point $x$ (say) in $X$. Again, since the space $(X, \tau)$ is first countable, it follows, by Theorem 17, page 202, that there exists a sub-sequence $\{x_{k_i} : i = 1, 2, \ldots\}$ of the sequence $\{x_i : i = 1, 2, \ldots\}$, such that $\lim x_{k_i} = x$. Thus the sequence $\{x_i : i = 1, 2, \ldots\}$ contains a convergent subsequence $\{x_{k_i} : i = 1, 2, \ldots\}$. Hence the space $(X, \tau)$ is sequentially compact.//

**Note :** In proving Theorem 16 (*b*), we have merely used the property that $(X, \tau)$ is Frechet compact (in place of its countable compactness). Hence, *every Frechet compact, first countable space is sequentially compact.*

In view of the fact that a second countable space is first countable and also a Lindeloff space, it follows from Theorems 14, 15 and 16 that:

**Theorem 17 :** *For a second countable $T_1$-space, any one of the four properties,* (*i*) *compactness,* (*ii*) *countable compactness,* (*iii*) *sequential compactness, and* (*iv*) *Frechet compactness, implies the other three.*

It can be shown by constructing suitable counter-examples that no other direct implications exist between the Lindeloff property and the four compactness properties.

**Example 4.** Example of a second countable (and hence Lindeloff), Frechet compact space that is not countably compact.

**Solution :** Let $(\mathbf{N}, \tau)$ be the topological space, defined in Example 7 (*e*), page 103. Thus $\mathbf{N}$ is the set of natural numbers, and $\tau$ is the odd-even topology on $\mathbf{N}$. [The topology $\tau$ is generated by the base

$$\mathbf{B} = \phi \cup \{(2n-1,\ 2n) : n = 1, 2, \ldots\}].$$

The space $(\mathbf{N}, \tau)$ is second countable, since the open base $\mathbf{B}$ of $\tau$ is countable. Also $\mathbf{B}$ forms a countable open covering of $\mathbf{N}$, for which there is no finite subcovering, hence $(\mathbf{N}, \tau)$ is not countably compact.

Let $P$ be an infinite subset of **N**, and let $p \in P$. Let now, $x = p + 1$ if $p$ is odd, and $x = p - 1$ if $p$ is even. Then every open set, containing $x$, also contains $p$; hence $x$ is an accumulation point of $P$ in **N**. Consequently, the space $(\mathbf{N}, \tau)$ is Frechet compact.

**Example 5.** Example of a compact Hausdorff space that is not sequentially compact.

**Solution :** Let **I** denote the closed unit interval [0, 1] with the subspace topology $\sigma_{\mathbf{I}}$, induced by the usual topology $\sigma$ of the real number space $(\mathbf{R}, \sigma)$. Let $(X, \tau) = \pi \{I_r : I_r = I, r \in \mathbf{R}\}$. Thus $X = \mathbf{I}^{\mathbf{I}}$ is the uncountable product of **I**. Hence $X$ is compact and $T_2$, since $I$ is so. Again $X$ is not sequentially compact, since the sequence of function $f_n \in X$ defined by $f_n(x) =$ the $n$th digit in the binary expansion of $x$, has no convergent subsequence.

For, suppose $\{f_{n_k}\}$ is a subsequence which converges to a point $f \in X$. Then, for each $x \in \mathbf{I}$, $f_{n_k}(x)$ converges in **I** to $f(x)$. Let $p \in \mathbf{I}$ have the property that $\alpha_{n_k}(p) = 0$ or 1 according as whether $k$ is odd or even. Then $\{\alpha_{n_k}(p)\}$ is 0, 1, 0, 1, ... which cannot converge.

*Chapter 7*

# CONNECTEDNESS

## Art. 1 INTRODUCTION

Since the inception of the concept of topology in Chapter 2, we have seen that the subject excellently generalises some concepts and results of the Real line. In calculus we have come across the following theorem :

Intermediate value Theorem : If $f: [a, b] \to \mathbf{R}$ is continuous and if $r$ is a real number between $f(a)$ and $f(b)$, then there is an element $c \in [a, b]$ such that $f(c) = r$.

The question is now as follows :

How is the said theorem generalized in topology? Actually the theorem depends not only on the continuity of the function but also on some property of $[a, b]$, other than compactness. The property is known to be connectedness whose definition in topological setting would be as follows :

**Definition 1.** *Let $(X, T)$ be a topological space. It is called connected if it cannot be expressed as a disjoint union of non-empty sets A and B satisfying Hausdorff-Lennaes condition as in Theorem 2, Chapter 3 i.e., $(\overline{A} \cap B) \cup (A \cap \overline{B}) = \phi$. A subset A of $(X, T)$ is connected if it is connected as a topological space $(A, T_A)$. $(A, B)$ is said to determine a separation for X.*

**Theorem 1.** *$(X, T)$ is connected iff X cannot be expressed as $U \cup V$ where U and V are disjoint open sets in X.*

**Proof :** If $(X, T)$ is connected and $X = U \cup V$ with $U \cap V = \phi$, $U$ and $V$ are open then obviously $\overline{U} \cap V = \overline{V} \cap U = \phi$ and we are done.

Conversely, if $(X, T)$ be not connected, then there exists disjoint sets $A$ and $B$ such that $X = A \cup B$ where $\overline{A} \cap B = \overline{B} \cap A = \phi$. Now $(B = X - \overline{A})$ $\Rightarrow B$ is open; similarly $A$ is open as well and $X$ can be expressed as the union of two disjoint open sets.

**Theorem 2.** *A topological space $(X, T)$ is connected iff the only subsets of X that are both open and closed in X are the empty set and X itself.*

Proof is obvious.

**Definition 2 :** *A topological space is said to be disconnected if it is not connected.*

**Example 1.** The set **Q** of rationals, with the subspace topology is not connected : in fact, if $r$ is any irrational number $\mathbf{Q} \cap (-\infty, r)$ and $\mathbf{Q} \cap (r, \infty)$ are both open in **Q** and their union constitutes **Q**.

**Theorem 3 :** *A subspace A of a topological space (X, T) is connected iff there exists no continuous mapping $f: A \to R$ such that $f(A)$ consists of exactly two points.*

**Proof :** Let $A$ be connected; if possible, let there be a continuous $f: A \to R$ such that $f(A) = \{a, b\}$ where $a, b \in R$; then $f^{-1}(\{a\})$ and $f^{-1}(\{b\})$ are disjoint closed in $A$; further $A = f^{-1}(\{a\}) \cup f^{-1}(\{b\})$ giving $A$ to be disconnected—a contradiction.

If $A$ be not connected then $A = A_1 \cup A_2$ where $A_1$ and $A_2$ are disjoint closed sets in $A$; then $\chi_{A_1}$ is a continuous function on $A$ having two values 0 and 1 in $R$; if $K$ is a closed set in $R$ containing 0 but not 1 then $\chi_{A_1}^{-1}(K)$ is $A_2$ which is closed in $A$; if $K$ contains 1 but not 0 then $\chi_{A_1}^{-1}(K)$ is $A_2$ which is closed in $A$. If $K$ contains none of 0 and 1 then $\chi_{A_1}^{-1}(K)$ is $\phi$ in $A$ where as, if it contains both 0 and 1, then $\chi_{A_1}^{-1}(K)$ $= A$. But both $\phi$ and $A$ are closed in $A$ and thus $\chi_{A_1}$ is continuous on $A$.

**Theorem 4 :** *The continuous image of a connected set (X, T) is connected i.e., if (X, T) is connected and $f: (X, T) \to (Y, T')$ is continuous, then $(f(X), T'_{f(X)})$ is connected.*

**Proof :** If $(f(X), T'_{f(X)})$ is not connected then there is a continuous function (by Theorem 3) $g: f(X) \to R$ such that $g(f(X))$ consists of two points in $R$. But $gof$ is continuous on $X$ having two pointic image in $R$. By Theorem 3, $X$ is disconnected—a contradiction.

**Corollary 1 :** *If $f: X \to Y$ determines a homeomorphic correspondence between (X, T) and (Y, T′) and further, (X, T) is connected, then so is (Y, T′). Thus, connectedness is a topological property.*

**Theorem 5 :** *Let (X, T) be a topological space. The union of any family of connected subsets, of X, having atleast one point in common, is connected.*

**Proof :** Let $C = \bigcup_{\alpha} A_\alpha$, $y_0 \in \bigcap_{\alpha} A_\alpha$, let 2 denote a two point space in $R$ (obviously, its topology is the discrete topology).

If possible, let there be a continuous $f: C \to 2$. Since each $A_\alpha$ is connected, no $f/A_\alpha$ is surjective and because $y_0 \in A_\alpha$ for each $\alpha$, $f(y) = f(y_0)$ for all $y \in A_\alpha$ and all $\alpha$. Thus $f$ cannot be surjective giving $C$ to be connected.

**Lemma 1 :** *If the sets $C$ and $D$ form a separation of $(X, T)$ and if $Y$ is a connected subset of $X$, then $Y$ lies wholly within either $C$ or $D$.*

**Proof :** Since $C \in T$ and $D \in T$ as well, $C \cap Y \in T_Y$ and $D \cap Y \in T_Y$; obviously, $Y = (C \cap Y) \cup (D \cap Y)$; if both of $C \cap Y$ and $(D \cap Y)$ are non-empty $Y$ would be disconnected—a contradiction.

If $C \cap Y = \phi$, $Y \subset D$ and if $D \cap Y = \phi$, then $Y \subset C$.

**Theorem 5 :** *Let $A$ be a connected subset in $(X, T)$. If $A \subset B \subset \overline{A}$, then $B$ is also connected.*

**Proof :** If possible, let $B$ be not connected and $B = C \cup D$ where $(C, D)$ determine a separation for $B$. By Lemma 1 either $A \subset C$ or $A \subset D$ since $A$ is connected. Suppose $A \subset C$; then $\overline{A} \subset \overline{C}$; since $\overline{C} \cap D = \phi$, $B$ cannot intersect $D$. This contradicts the fact that $D$ is a non-empty subset of $B$.

**Corollary 2 :** *It is obvious from Theorem 5 that if $A$ be connected then so is $\overline{A}$.*

**Theorem 6 :** *Let $(X, T)$ and $(Y, T')$ be topological spaces then :*

(*i*) *if $(X \times Y, T \times T')$ is connected than so are $X$ and $Y$;*

(*ii*) *if $X$ and $Y$ are both connected then so is $X \times Y$.*

**Proof :** (*i*) Let $p_x$, $p_y$ denote, as usual, the projection maps :

$$\left.\begin{array}{l} p_x : X \times Y \to X \\ (p, q) \to p \end{array}\right\}; \quad \left.\begin{array}{l} p_y : X \times Y \to Y \\ (p, q) \to q \end{array}\right\};$$

we know that both $p_x$ and $p_y$ are continuous; by Theorem 4,

$$p_x(X \times Y) = X \text{ and } p_y(X \times Y) = Y$$

are both connected.

(*ii*) For the other part, let $p \in X$ and let $A_p = \{p\} \times Y$; the mapping $\left.\begin{array}{l} h : Y \to A_p \\ q \to (p, q) \end{array}\right\}$ is obviously a homeomorphism; hence, $Y$ being connected, so is $A_p$. In the same fashion, we can say that, if $q_0 \in Y$, the set $B_{q_0} = X \times \{q_0\}$ is connected. It is obvious that $(p, q_0) \in A_p \cap B_{q_0}$ and thus, by Theorem 5, $H_p = B_{q_0} \cup A_p$ is a connected set for each $p \in X$. Take $H = \bigcup_{p \in X} H_p$; obviously, $B_{q_0} \subset \bigcap H_p$ and by Theorem 5, $H$ is connected. But $H = X \times Y$ and we are done.

**Note 1 :** The proof for finite product of connected spaces follows by induction; of course, this process requires the obvious fact that $(X_1 X \ldots \times X_{n-1}) \times X_n$ is homeomorphic to $X_1 \times X_2 \times \ldots \times X_n$.

**Theorem 7 :** *Let* $\{X_\alpha \mid \alpha \in \Lambda\}$ *be a collection of topological spaces. Then :*

(*i*) *If* $\left( \prod_{\alpha \in \Lambda} X_\alpha, T \right)$ *where* $T$ *is the product topology, is connected, then so is* $X_\alpha$ *for each* $\alpha$;

(*ii*) *if each* $X_\alpha$ *is connected for each* $\alpha$, *then so is* $\left( \prod_{\alpha \in \Lambda} X_\alpha, T \right)$.

**Proof :** Proof of (*i*) is the same as (*i*) in Theorem 6.

(*ii*) For the other part, let $\hat{b} = (b_\alpha)$ be an arbitrary but fixed point of $\prod_{\alpha \in \Lambda} X_\alpha$; let $F = \{\alpha_1, \ldots, \alpha_n\} \subset \Lambda$; let us collect all points $x = (x_\alpha)_{\alpha \in \Lambda}$ from $X$ such that $x_\alpha = b_\alpha$ for $\alpha \neq \alpha_1, \ldots, \alpha_n$ and name the set as $X_F$. We consider a mapping

$$f : X_{\alpha_1} \times X_{\alpha_2} \times \ldots X_{\alpha_n} \to X_F : (x_{\alpha_1}, \ldots, x_{\alpha_n}) \to (y_\alpha)_{\alpha \in \Lambda}$$

where $y_\alpha = x_\alpha$ for $\alpha = \alpha_1, \ldots, \alpha_n$ and $y_\alpha = b_\alpha$ for all other values of $\alpha$. Obviously $f$ is a bijection, carries a basis element for $X_{\alpha_1} \times \ldots, \times X_{\alpha_n}$ to a basis element for $X_F$; $f$ is continuous as well and is thus a homeomorphism.

Let $Y = \cup \{X_F \mid F$ is any finite subset of $\Lambda\}$; obviously, $X_{\alpha_1} \times \ldots \times X_{\alpha_n}$ is connected by Note 1 and so is $X_F$ by Corollary 1 and $Y$ is connected by Theorem 5 since $\hat{b}$ is a member of each $X_F$.

We show that $\overline{Y} = X$ which would assert by Corollary 2 that $X$ is connected.

Let $x = (x_\alpha)_{\alpha \in \Lambda} \in X$ and let $U = \prod U_\alpha$ be an arbitrary basic open set containing $x$; by definition, $U_\alpha = X_\alpha$ except for a finite number of indices, say $\alpha_1, \ldots, \alpha_n$. Let $y_\alpha = \begin{cases} x_\alpha \text{ for } \alpha = \alpha_1, \ldots, \alpha_n \\ b_\alpha \text{ for all } \alpha \in \Lambda / \{\alpha_1, \ldots, \alpha_n\} \end{cases}$; obviously, $(y_\alpha) \in Y$

as because it belongs to the space $X_F$ where $F = \{\alpha_1, ..., \alpha_n\}$; $(y_\alpha)$ is a point of $Y$ as well, since $y_\alpha = x_\alpha \in U_\alpha$ for $\alpha = \alpha_1, ..., \alpha_n$ and $y_\alpha = b_\alpha \in X_\alpha$ for all other values of $\alpha$; then $(y_\alpha) \in U \cap Y$ and thus $(x_\alpha) \in \overline{Y}$.

## Art. 2

In Art. 1, in different theorems, we have come across different processes for constructing a connected space from a collection of connected spaces or connected subsets. The question comes automatically :

(*i*) Which among our best known topological spaces are connected? What about the connectedness of their subsets?

To start with, we refer to Theorem 6 of Chapter 2 in which we have the order topology for a linearly ordered set. Since the real line **R** is a linearly ordered set and the order topology is the usual topology, we investigate the problem of connectedness for such a linearly ordered set $X$ with its order topology?

**Theorem 8 :** *Let X be a linearly ordered set with two properties:*

(*i*) *any subset of X, bounded above has l.u.b.;*

(*ii*) *if $x < y$ where $x, y \in X$, then there exists $z \in X$ such that $x < z < y$.*

Consider $T$ to be the order-topology on $X$, then $(X, T)$ is connected and so is every interval and ray (as defined in 21, Chapter 2).

**Proof :** Let us start with a subset $A \subset X$ having the property that if $a < b$, $(a, b) \in A \times A$, then $[a, b] \subset A$. If $A = X$, it obviously possesses the property as required.

If possible, let $A = P \cup Q$ ...(1)

where $P$ and $Q$ are disjoint open sets in $A$.

Let $a \in P$ and $b \in Q$ be such points as $a < b$; by assumption $[a, b] \subset A$. We shall find a point $p \in [a, b]$ which does not belong either to $P$ or to $Q$ and which will fetch a contradiction in assumption (1) thereby ensuring the connectedness of $A$.

Consider $P_0 = P \cap [a, b]$ and $Q_0 = Q \cap [a, b]$.

By definition, $P_0$ and $Q_0$ are open in $[a, b]$ in the subspace topology. Obviously, $P_0$ is bounded above and has an *l.u.b.* by assumption, say, $p$. We ensure that $p$ does belong neither to $P_0$ nor to $Q_0$.

If possible, let $p \in Q_0$. Then $p \neq a$ since $a \in P_0$. By the definition of $Q_0$, either $p = b$ or $a < p < b$. ($p < a$ is not possible by the definition of $p$ itself).

We shall reach a contradiction by showing that there exists an upper bound $d < p$ for $P_0$. To establish the fact that such a '$d$' exists we show that $(d, b] \cap P_0 = \phi$ which would ensure that no element of $P_0$ can be greater than $d$ in the given ordering.

Since $p \in Q_0$ and $Q_0$ is open there is an open interval $(d, p) \subset Q_0$; if $p = b$, $(d, b] \subset Q_0$, and thus $(d, b] \cap P_0 = \phi$, so that there is an upper bound $d$ of $P_0$ less than $p$—a contradiction. If $p < b$, $(p, b]$ has null intersection with $P_0$ by definition of $p$ and $(d, p] \cap P_0 = \phi$; thus $(d, b] \cap P_0 = \phi$ since $(d, b] = (d, p] \cup (p, b]$.

If $p \in P_0$, then $p \neq b$ and thereby, either $p = a$ or $a < p < b$. Since $P_0$ is open in $[a, b]$, there must be same interval of the form $[p, e)$ contained in $P_0$; by hypothesis, we can choose a point $z$ of $X$ such that $p < z < e$. Then $z \in P_0$, contrary to the fact that $p = l.u.b.\ P_0$.

**Corollary 3:** *Since* **R** *with its usual ordering has the properties of X and its usual topology is the order topology,* **R** *is connected and so is every interval and ray of R.*

**Note 2:** A linearly ordered set $X$ with the properties (*i*) and (*ii*) as in Theorem 2.1 is called a linear continuum.

**Note 3:** We promised a generalisation of the Intermediate value Theorem of Calculus in the introduction. We now give the statement and proof of the generalized version of the said theorem.

**Theorem 9.** *Let* $(X, T)$ *be a connected topological space and let* $Y$ *be a linearly ordered set with the order topology* $T'$. *If* $f : (X, T) \to (Y, T')$ *is continuous, a and b are two points in X with* $f(a) < r < f(b)$, *then there exists* $c \in X$ *such that* $f(c) = r$.

**Proof:** Since $f$ is continuous and $X$ is connected so is $f(X)$ with the subspace topology inherited from $T'$; now, if there would be no $c$ such that $f(c) = r$ then $f(X) = (f(X) \cap (-\infty, r)) \cup (f(X) \cap (r + \infty)) = A \cup B$, where $A$ and $B$ both belong to $T_{f(X)}$ and are disjoint open sets in $f(X)$. Then $f(X)$ becomes disconnected a contradiction.

**Note 4:** $X = [a, b]$ and $Y = \mathbf{R}$, with their usual topologies in the above Theorem, would give the I.V. Theorem of Calculus.

**Note 5:** We can think of the complex plane with the lexicographic order and it would turn into a linear continuum. But the order topology would not be same as the usual topology of the complex plane.

**Example 1.** ***We order C lexicographically as follows :*** $(x, y) < (u, v)$ ***iff either*** $x < u$, ***or*** $(x = u$ ***and*** $y < v)$ ***where*** $x, y, u, v$ ***are real numbers.***

The condition (*ii*) as in Theorem 2.1 is obvious for such a system but condition (*i*) needs verification. Consider a bounded set $A = \{(x, y) : x \leq m, y \leq n\}$. Let $\alpha = l.u.b. \{(x, 0) \mid (x, y) \in A$ for some $y\}$. This *l.u.b.* exists since the set is a subset of the real line and is bounded above.

Now there are two possibilities :

**Case I :** $A \cap \{(\alpha, y) \mid y \in \boldsymbol{R}\} = \phi$; *in this case* $(\alpha, 0)$ *is the l.u.b. for A.*

**Case II :** $A \cap \{(\alpha, y) \mid y \in \boldsymbol{R}\} \neq \phi$; *then l.u.b.*

$$\{(\alpha, y) \in A \mid y \leq n\} = l.u.b.\, A.$$

Thus $C$ is a linear continuum.

If we name the order topology on $C$ to be $\sigma$ and the usual topology as $T$ then, for any open sphere $S((x, y), \varepsilon)$ in the usual topology where

$$* \quad S((x, y), \varepsilon) = \{(x', y') \in C / |(x, y) - (x', y)| < \varepsilon\},$$

$$U^{\varepsilon}_{(x, y)} = \left\{ (x, y) \,\middle|\, -\frac{\varepsilon}{2} < y < \frac{\varepsilon}{2} \right\}$$

contains $(x, y)$ and is contained in $S((x, y), \varepsilon)$ and thus $T \subset \sigma$. But if we consider the *nbd* : $U^{\varepsilon}_{(x, y)}$ of $(x, y)$, then no open sphere with centre $(x, y)$ in $T$ would be contained in $U^{\varepsilon}_{(x, y)}$ and thus $\sigma$ is strictly finer than $T$.

## Art. 3 COMPONENTS AND QUASI COMPONENTS IN A TOPOLOGICAL SPACE

**Definition 3 :** *Let* $(X, T)$ *be a topological space.*

(*a*) *A subset* $A \subset X$ *is called a component of X if it is connected and not strictly included in any connected set B in X. If* $x \in X$, *a component at x is always that component in the space X that contains the point x.*

(*b*) *A quasi-component at a point x in X is the intersection of all clopen (i.e., both open and closed) sets in X containing the given point.*

(*c*) *X is called totally disconnected if every quasi-component of this space contains only one point.*

**Theorem 10 :** *A component C in a topological space* $(X, T)$ *is closed in X; any two components of X are either identical or disjoint.*

**Proof :** Since $C$ is a component, it is a maximal connected set in $X$; now $\overline{C}$ is connected by Corollary 2; hence $C = \overline{C}$ giving $C$ to be closed.

For the other assertion, let $A$ and $B$ be two components of $X$; if $A \cap B \neq \phi$ then we consider $A \cup B$. By Theorem 5 $A \cup B$ is connected—a contradiction since $A$ and $B$ are both maximal connected sets.

**Theorem 11 :** *Let $X$ be a topological space, $x$ a point of $X$ and $B^x$ the family of all connected subsets of $X$ containing $x$. Then*

$$K^x = \cup \{P : P \in B^x\}$$

*is a component of $X$.*

**Proof :** Since $\{x\}$ is connected and $x \in \{x\}$, $\{x\} \in B^x$ and $B^x \neq \phi$. Obviously $x \in \cap \{P : P \in B^x\}$; hence, by Theorem 5, $K^x$ is connected. If $Q \supset K^x$ and $Q$ is connected, then $x \in Q$ and hence $Q \in B^x$. Consequently, $Q \subset K^x$. Thus $K^x$ is a maximal connected set in $X$.

**Example 1.** ***Given a topological space $(X, T)$, let, for two points $x, y$ belonging to $X$, $x \sim y$ if there is a connected subset of $X$ containing both $x$ and $y$; show that :***

***(i)*** ***'~' is an equivalence relation;***

***(ii)*** ***the disjoint equivalence classes constitute all the components of the space $X$.***

**Solution :** (*i*) Reflexivity and symmetry for '~' are obvious. For transitivity, let $x \sim y$ and $y \sim z$ *i.e.*, there exists a connected set $A$ containing $x$ and $y$ and a connected set $B$ containing $y$ and $z$. Thus $A \cup B$ is a connected set ($\because\ y \in A \cap B$) containing $x$ and $z$. Thus $x \sim z$.

(*ii*) For the other part, let $C$ be an equivalence class. Choose $x_0 \in C$; for each $x \in C$, we know that $x \sim x_0$ and so there exists a connected set $A_x$ containing $x_0$ and $x$.

If possible, let $A_x \cap D \neq \phi$ where $D$ is an equivalence class other than $C$; but $A_x \cap C \neq \phi$ as well. Let $x_1 \in A_x \cap D$ and $x_2 \in A_x \cap C$; then $x_1$ and $x_2$ belong to the same connected set $A_x$ and thus $x_1 \sim x_2$. But this leads to a contradiction as because $x_1$ and $x_2$ belong to two different equivalence classes.

Hence $A_x \cap D = \phi$ for any equivalence class $D$ other than $C$. Thus $A_x \subset C$. Obviously $C = \underset{x \in C}{\cup} A_x$.

By Theorem 5, $C$ is connected since $x_0 \in \cap \{A_x \mid x \in C\}$; $C$ is also maximal connected : in fact, if $D$ is another connected set containing

$C$ then $D \cap C \neq \phi$ and thus, by argument as above, $D$ cannot intersect any other equivalence class and hence $D = C$. Thus every equivalence class is a component.

That every component is an equivalence class (for '~') is obvious from definition.

**Example 2.** ***Give an example of a space in which a quasi-component at a point is different from its component.***

**Solution :** Let

$$X = \left\{ (0,\ 0),\ (1,\ 0),\ (x,\ y) : 0 \leq x \leq 1,\ y = \frac{1}{n},\ n = 1, 2, \ldots \right\}$$

we consider the topology $T$ to be the subspace topology inherited from $\mathbf{R}^2$. The picture shows that $X$ consists of some horizontal lines which do not intersect. The lines are closer to themselves when they are close to the $x$-axis but the line segment from (0, 0) to (1, 0) is not included in $X$ nor is the line segment from (0, 0) to (0, 1); thus there is no connected subset of $X$ containing (0, 0). Hence the component at (0, 0) in $X$ consists of one point *i.e.*, (0, 0) itself.

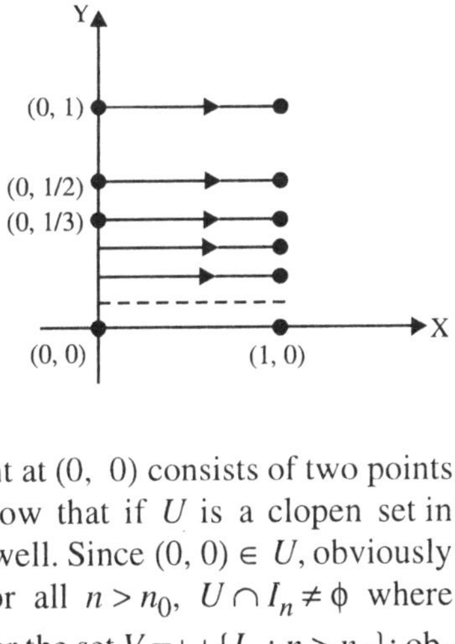


We now prove that the quasi-component at (0, 0) consists of two points (0, 0) and (1, 0); for this, it suffices to show that if $U$ is a clopen set in $X$ containing (0, 0) then it contains (1, 0) as well. Since $(0, 0) \in U$, obviously there is a positive integer $n_0$ such that for all $n > n_0$, $U \cap I_n \neq \phi$ where $I_n = \{(x,\ y) : 0 \leq x \leq 1,\ y = 1/n\}$. We consider the set $V = \cup \{I_n : n > n_0\}$; obviously, (1, 0) is an accumulation point of $V$; now that $U$ is a clopen set and $U \cap I_n \neq \phi$, $U \supset I_n$ : otherwise, $I_n$ would have a clopen subset which is not possible since each $I_n$ is connected; thus $U \supset V$ and (1, 0) is an accumulation point of $V$ and hence of $U$ as well. Thus $(1, 0) \in U$ since $U$ is closed.

## Art. 4 LOCAL CONNECTIVITY AND PATH CONNECTIVITY

**Definition 4 :** *A space* $(X, T)$ *is said to be locally connected at a point* $p$ *iff, given any neighbourhood* $U$ *of* $p$*, there exists a connected neighbourhood*

*V of p such that $V \subset U$. The space X is said to be locally connected iff it is locally connected at each of its points.*

**Theorem 12 :** *Let (X, T) be a topological space. Then, the following are equivalent :*

(*i*) *the component of an open set is an open set;*

(*ii*) *(X, T) is locally connected.*

**Proof :** Let $x \in X$ and let $O_x$ be a neighbourhood of $x$; we consider a component $V$ of the open set $O_x$, containing the point $x$. By (*i*), $V$ is open and as such given a neighbourhood $O_x$ of $x$, there exists a connected neighbourhood $V$ of $x$ such that $x \in V \subset O_x$. Thus $(X, T)$ is locally connected.

Conversely, let $x \in W$ where $W$ is a component of the open set $V$ containing $x$. By (*ii*) there is a connected set $U$ such that $x \in \text{int}\, U \subset U \subset V$; since $x \in W \cap U$ and both of $W$ and $U$ are connected, we have $W \cup U$ to be connected and contained in $V$; since $W$ is a component we conclude that $W \cup U = W$ *i.e.*, $U \subset W$. Finally, $x \in \text{int}\, U \subset W$ giving $W$ to be open.

**Theorem 13 :** *Let (X, T) and (Y, T′) be topological spaces and let $f : X \to Y$ be a continuous mapping which is closed. Then, if X is locally connected, then so is f (X).*

**Proof :** To prove that $f(X)$, (of course with its subspace topology) is locally connected, it would suffice, by Theorem 12, to show that if $U$ is open in $f(X)$ and if $C$ is any component in $U$, then $C$ is open in $f(X)$; since $f : (X, T) \to (Y, T')$ is continuous, $f : (X, T) \to (f(X), T_{f(X)})$ is continuous as well. Now $U$ is open in $f(X)$ implies that $f^{-1}(U)$ is open in $X$. Let $A$ be any component of $f^{-1}(U)$; since $X$ is locally connected $A$ is open in $X$. Since $A$ is connected, so is $f(A)$; since $C$ is a component of $U$ and $f(A) \subset U$, it must be true that either $f(A) \cap C = \phi$ or $f(A) \subset C$. Hence, we can deduce that $f^{-1}(C)$ is the union of a collection of components of $f^{-1}(U)$; but $X$ being locally connected every component of $f^{-1}(U)$ is open and so $f^{-1}(C)$ is open in $X$ consequently; $X - f^{-1}(C)$ is closed. Since $f$ is a closed mapping, $f(X - f^{-1}(C)) = f(X) - C$ is closed. Thus $C$ is open and $f(X)$ is locally connected.

**Corollary 5 :** *From Theorem 13, it obviously follows that local connectivity is a topological property.*

**Theorem 14 :** *Let $(X_a, T_a)_{\alpha \to \Lambda}$ be a collection of topological spaces; let $X = \prod_{\alpha \to \Lambda} X_\alpha$ and let T be the product topology on X. Let (X, T) be locally connected. Then :*

(*i*) *each* $X_\alpha$ *is locally connected;*

(*ii*) *barring a finite number of* $\alpha$'*s, each* $X_\alpha$ *is connected.*

**Proof :** (*i*) Let $x_\alpha \in X_\alpha$; let $U_\alpha$ be an open set in $X_\alpha$ containing $x_\alpha$; then $U = U_\alpha \times \prod_{\substack{\beta \neq \alpha \\ \beta \in \Lambda}} X_\beta$ is an open set in $X$; since it is an open set containing a point $x$ whose $\alpha$'th coordinate is $x_\alpha$ we have $x \in V \subset U$ when $V$ is a connected neighbourhood of $x$; but $p_\alpha$—the projection map into $X_\alpha$—being an open map, $p_\alpha(x) \in p_\alpha(V) \subset p_\alpha(U)$ giving $p_\alpha(V)$ to be a connected neighbourhood of $p_\alpha(x) = x_\alpha$ and contained in $p_\alpha(U) = U_\alpha$; thus $X_\alpha$ is locally connected.

(*ii*) Let $V$ be any connected open set in $X$; we have $p_\alpha(V) = X_\alpha$ except for a finite number of indices $\alpha$ and thus each $X_\alpha$ is connected except for a finite number of indices.

**Theorem 15 :** *Let* $(X_\alpha, T_\alpha)_{\alpha \in \Lambda}$ *be a collection of topological spaces; let each* $X_\alpha$ *be locally connected and let, barring a finite number of* $\alpha$'*s, all the* $X_\alpha$'*s be connected. Then* $X = \prod_{\alpha \in \Lambda} X_\alpha$ *is locally connected with the product topology* $T$.

**Proof :** Let $F = \{\alpha_1, ..., \alpha_n\} \subset \Lambda$ be a finite number of indices for which $X_\alpha$'s are not connected. Let $V \in T$ and $x \in V$; then $x \in \prod_{\beta \in \Lambda} U_\beta \subset V$ where $U_\beta = X_\beta$ except for (say), $\beta_1, ..., \beta_n$. For each $\beta_i$, let $V_{\beta_i}$ be a connected neighbourhood of $p_{\beta_i}(x)$ such that $V_{\beta_i} \subset U_{\beta_i}$. Now, for each $\gamma_1, ..., \gamma_k \in F - \{\beta_1, ..., \beta_n\}$, let $V_{\gamma_i}$ be a connected neighbourhood of $p_{\gamma_i}(x)$. Then

$$V_{\beta_1} \times \ldots V_{\beta_n} \times V_{\gamma_1} \times \ldots \times V_{\gamma_k} \times \prod_{\substack{\alpha \neq \beta_1, ..., \beta_n \\ \alpha \neq \gamma_1, ..., \gamma_k}} X_\alpha$$

is a connected neighbourhood of $x$ (by Theorem 7 (*ii*)) and is contained in $\prod_{\substack{\beta \in \Lambda \\ \beta \neq \beta_1, ..., \beta_n}} X_\beta \times U_{\beta_1} \times \ldots \times U_{\beta_n} . \subset V$. Thus $\prod_{\alpha \in \Lambda} X_\alpha$ is locally connected.

**Example 1.** ***The Cartesian product*** $\prod_{\alpha \in \Lambda} X_\alpha$ ***where*** $\Lambda$ ***is an infinite set need not be locally connected even if each*** $X_\alpha$ ***is locally connected.***

Let us consider $X_\alpha = \{1, 2\}$, with the discrete topology, for each $\alpha$. Obviously, the product topology on $X = \prod_{\alpha \in \Lambda} X_\alpha$ cannot be discrete topology as because any one-pointic set in $\prod_{\alpha \in \Lambda} X_\alpha$ cannot be open by definition of product topology.

Obviously, each $X_\alpha$ is locally connected since both $\{1\}$ and $\{2\}$ are connected and open as well.

In $X$, let us consider a point $y = (y_\alpha)$; let $C(y)$ denote the component of the point $y$ in $X$; we show that $C(y)$ is an one-point set : if not, *i.e.*, if $C(y) \neq \{y\}$, select $\hat{y} \in C(y) - \{y\}$ *i.e.*, $\hat{y}_\alpha \neq y_\alpha$ at least for same $\alpha \in \Lambda$; now $p_\alpha(C(y))$ is a connected set in $X_\alpha$ and contains $\hat{y}_\alpha$ which is different from $y_\alpha$—a contradiction as because, in $X_\alpha$, the only connected sets are one pointic sets. Thus $C(y) = \{y\}$.

It now follows, that in $X$, the only connected sets are one-pointic sets and none of them is open. Thus $X$ cannot be locally connected.

**Example 2.** ***The continuous image of a locally connected space need not be locally connected.***

Let $X = \{0, 1, 2, \ldots\}$ be endowed with discrete topology; let $Y = \{0\} \cup \{1/n : n = 1, 2, \ldots\}$ be considered with the subspace topology from the real line $R$; obviously, each $\{1/n\}$, for $n = 1, 2, \ldots$ is an open set and the open set containing '0' contains all but finitely many elements of the sequence $\{1/n\}$. The topology in $Y$ is not the discrete topology but each one-pointic set in $Y$ is connected and, since $\{0\}$ is not an open set, '0' does not have any connected open neighbourhood. Thus $Y$ is not locally connected.

Consider $\left.\begin{array}{r} f : X \to Y \\ f(0) = 0 \\ f(n) = 1/n \text{ for } n = 1, 2, \ldots \end{array}\right\}$ Obviously, $f$ is continuous bijection, $X$ is locally connected, but $f(X)$ is not locally connected.

**Note 6.** Next, we are interested in the Example of a connected space which is not locally connected; for this we need the following definition and a Theorem related to it.

**Definition 6 :** *Let $X$ and $Y$ be two sets and let $f : X \to Y$. The graph of $f$ is the subset $F$ of the Cartesian product $X \times Y$ consisting of all points $(x, f(x))$.*

**Theorem 16 :** *Let X and Y be topological spaces and $f: X \to Y$; let F be the graph of f. Then* $\left.\begin{array}{c} g: X \to F \\ g(a) = (a, f(a)) \end{array}\right\}$ *is a homeomorphism iff f is continuous.*

**Proof :** Let $f$ be continuous. Let $W$ be an open set in $F$ containing $(a, g(a))$; obviously, $W = G \cap F$ where $G$ is open in $X \times Y$ (as because $F$ inherits the topology from $X \times Y$). By definition of product topology there are open sets $U$ and $V$ in $X$ and $Y$ respectively such that $(a, f(a)) \in u \times V \subset G$; now, $V$ is an open neighbourhood of $f(a)$ in $Y$ and $f$ is continuous which is why $f^{-1}(V)$ is an open neighbourhood of $a$ in $X$; furthermore, $g(U \cap f^{-1}(V)) \subset (U \times V) \cap F \subset G \cap F = W$. Thus $g$ is continuous at $a$. That $g$ is a bijection is obvious; so, it remains to show that $g$ is open or in other words, $g^{-1}$ is continuous.

Now $\left.\begin{array}{c} g^{-1}: F \to X \\ (a, f(a)) \to a \end{array}\right\}$; we know that $\left.\begin{array}{c} p_1: X \times Y \to X \\ (x, y) \to x \end{array}\right\}$ is continuous and $g^{-1} = p/F$; thus $g^{-1}$ is continuous and $g$ becomes a homeomorphism.

Conversely, suppose that $g$ is a homeomorphism; then, if $G$ be open in $F$, then so is $g^{-1}(G)$ in $X$. Let $V$ be open in $Y$, then $(X \times V)$ is open in $X \times Y$ and $(X \times V) \cap F$ is open in $F$; then, $g^{-1}[(X \times V) \cap F]$ is open in $X$. But $g^{-1}[(X \times V) \cap F] = f^{-1}(V)$ and thus $f$ is continuous.

**Example 3.** ***Example of a non-locally connected space which is connected.***

Let $\left.\begin{array}{c} f: [0, 1] = I \to \mathbf{R} \\ f(0) = 0 \\ f(x) = \sin \pi/x \quad 0 < x \leq 1 \end{array}\right\}$; obviously the function is continuous on $(0, 1]$.

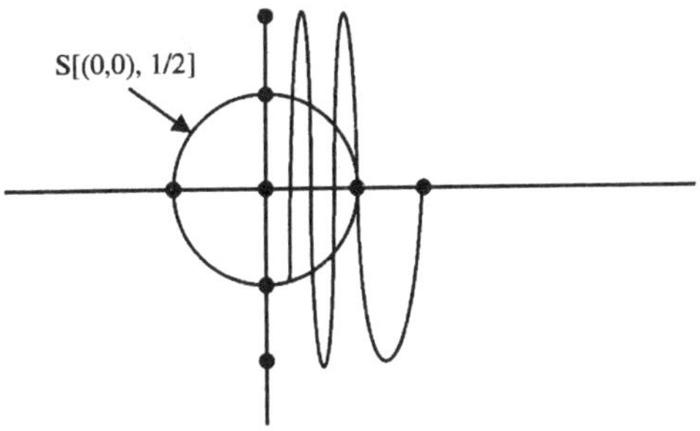

We consider its graph $F$ as is drawn in the figure.

Since $(0, 1]$ is connected and $f/(0, 1]$ is continuous, the graph $G$ of $f/(0, 1]$ is connected (by Theorem 16). Obviously, $F - \{(0, 0)\} = G$ and $(0, 0)$ is a limit point of $G$. By Theorem 5, $F$ is connected.

We now show that $F$ is not locally connected at the point $(0, 0)$; let $S[(0, 0), 1/2] = W$ [see * in Art. 2] and let $V = W \cap F$; then $V$ is a neighbourhood of $(0, 0)$ in $F$. We can prove that $F$ is not locally connected by showing that $V$ can not contain any connected neighbourhood of the origin.

If possible, let $U$ be any neighbourhood of $(0, 0)$ in $F$ contained in $V$; it is easy to see that, for every integer $n$, the point $(1/n, 0) \in F$. Let $m$ be an integer, large enough, such that $(1/n, 0) \in U$ for all $n \geq m$.

Let $\left.\begin{array}{l} g : U \to R \\ (x, y) \to x \end{array}\right\}$; since $g = p_1/U$ where $\left.\begin{array}{l} p_1 : \mathbf{R}^2 \to \mathbf{R} \\ (x, y) \to x \end{array}\right\}$ is continuous,

$g$ is continuous; obviously $U$ contains the point $(0, 0)$ and the point $\left(\dfrac{1}{m+1}, 0\right)$, but does not contain the point $\left(\dfrac{2}{(2m+3)}, q\right)$ for any

$q \in [-1, 1]$. Since $0 < \dfrac{2}{2m+3} < \dfrac{1}{m+1}$ and for no point $(p, q)$ in $U$,

$g((p, q)) = \dfrac{2}{2m+3}$ we have $U$ to be non-connected (by Theorem 9).

Thus $F$ is not locally connected (of course with its subspace topology inherited from $\mathbf{R}^2$).

**Example 4.** ***Example of a locally connected space which is not connected.***

We consider $X = \{0, 1\}$ with its discrete topology. Obviously, $\{0\}$ and $\{1\}$ are both connected and open as well and thus $X$ is locally connected. But $X$ is not connected as it is the union of two disjoint open sets.

**Definition 7 :** *(a) Let $(X, T)$ be a topological space; by a path in $X$, we mean a continuous map $f : I \to X$ where $I = [0, 1]$ is considered with its subspace topology inherited from $R$. The points $f(0)$ and $f(1)$ are called the initial point and the final point of the path respectively.*

(b) *In a given topological space $(X, T)$. Let us define an equivalence relation '$\sim$' as follows : if $a$ and $b$ are two points of $X$, then $a \sim b$ iff there exists a path $f : I \to X$. Such that $f(0) = a$ and $f(1) = b$.*

'$\sim$' is obviously an equivalence relation : in fact, let $a \sim b$ and $b \sim c$; then there exists continuous functions $f_1$ and $f_2$, each on $I$, having values in $X$, such that $f_1(0) = a$, $f_1(1) = b$, $f_2(0) = b$, $f_2(1) = c$; let $h : I \to X$ be such that $\left.\begin{array}{l} h(t) = f_1(2t) : 0 \le t \le 1/2 \\ \quad\; = f_2(2t-1) : 1/2 \le t \le 1 \end{array}\right\}$; obviously, $h$ is continuous (why?),

$h(0) = a$ and $h(1) = c$; thus $a \sim c$.

This equivalence relation divides $X$ into disjoint classes called the path-components of $X$. If it so happens that $X$ has only one path-component, then $X$ is called path connected; in other words, $X$ is path connected iff for any two points $x$, $y$ in $X$, there exists a continuous function $f : \mathbf{I} \to X$ such that $f(0) = x$, $f(1) = y$.

Whenever we call a set $A \subset X$ path connected, we mean that $(A, T_A)$ is path connected.

**Theorem 16 :** *Every path connected space is connected.*

**Proof :** Let $(X, T)$ be a path connected topological space and let '$a$' be considered as a fixed point of $X$; let $x \in X$; then there exists a continuous function $f_x : \mathbf{I} \to X$ such that $f_x(0) = a$ and $f_x(1) = x$; obviously $f_x(\mathbf{I})$ is a connected subset of $X$ containing $a$ and $x$; if we vary $x$ over $X$, each $f_x(\mathbf{I})$, however, contains the point $a$. Thus $X = \bigcup_{x \in X} f_x(\mathbf{I})$ and so $X$ is connected by Theorem 5.

**Theorem 17 :** *Every continuous image of a path-connected space is connected.*

**Proof :** Without loss of generality we could take a continuous surjection $f$ mapping a topological space $(X, T)$, which is path connected, onto a topological space $(Y, T')$; we have to show that $(Y, T')$ is path connected.

Let $x$, $y$ be two points in $Y$; since $f$ is a surjection, $f$ points '$a$' and '$b$' in $X$ such that $f(a) = x$ and $f(b) = y$; since $X$ is path-connected, $\exists$ a continuous $g : \mathbf{I} \to X$ such that $g(0) = a$ and $g(1) = b$; consider $fog : \mathbf{I} \to Y$; obviously $fog$, being the composite of two continuous maps is continuous and $fog(0) = x$ and $fog(1) = y$. Thus $Y$ is path connected.

**Theorem 18 :** *Let $(X_\alpha, T_\alpha)_{\alpha \in \Lambda}$ be a collection of path connected topological spaces and let $X = \prod_{\alpha \in \Lambda} X_\alpha$ be considered with the product topology $T$. Then $(X, T)$ is path connected.*

**Proof :** Let $x = (x_\alpha)$ and $y = (y_\alpha)$ be two points in $X$; for each $\alpha$, $\exists$ a continuous $\sigma_\alpha : \mathbf{I} \to X_\alpha$ such that $\sigma_\alpha(0) = x_\alpha$ and $\sigma_\alpha(1) = y_\alpha$. Let $\sigma : \mathbf{I} \to X$ be defined as $\sigma(t) = (\sigma_\alpha(t))$ for any $t \in \mathbf{I}$. If $p_\alpha : X \to X_\alpha$ is the $\alpha$'th projection map then $p_\alpha \, o \, \sigma = \sigma_\alpha$ for every $\alpha$; since $\sigma_\alpha$ is continuous for each $\alpha$, so is $p_\alpha o \sigma$ for each $\alpha$; by Theorem 63, Chapter 2, $\sigma$ is continuous; $\sigma(0) = (\sigma_\alpha \; (0)) = (x_\alpha) = x$ and $\sigma(1) = (\sigma_\alpha \; (1)) = (y_\alpha) = y$ and thus $X$ is path connected.

## Art. 5 EXERCISES

**Example 1.** ***Let (X, T) be a connected topological space and let $C \subset X$ be connected; if (A, B) is a separation of $X \setminus C$; then show that each of the sets $A \cup C$ and $B \cup C$ is connected.***

**Solution :** Suppose that $A \cup C$ is not connected; hence there is a separation $(E, F)$ such that $A \cup C = E \cup F$. By Lemma 1, $C$ being connected, is completely contained either in $E$ or in $F$ say $F$; obviously, $E \subset A$. We claim that $(E, F \cup B)$ forms a separation for $X$ leading to a contradiction. Clearly, both $E$ and $F \cup B$ are non void, their intersection is void and their union is $X$; we claim that both are closed in $X$. Let $x$ be a limit point of $E$.

It cannot be in $F$, since $(E, F)$ is a separation of $A \cup C$; $x$ cannot be in $B$ as well, since $(A, B)$ is a separation. Hence $x \in E$ and $E$ becomes closed; the limit points of $F \cup B$ are either limit points of $B$ or of $F$. No limit point of $F$ is in $E$, since $(E, F)$ is a separation; also, no limit point of $B$ is in $E$, since $E \subset A$ and $(A, B)$ is a separation. Thus $(E, F \cup B)$ is a separation and $E \cup F \cup B = X$.

**Example 2.** ***Let (X, T) be a topological space and let A, B be subsets of X. If each of the sets A, B is closed in their union and if $A \cup B$ and $A \cap B$ are connected, then show that A is connected and B is connected.***

**Solution :** We have nothing to prove if $A \subset B$ or $B \subset A$. Otherwise, we may assume that each of the sets $A - B$ and $B - A$ is non-empty. Since both of $A$ and $B$ are closed in $A \cup B$ we have $A \cup B \setminus B = A \setminus B$ is open in $A \cup B$ and so is $B \setminus A$ in $A \cup B$. Consequently, $(A \setminus B, \; B \setminus A)$ forms a separation and further $(A \cup B) \setminus (A \cap B) = (A \setminus B) \cup (B \setminus A)$ where both of $A \cup B$ and $A \cap B$ are connected. By Example $A = (A \setminus B) \cup (A \cap B)$ and

$$B = (B \setminus A) \cup (A \cap B)$$

are connected.

**Example 3.** ***Let (X, T) be a connected space and let $x \in S$ with $S - \{x\} = A \cup B$, where (A, B) is a separation. Then show that $A \cup \{x\}$ and $B \cup \{x\}$ are each connected. Moreover, if $\{x\}$ is a closed set in X, then show that $\overline{A} = A \cup \{x\}$ and $\overline{B} = B \cup \{x\}$.***

**Example 4.** *Let $\{A_\alpha\}$ be such a collection of connected subsets of a topological space X such that intersection of every pair of sets of the collection $\{A_\alpha\}$ is non empty. Then show that $\bigcup_\alpha A_\alpha$ is a connected set.*

**Example 5.** *Let A and B be connected subsets of a topological space X such that $\bar{A} \cap B \neq \phi$; show that $A \cup B$ is connected.*

**Example 6.** *Is it true that a connected set consisting of more than one point is dense in itself? Justify your answer.*

**Example 7.** *If a space contains a connected, every where dense subset, then show that the space is connected.*

**Example 8.** *Let (X, T) and (X, σ) be two connected spaces. Prove or disprove : $(X, T \cap \sigma)$ and $(X, T \vee \sigma)$ are connected.*

**Example 9.** *Let (X, T) and (X, σ) be two topological spaces such that $T \subset \sigma$. Does there exist any relation between the components of X with respect to the two topologies? Justify your answer.*

**Example 10.** *Let K denote the set $\{1/n \mid n \in Z_+\}$, and define*

$$C = ([0,\ 1] \times 0) \cup (K \times [0,\ 1] \cup (\{0\} \times [0,\ 1]).$$

The space $C$ is called the comb space. The $D$ obtained by deleting from $C$ the points of the set $\{0\} \times (0,\ 1)$ is called the deleted comb space. Let $T$ denote the usual topology on **R**. Show that $(C, T_C)$ is path-connected but $(D, T_D)$ is not so. Further show that $(D, T_D)$ is connected.

**Example 11.** *If A is a connected set in a topological space is it true that Interior A and Frontier A are both connected? Justify your answer. If both of Interior A and Frontier A of a set A are connected, is it true that A is connected? Justify your answer.*

**Example 12.** *Let A be a countable subset of $R^2$. If $R^2$ be considered with its usual topology show that $R^2 - A$ is path connected.*

**Example 13.** *Show that $R^2$ and R are not homeomorphic.*

**Example 14.** *If $\{A_\alpha\}$ is a collection of path-connected subsets of X and if $\cap A_\alpha \neq \phi$ is $\cup A_\alpha$ necessarily path-connected? Justify your answer.*

**Example 15.** *If $A \subset X$, A is path connected, is $\bar{A}$ necessarily path-connected? Justify your answer.*

**Example 16.** *If X is locally connected show that the quasi components of X are the same as components of X.*

**Example 17.** *A space X is said to be connected im Kleinen at x if for every neighbourhood U of x, there is a connected subset A of U that contains a neighbourhood of x. Show that if X is connected im Kleinen at each of its points, then X is locally connected.*

**Example 18.** *Determine the path components of $(D, T_D)$ as in Example 10.*

*Chapter 8*

# METRIC SPACES

## INTRODUCTION

We consider the usual topology on the real line **R** as discussed in 211. The base for the topology is given by the collection of all open intervals $(a, b)$ where $a$ and $b$ are finite real numbers. If we now consider the collection $\{(a-\varepsilon, a+\varepsilon) : a \in \mathbf{R}, \varepsilon > 0\}$, then again, this collection gives the usual topology on $\mathbf{R}$ : in fact, for any point in $p$ in $(a, b)$ we can find an $\varepsilon > 0$ such that $p \in (p-\varepsilon, p+\varepsilon) \subset (a, b)$ and as such the two topologies generated by the two collections are actually the same. If we take $d(x, y) = |x - y|$ for any two real numbers $x$ and $y$, then

$$(a-\varepsilon, a+\varepsilon) = \{x \in \mathbf{R} : d(x, a) < \varepsilon\}.$$

We denote $(a-\varepsilon, a+\varepsilon)$ by $S(a, \varepsilon)$. '$d$' is obviously written according to the standard Euclidean formula.

In example 2 of Art. 6.3 we have given the usual topology of the Euclidean plane $\mathbf{R}^2$. In fact, the base for such a topology has been constructed via the sets $S(a, \varepsilon)$ where

$$S(a, \varepsilon) = \{x \in \mathbf{R}^2 \mid d(a, x) < \varepsilon, a \in \mathbf{R}^2, \varepsilon > 0\}$$

and '$d$' is determined by the standard Euclidean formula *i.e.*, for any two points $(x_1, y_1)$, $(x_2, y_2)$ in $\mathbf{R}^2$,

$$d[(x_1, y_1), (x_2, y_2)] = \sqrt{(x_1 - x_2)^2 + (y_1 - y_2)^2}.$$

We called $S(a, \varepsilon)$ the open sphere around $a$ of radius $\varepsilon$. In this chapter we look at a natural generalization of the Euclidean plane, namely any space whose structure is specified by a distance function.

## Art. 1 METRIC

**Definition 1 : Metric space.** A non-empty set $X$ has a metric or distance function if for any pair of points $x$, $y$ in $X$, there is a real number $d(x, y)$ with the following properties :

1. $d(x, y) \geq 0$, equality occurring when and only when $x = y$.
2. $d(x, y) = d(y, x)$ (symmetry).

3. $d(x, y) + d(y, z) \geq d(x, z)$ (Triangle inequality), where $z \in X$.

The real number $d(x, y)$ is called the distance between $x$ and $y$. The combination $(X, d)$ is a metric space.

**Example 1.** *Define $d(x, y) = |x - y|$ for $x, y \in R$. Then $(R, d)$ is a metric space.*

**Example 2.** *Define $d((x, y), (x', y')) = \sqrt{(x - x')^2 + (y - y')^2}$ for $(x, y), (x', y') \in R^2$. Then $(R^2, d)$ is a metric space.*

**Example 3.** *Let $X$ be a non void set; let us define*

$$d(x, x) = 0 \ (x \in X)$$
$$d(x, y) = 1 \ (x, y \in X;\ x \neq y)$$

It is easy to verify that $(X, d)$ is a metric space. '$d$' is called the discrete metric.

**Example 4.** *For any two points $x = (x_1, ..., x_n)$ and $y = (y_1, y_2, ..., y_n)$ in $R^n$ we define $d(x, y) = \sqrt{\sum_{k=1}^{n} (x_k - y_k)^2}$ ; obviously, for $n = 1$ or 2, $d(x, y)$ gives the standard Euclidean formula. The verification of the properties (1) and (2) as in definition 1 are rather obvious. We only verify the triangle inequality : Let $z = (z_1, z_2, ..., z_n)$. For $k = 1, 2, ..., n$, let $a_k = x_k - z_k$ and $b_k = z_k - y_k$. Then*

$$d(x, z) = \sqrt{\sum_{k=1}^{n} a_k^2} \text{ and } d(z, y) = \sqrt{\sum_{k=1}^{n} b_k^2}$$

and $$d(x, y) = \sqrt{\sum_{k=1}^{n} (a_k + b_k)^2}$$

Finally, what we have to show is the following :

$$\sqrt{\sum_{k=1}^{n} (a_k + b_k)^2} \leq \sqrt{\sum_{k=1}^{n} a_k^2} + \sqrt{\sum_{k=1}^{n} b_k^2} \qquad ...(1)$$

For this, we first prove an equality named after Cauchy : Let

$$a = (a_1, ..., a_n) \text{ and } b = (b_1, ..., b_n)$$

be two $n$-tuples of real or complex numbers. Then

$$\sum_{i=1}^{n} |a_i\ b_i| \le \left(\sum_{i=1}^{n} |a_i|^2\right)^{1/2} \left(\sum_{i=1}^{n} |b_i|^2\right)^{1/2} : \qquad ...(2)$$

in fact, we first note that if $x$ and $y$ are any two non-negative real numbers then

$$x^{1/2}\ y^{1/2} \le \frac{x+y}{2}; \qquad ...(3)$$

the validity is obvious by squaring both sides, rearranging and finding its equivalence with $0 \le (x-y)^2$, which is evidently true. If $x = 0$ or $y = 0$, the assertion (2) is obvious. We therefore assume that $x \ne 0$, $y \ne 0$. We define $x_i$ and $y_i$ by

$$x_i = \left(\frac{|a_i|}{\|a\|}\right)^2, \quad y_i = \left(\frac{|b_i|}{\|b\|}\right)^2$$

where $$\|a\| = \left(\sum_{i=1}^{n} |a_i|^2\right)^{1/2} \text{ and } \|b\| = \left(\sum_{i=1}^{n} |b_i|^2\right)^{1/2};$$

by (3), $$\frac{|a_i\ b_i|}{\|a\|\ \|b\|} \le \frac{|a_i|^2/\|a\|^2 + |b_i|^2/\|b\|^2}{2}$$

for each $i$; summing these inequalities as $i$ varies from 1 to $n$ yields.

$$\frac{\sum_{i=1}^{n} |a_i\ b_i|}{\|a\|\ \|b\|} \le \frac{1+1}{2} = 1. \text{ from which (2) follows}$$

Using (2) we now have $\|a+b\|^2 = \sum_{i=1}^{n} |a_i + b_i|\ |a_i + b_i|$

$$\le \sum_{i=1}^{n} |a_i + b_i|\ (|a_i| + |b_i|) = \sum_{i=1}^{n} |a_i + b_i|\ |a_i| + \sum_{i=1}^{n} |a_i + b_i|\ |b_i|$$

$$\le \|a+b\|\ \|a\| + \|a+b\|\ \|b\| = \|a+b\|\ (\|a\| + \|b\|)$$

and thus, changing the notation we have the inequality (1).

The metric space $(\mathbf{R}^n, d)$ is called Euclidean $n$-space giving the Euclidean plane when $n = 2$. The metric '$d$' is termed as the Euclidean metric.

**Example 5.** ***Let 'm' denote the set of all bounded sequences of real number. If $a = \{a_n\}_{n=1}^{\infty}$ and $b = \{b_n\}_{n=1}^{\infty}$ are points in m, define***

$$d(a, b) = \sup_{i \le n < \infty} |a_n - b_n|$$

It is obvious that requirements (1) and (2) from definition 1 are satisfied. To demonstrate the triangle inequality, let $c = \{c_n\}_{n=1}^{\infty}$ also be a point in $m$. For any natural number, we have

$$|a_k - b_k| = |a_k - c_k + c_k - b_k| \le |a_k - c_k| + |c_k - b_k|$$

$$\le \sup_{i \le n < \infty} |a_n - c_n| + \sup_{i \le n < \infty} |c_n - b_n|$$

and so $|a_k - b_k| \le d(a, c) + d(c, b)$ ($k$ is a natural number).

From this, it follows that $\sup_{i \le k < \infty} |a_k - b_k| \le d(a, c) + d(c, b)$ and hence the validity of '$d$' as a metric.

## Art. 2 THE METRIC TOPOLOGY

As in the introduction we introduce the concept of open sphere $S(x, \varepsilon)$ (or $S_d(x, \varepsilon)$) in an arbitrary metric spaces $(X, d)$ as follows :

$$S(x, \varepsilon) = \{y \in X \mid d(x, y) < \varepsilon\}.$$

**Theorem 1 :** *In a metric space $(X, d)$ the collection*

$$\mathcal{B} = \{S(x, \varepsilon) : x \in X;\ \varepsilon > 0\}$$

*alongwith $\phi$ is a base for same topology on X.*

**Proof :** The first condition of Theorem 3, Chapter 1 is obviously satisfied by the inclusion of the null set $\phi$. The 2nd condition of theorem 3 is obvious since $x \in S(x, \varepsilon)$ for any $\varepsilon > 0$. Before checking the 3rd condition for a basis, we show that if $y$ is a point of the element $S(x, \varepsilon)$, then there is a $\delta > 0$ such that $y \in S(y, \delta) \subset S(x, \varepsilon)$; in fact; if, $\delta = \varepsilon - d(x, y)$ then $y \in S(y, \delta) \subset S(x, \varepsilon)$ since, if $z \in S(y, \delta)$, then $d(y, z) < \varepsilon - d(x, y)$ from which we conclude that $d(x, z) \le d(x, y) + d(y, z) < \varepsilon$.

Now to check the 3rd condition for a basis, let $B_1, B_2 \in \mathcal{B}$ and let $y \in B_1 \cap B_2$; we have just seen that we can choose positive numbers $\delta_1$ and $\delta_2$ so that $S(y, \delta_1) \subset B_1$ and $S(y, \delta_2) \subset B_2$. Let $\delta = \min\{\delta_1, \delta_2\}$; then

$y \in S(y, \delta) \subset B_1 \cap B_2$ and $B_1 \cap B_2$ is expressible as union of some sets belonging to $\mathcal{B}$.

Hence $\mathcal{B}$ is a base for some topology on $X$ which we call the metric topology on $X$ and denote it by $T(d)$.

**Note 1 :** Now that we have been able to introduce topology via a metric there are many questions which arise almost, immediately :

1. Can it so happen that in a metric space $(X, d)$ there is another metric 6 such that $T(d) = T(6)$ ?
2. If $(X, T)$ is a topological space, is it always true that there is a metric $d$ in $X$ such that $T(d) = T$ ?
3. Do the properties that involve a specific metric for $X$ depend in general, on the topology of $X$ ?

**Note 2 :** We first show that in a metric space $(X, d)$ there is always another metric $e$ such that $T(e) = T(d)$ : in fact, let us take

$$e(x, y) = \min \{d(x, y), 1\}.$$

We first check that '$e$' is a metric in $X$. The first two conditions being obvious, we just check the triangle inequality.

$$e(x, z) \leq e(x, y) + e(y, z.)$$

Now if either $d(x, y) \geq 1$ or $d(y, z) \geq 1$, then the right side of this inequality is at least 1; the left hand side is atmost 1 by definition and the inequality holds.

If $d(x, y) < 1$ and $d(y, z) < 1$ then

$$d(x, z) \leq d(x, y) + d(y, z) = e(x, y) + e(y, z).$$

Since $e(x, z) \leq d(x, z)$ by definition, the triangle inequality holds for '$e$' as well. That $T(e) = T(d)$ is obviously by application of Theorem 5 (Chapter 1) and the fact that $S_d(x, \varepsilon) \subset S_e(x, \varepsilon)$ and $S_e x, \delta) \subset S_d(x, \varepsilon)$ where

$$\delta = \min \{\varepsilon, 1\}.$$

**Note 3 :** Before answering Question 2 above we first mention that a topological space $(X, T)$ in which there is a metric with $T = T(d)$ is said to be metrizable. The answer to the question is negative since every metric space is first countable and there are topological spaces which are not first countable (For example, Example 1, Chapter 5).

**Note 4 :** With the help of Theorem 1 (Chapter 1) and Theorem 1, we can now assert that if $U \in T(d)$ then for $p \in U$, there exists $\delta > 0$ such that $p \in S(p, \delta) \subset U$.

**Theorem 2 :** *Every metrizable space $(X, T)$ is first countable.*

**Proof :** Obviously, we have metric $d$ in $X$ such that $T(d) = T$. Consider $v = \{S(x, 1/n) : x \in X, n \in N\}$ where $N$ is the set of natural numbers; obviously $v \subset B$ where $\mathcal{B} = \{S(x, \varepsilon) : x \in X, \varepsilon > 0\}$ gives the topology $T(d)$. To show that $v$ is a base for $T(d)$ we just observe that for $U \in T(d)$ and $x \in U$, there is an $\varepsilon > 0$ such that $x \in S(x, \varepsilon) \subset U$; choose $n \in N$ such that $\frac{1}{n} < \varepsilon$. Then $x \in S\left(x, \frac{1}{n}\right) \subset S(x, \varepsilon) \subset U$; by Theorem 1 (Chapter 1), $v$ is a base for $T(d)$. For any point $x \in X$, $\left\{S\left(x, \frac{1}{n}\right) : n \in N\right\}$ is a countable set giving a countable base at the point $x$. Thus $(X, T)$ is first countable.

**Note 5 :** To answer Question (3) in the negative we first define a property that involves a specific metric $d$ in $X$.

**Definition 2 :** *Let $(X, d)$ be a metric space, A subset $Y \subset X$ is said to be bounded if there is same number $\varepsilon > 0$ such that $d(x, y) \le \varepsilon$ for every pair of points $x$, $y$ in $Y$. If $Y$ is bounded we call sup $\{d(x, y) \mid x, y \in Y\}$ the diameter of $Y$ and denote it by diam $Y$ or $\delta(Y)$.*

**Note 6 :** It can be easily seen that boundedness depends only on the metric $d$ and not on the topology. Consider $[0, \infty)$ in $\mathbf{R}$ with the usual metric $d$; it is not a bounded set; but if we consider $e$ in $\mathbf{R}$ (which gives the same topology as $d$), $[0, \infty)$ is a bounded set, by definition of $e$ (in note 2).

**Note 7 :** Coming back to Note 2 we can say that $d$ and $e$ are equivalent; in fact, for any two metrics $d$ and $e$ on the same set, $d$ is equivalent to $e$ (notation $d \sim e$) iff $T(d) = T(e)$. We can assert something more than what we did in Note 2.

**Note 8 :** If $(X, T(d))$ is a topological space. Then for each $M > 0$ there is a metric $d_M \sim d$ such that $d_M(x, y) \le M$ for all $(x, y)$.

**Proof :** Given $M > 0$, let us define $d_M(x, y) = \min[M, d(x, y)]$; as in Note 2, $d_M$ is a metric and $T(d_M) = T(d)$.

**Note 9 :** It is obvious that in Notes 2 and 3 we have proved that each metric space is homeomorphic to a bounded metric space with boundedness as defined in Definition 2. $d_M$ is, in fact, called a bounded metric for obvious reasons.

## Art. 3

Metric '$d$' as we have defined in a set $X$ is actually a function on $X \times X$ having values in $\mathbf{R}$. With metric $d$, $X$ has now been topologized with $T(d)$ and $X \times X$ has been topologized with the product topology $T(d) \times T(d)$. [Art. 3 in Chapter 2]; $\mathbf{R}$ has usual topology. So the question of continuity arises for $d : X \times X \to \mathbf{R}$. Before tackling the question we further define some more 'metric like' concepts :

**Definition 3 :** *Let $(X, d)$ be a metric space and $Y \subset X$.*

(*a*) *For $x \in X$, we define $d(x, Y) = inf\{d(x, y) \mid y \in Y\}$.*

(*b*) *For non empty sets A and B in X, we define*

$$d(A, B) = inf\{d(a, b) \mid a \in A, b \in B\}.$$

**Example 1.** ***$d(x, A) = 0$ iff $x \in \overline{A}$; thus $\overline{A} = \{x \mid d(x, A) = 0\}$.***

**Proof :** Remembering that the open spheres like $S(x, \varepsilon)$ for $x \in X$ and $\varepsilon > 0$ form the basis for the metric topology, we have $x \in \overline{A}$ iff for all $S(x, \varepsilon)$, $A \cap S(x, \varepsilon) \neq \phi$ *i.e.*, iff for all $\varepsilon > 0$, there exists $a_\varepsilon \in A$ such that $d(x, a_\varepsilon) < \varepsilon$ *i.e.*, iff $d(x, A) = 0$. We note further that

$$\overline{A} = \{x \mid d(x, A) = 0\}.$$

**Example 2.** ***$e \sim d$ iff: for each $A \subset X$, $d(x, A) = 0$ iff $e(x, A) = 0$.***

**Proof :** The last line in the previous example and Example 4 (*ii*) in Chapter 2 solve the problem.

**Theorem 3 :** *Let $(X, d)$ be a metric space and let $A \subset X$. Then the map $f : X \to R$ defined by $x \to d(x, A)$ is continuous.*

**Proof :** To find the continuity character of a real valued function $f$ we appeal to Theorem 1, Art. 1 in Chapter 4.

Let $x, y$ be any two elements of $X$. Then, for each $a \in A$, we have $d(x, a) \leq d(x, y) + d(y, a)$, so that

$$d(x, A) = \inf_a d(x, a) \leq d(x, y) + \inf_a d(y, a) = d(x, y) + d(y, A)$$

which shows that $d(x, A) - d(y, A) \leq d(x, y)$. Interchanging the roles of $x$ and $y$, we obtain $|d(x, A) - d(y, A)| \leq d(x, y)$. Thus, corresponding to $\varepsilon > 0$ we can find *a nbd* $S(x, \varepsilon)$ of $x$ such that whenever $y \in S(x, \varepsilon)$ *i.e.*, $d(x, y) < \varepsilon$, we have $|d(x, A) - d(y, A)| < \varepsilon$. Hence the function $x \to d(x, A)$ is continuous.

**Corollary 1 :** *If $A = \{a\}$ in Theorem 3. Then the function*

$$\left.\begin{array}{l} f : X \to \mathbf{R} \\ x \to d(x, a) \end{array}\right\}$$

*is a continuous function.*

**Note 10 :** From the definition 3 (*b*), it is obvious that $d(A, B) \neq 0$ implies $A \cap B = \phi$; but the converse is however not true in general : in fact, let us take $A = \{n \mid n \in N\}$ and $B = \left\{ n + \frac{1}{2n} \,\middle|\, n \in \mathbf{N} \right\}$ obviously, both A and $B$ are closed sets and $A \cap B = \phi$. But $d(A, B) = 0$.

If $A = \phi$ or $B = \phi$ we have $A \cap B = \phi$; the converse implication can however be had if we take $d(A, \phi) = d(\phi, B) = \infty$.

**Note 11 :** To correlate the concept of diameter $\delta(A \cup B)$ for $A, B \subset X$ and $d(A, B)$ we have the following obvious results :

(*i*) If $x \in A$, $y \in B$, then $0 \le d(A, B) \le d(x, y) \le \delta(A \cup B)$.

(*ii*) $\delta(A \cup B) \le \delta(A) + d(A, B) + \delta(B)$.

The proof of (*i*) is very easy; to prove (*ii*) we take two arbitrary but fixed points $a$ and $b$ from $A$ and $B$; then

$$d(x, y) \le d(x, a) + d(a, b) + d(b, y)$$

where $x \in A$ and $y \in B$; obviously

$$d(x, y) \le \delta(A) + d(a, b) + \delta(B); \quad \text{...(1)}$$

since $a$ and $b$ were chosen arbitrarily, the relation holds for any

$$(a, b) \in A \times B.$$

If possible, let $\delta(A \cup B) > \delta(A) + d(A, B) + \delta(B)$; then, by the definition of 'sup', $\exists$ at least two points $p$, $q$ from $A \cup B$ such that

$$d(p, q) > \delta(A) + d(A, B) + \delta(B);$$

by the definition of a diameter, it cannot so happen that

$$(p, q) \in A \times A \text{ or } (p, q) \in B \times B;$$

thus $p \in A$, and $q \in B$. Thus $d(p, q) - \delta(A) - \delta(B) > d(A, B)$; by definition of 'inf', there is $(s, t) \in A \times B$ such that $d(s, t) < d(p, q) - \delta(A) - \delta(B)$ *i.e.*, $d(p, q) > d(s, t) + \delta(A) + \delta(B)$ contradicting (1).

## Art. 4 EXERCISES

**Example 3.** ***Let (X, d) be a metric space. Let $X \times X$ be treated with the product topology $T(d) \times T(d)$. Then*** $\left.\begin{array}{l} d : X \times X \to R \\ (x, y) \to d(x, y) \end{array}\right\}$ ***is a continuous function.***

**Proof :** Let $(d(x, y) - \varepsilon, \; d(x, y) + \varepsilon)$ be a *nbd* of $d(x, y)$ in the usual topology of $R$. Consider $S(x, \varepsilon/2) \times S(y, \varepsilon/2)$ to be a basic open *nbd* of $(x, y)$ in $X \times X$.

Then $(x_1, y_1) \in S(x, \varepsilon/2) \times S(y, \varepsilon/2)$ implies that $d(x, x_1) < \varepsilon/2$ and $d(y, y_1) < \varepsilon/2$.

Obviously, $d(x, y) \le d(x, x_1) + d(x_1, y_1) + d(y_1, y)$

*i.e.,* $d(x, y) - d(x_1, y_1) \le d(x, x_1) + d(y, y_1)$

*i.e.,* $|d(x, y) - d(x_1, y_1)| \le d(x, x_1) + d(y, y_1) < \varepsilon/2 + \varepsilon/2 = \varepsilon$

*i.e.,* $d(x_1, y_1) \in (d(x, y) - \varepsilon, d(x, y) + \varepsilon)$

and thus $d$ is continuous.

**Example 4.** ***If, for any two elements $x = (x_1, ..., x_n)$ and $y = (y_1, ..., y_n)$ of $R^n$ we write $e(x, y) = \max\{|x_1 - y_1|, ..., |x_n - y_n|\}$, then the usual Euclidean metric 'd' is equivalent to 'e'.***

**Proof :** That '$e$' is a metric is obvious; however we verify the triangle inequality; if $z = (z_1, ..., z_n) \in \mathbf{R}^n$, than

$$|x_k - z_k| \le |x_k - y_k| + |y_k - z_k|$$

*i.e.,* $|x_k - z_k| \le e(x, y) + e(y, z)$ and finally,

$$e(x, z) = \max_{1 \le k \le n} \{|x_k - z_k|\} \le e(x, y) + e(y, z)$$

By definition of $d$ and $e$, $e(x, y) \le d(x, y) \le \sqrt{n}\, e(x, y)$ for any two points $x, y \in \mathbf{R}^n$. ...(1)

Let $x \in S_e(x, \varepsilon)$; we claim that $x \in S_d(x, \varepsilon) \subset S_e(x, \varepsilon)$ and this is obvious from $e(x, y) \le d(x, y)$ as in (1).

Similarly, if $x \in S_d(x, \varepsilon)$, then $x \in S_e(x, \varepsilon/\sqrt{n}) \subset S_d(x, \varepsilon)$; in fact, $y \in S_e(x, \varepsilon/\sqrt{n}) \Rightarrow e(x, y) < \varepsilon/\sqrt{n} \Rightarrow d(x, y) < \varepsilon$ from (1).

By Theorem 5, (Chapter 1), $T(d) = T(e)$.

**Example 5.** ***Let $(X, d)$ be a metric space; we define the mapping***

$$e : X \times X \to R$$

$$(x, y) \to \frac{d(x, y)}{1 + d(x, y)}$$

Then $e$ is a metric on $X$ which is equivalent to $d$.

**Proof :** We only verify the triangle inequality, the other conditions being obvious; let $x, y, z \in X$ and suppose $x \ne z$. We have the following:

$$e(x, y) + e(y, z) = \frac{d(x, y)}{1 + d(x, y)} + \frac{d(y, z)}{1 + d(y, z)} \ge \frac{d(x, y) + d(y, z)}{1 + d(x, y) + d(y, z)}$$

$$\ge \frac{1}{1 + \dfrac{1}{d(x, y) + d(y, z)}} \ge \frac{1}{1 + \dfrac{1}{d(x, z)}} = e(x, z)$$

Thus, for any three points $x$, $y$, $z$ of $X$, with $x \neq z$, we have

$$e(x, z) \leq e(x, y) + e(y, z).$$

If $x = z$, the inequality is immediate.

To complete the proof we need verify that $T(e) = T(d)$; in fact, it is easy to see that $y \in S_d(x, \varepsilon)$ iff $d(x, y) < \varepsilon$ *i.e.*, iff $\dfrac{d(x, y)}{1 + d(x, y)} < \dfrac{\varepsilon}{1+\varepsilon}$ *i.e.*, iff $e(x, y) < \dfrac{\varepsilon}{1+\varepsilon}$ *i.e.*, if $y \in S_e\left(x, \dfrac{\varepsilon}{1+\varepsilon}\right)$.

Thus $S_d(x, \varepsilon) = S_e\left(x, \dfrac{\varepsilon}{1+\varepsilon}\right)$ completing the proof.

**Note 12 :** As per definition, '$e$' above is a bounded metric, also called a $B$-metric. Thus, given a metric $d$ for a space. $X$, there always exists a $B$-metric $e$ for $X$ such that $T(d) = T(e)$. The metric '$e$' as in Note 2 is also a bounded metric.

We now make a very important observation. In Corollary 2 of Theorem 3 in Chapter 6 we observed that every closed and bounded subset $X$ of the real number space is compact; the boundedness, though it was nor specially mentioned, was, with respect to the usual distance function

$$d(x, y) = |x - y|.$$

If, instead of '$d$', we have $e$ as in Example 4. Then **R** itself becomes bounded; it is also closed but still then it is not compact. The other part of Corollary 2 that every compact subset is closed and bounded in $R$ holds in any metrizable space $X$ with any metric giving its topology. In fact:

**Example 6.** ***Let X be a metrizable space, d any metric for X, and A any compact subset of X. Then :***

***(i) A is closed;*** ***(ii) A is bounded;***

***(iii)*** $\boldsymbol{\delta(A) < \infty}$;

***(iv) If A is non-empty, then there exist points a, b of A such that***

$$\boldsymbol{\delta(A) = d(a, b)}.$$

**Proof :** We first observe that $(X, T(d))$ is a $T_2$-space; for any two distinct points $x$ and $y$ in $X$, let $0 < 2\varepsilon < d(x, y)$. Obviously

$$S(x, \varepsilon) \cap S(y, \varepsilon) = \phi$$

giving a strong separation for $x$ and $y$. Thus $(X, d)$ is a $T_2$-space; $A$, being compact is thus closed by Theorem 8, Chapter 6 and thus we prove (1).

For the rest, let $A$ be a non-empty set and let $\hat{X} = X \times X$, $\hat{A} = A \times A$; let $f : \hat{X} \to \mathbf{R}$ be defined as follows : for $x = (x_1, x_2)$ where $x_1, x_2 \in X$, $f(x) = d(x_1, x_2)$. The mapping $f$ is continuous by Example 3; since $\hat{A}$ is a compact subset of $\hat{X}$, $f/\hat{A}$ is continuous and $f(\hat{A})$ must be a compact subset of $\mathbf{R}$. It follows that $f(\hat{A})$ is closed and bounded in $R$. Let $c = l.u.b.\, f(\hat{A})$. Then $c \in f(\hat{A})$; hence, there exists a point $p \in \hat{A}$ such that $f(p) = c$. Let $p = (a, b)$ where $a, b \in A$. Obviously, $d(a, b) = \delta(A)$. This proves (*iv*); (*ii*) and (*iii*) are obvious consequences of (*iv*).

**Example 7.** ***If X is a compact metrizable space, then every metric for X is a B-metric. Follows from Example 6.***

**Example 8.** ***Let (X, d) be a metric space. If the sets $A, B \subset X$ are non empty, compact and disjoint, then : (i) $d(A, B) > 0$; (ii) there exist points $a \in A$, $b \in B$ such that $d(A, B) = d(a, b)$.***

**Proof :** If possible, let $d(A, B) = 0$; then,

$$\inf \{d(a, b) : (a, b) \in A \times B\} = 0;$$

since $d/A \times B : A \times B \to \mathbf{R}$ is continuous and $A \times B$ is a compact set in $X \times X$, $d$ attains its minimum at some point $(a, b) \in A \times B$ *i.e.*, $d(a, b) = 0$ for some $(a, b) \in A \times B$; then $a \in \overline{B} = B$—a contradiction. Hence $d(A, B) > 0$ and as already mentioned, $\exists$ points $a \in A$ and $b \in B$ such that

$$d(a, b) = d(A, B).$$

**Example 9.** ***Let (X, d) be a metric space with the metric topology T (d); if $A \subset X$ be a non-empty subset and T (d)/A is the subspace topology on A then :***

(*i*) ***$d/A \times A$ is a metric on A and***

(*ii*) ***$T(d/A \times A) = T(d)/A$.***

**Proof :** If we put $d^* = d/A \times A$, it is evident that

$$S_{d*}(x, \varepsilon) = S_d(x, \varepsilon) \cap A$$

for any $x \in A$ and $\varepsilon > 0$. The proof follows from Theorem 58, Chapter 2.

**Example 10.** ***Let X = the real line* R *without the two reals 2 and 3; if d denotes the usual metric on the real line denote by $d^*$ the metric $d/X \times X$, in the metric space $(X, d^*)$ (which inherits the subspace topology from R by Example 9), consider $B = \{3 + 1/n : n \in N$ where N is the set of natural numbers$\}$ and $A = \{2 - 1/n : n \in N\}$. Show that $d^*(A, B) = 1$ and there do not exist points $a \in A$ and $b \in B$ such that $d^*(a, b) = 1$.***

The proof is obvious; further the example shows that without the properties of compactness (*ii*) of Example 8 is not valid; infact $A$ and $B$ are closed and bounded in $(X, d^*)$ but they are nor compact.

## Art. 5 CARTESIAN PRODUCT OF METRIZABLE SPACES

In Art. 6 of Chapter 2 we have discussed Cartesian product of metrisable spaces in the heading 'Construction of Topologies'.

With the property of metrisability, it is now pertinent to see if the product $\prod_{\alpha \in A} Y_\alpha$ where each $Y_\alpha$ is metrisable is metrisable with its product topology.

We start with a lemma :

**Lemma 1 :** *Let $(X_n, d_n)_{n \in N}$ be a countable family of metrizable spaces. Let $\delta_n (X_n) \leq M$ for all large n and let $\delta_n (X_n) \to 0$ as $n \to \infty$. Take $e (x, y) = \sup \{d_n (x_n, y_n)/n \in N\}$. Then $T(e)$ is the product topology of $\prod_{n \in N} (X_n, T(d_n)$ [$\delta_n$ denotes the diameter of $X_n$].*

**Proof :** Let, for $n_0 \in N$, $\delta_n (X_n) \leq M$ for all $n \geq n_0$; obviously '$e$' is well-defined; that '$e$' is a metric requires the verification of triangle inequality; in fact

$$e(x, z) = \sup_n d(x_n, z_n) \leq \sup_n [d_n (x_n, y_n) + d_n (y_n, z_n)]$$

$$\leq \sup_n d_n (x_n, y_n) + \sup_n d_n (y_n, z_n) = e(x, y) + e(y, z)$$

To show that the product topology is given by the metric $e$, let

$$x = (x_n) \in S(x_1, \varepsilon_1) \times \ldots \times S(x_n, \varepsilon_n) \times \prod_{n+1}^{\infty} X_n = U \text{ (say)}$$

Choose $\varepsilon = \min (\varepsilon_1, ..., \varepsilon_n)$; then $\varepsilon > 0$ and we would have

$$x \in S_e (x, \varepsilon) \subset U :$$

infact,

$$y \in S_e (x, \varepsilon) \Rightarrow e(x, y) < \varepsilon \Rightarrow \sup_n d_n (x_n, y_n) < \varepsilon$$

$$\Rightarrow \qquad d_i (x_i, y_i) < \varepsilon_n \text{ for } 1 \leq i \leq n$$

and hence $y = (y_n) \in U$ and thus $T \subset T(e)$

To see otherwise, let $x \in S_e(x, \varepsilon)$; since $\delta_n(X_n) \to 0$, there is an $n_0$ with $\delta_n(X_n) < \varepsilon/2$ for all $n \geq n_0$; let

$$u = S(x_1, \varepsilon/2) \times \ldots \times S(x_{n_0} \lambda, \varepsilon_2) \times \prod_{n_0+1}^{\infty} X_n;$$

we claim that $x \in U \subset S_e(x, \varepsilon)$ : in fact, let $y = (y_n) \in U$; obviously,

$$d_i(x_i, y_i) < \varepsilon/2 \text{ for } 1 \leq i \leq n_0$$

and since, this also does hold for $i \geq n_0$ (as because $\delta_n(X_n) < \varepsilon/2$ for all $n \geq n_0$), we have $y \in U \Rightarrow e(x, y) \leq \varepsilon/2 \Rightarrow y \in S_e(x, \varepsilon)$. Hence $T(e) \subset T$. Thus $T = T(e)$.

**Theorem 4 :** *Let $\{X_n \mid n \in \mathbf{N}\}$ be a family of metrizable spaces. Then $\prod_{n=1}^{\infty} X_n$, with the product topology T, is metrisable.*

**Proof :** Let $d_n$ be the metric for $X_n$; by Note 18, we have a metric $e^{1/n}$ for each $X_n$ ($e^{1/n}(x, y) = \min(1/n, d_n(x, y))$ for $(x, y) \in X_n \times X_n$) such that $T(e^{1/n}) = T(d_n)$ for each $n$. Obviously $\delta_n(X_n)$, when $X_n$ is endowed with the metric $e^{1/n}$, tends to zero as $n \to \infty$ and further, is bounded uniformly. By the Lemma above, if $e(x, y) = \sup_{n \in N} e^{1/n}(x_n, y_n)$, then $T(e) = T$.

**Theorem 5 :** *Let $\{X_\alpha \mid \alpha \in I\}$ be a family of spaces each of which has more than one point. If $\prod_{\alpha \in I} X_\alpha$ is metrisable then so is $X_\alpha$ for each $\alpha$.*

**Proof :** By Example 3, Chapter 2, each $X_\alpha$ is homeomorphic to a subspace of $\prod_{\alpha \in I} X_\alpha$ which is metrisable (Example 9). Hence each $X_\alpha$ is metrisable by Lemma 2 which is as follows:

**Lemma 2 :** Metrisability is invariant under homeomorphism.

**Proof :** Let $Y$ be endowed with a metric $d$ and let $(X, T)$ be homeomorphic to $(Y, T(d)$; then $X$ is metrisable : in fact, let $h : X \to Y$ be the homeomorphism and define $e(x, x') = d(h(x), h(x'))$; it is easy to verify that $e$ is a metric. To show that $T(e) = T$, let $U \in T$ and $x \in U$; obviously $h(x)$

$\in h(U)$ and $h(U)$ is open in $Y$; obviously, $\exists\, \varepsilon > 0$ such that $h(x) \in S_d(h(x), \varepsilon) \subset h(U)$. By definition of $e$ and since $h$ is a homeomorphism

$$h^{-1}(S_d(h(x), \varepsilon)) = S_e(x, \varepsilon)$$

and thus $x \in S_e(x, \varepsilon) \subset U$; thus $T \subset T(e)$ ...(1)

Since every basic open set $S_e(x, \varepsilon)$ is equal to $h^{-1}(S_d(h(x), \varepsilon))$ and $h$ is a homeomorphism, $S_e(x, \varepsilon) \in T$ and thus $T(e) \subset T$ ...(2)

Finally, from (1) and (2) $T(e) = T$.

**Note 13 :** We now show by an example that for an arbitrary index set I, $\prod_{\alpha \in I}(X_\alpha, T(d_\alpha))$ need not be metrisable. For this we first refer to Theorem 17, Chapter 5 which asserts that if $(X, T)$ is a first countable space and $A \subset X$ then $x \in \overline{A}$ implies that there exists a sequence $\{x_n\} \subset A$ converging to $x$; we then appeal to Theorem 2 and finally resolve that for a metrisable space $X$, if $A \subset X$ and $x \in \overline{A}$ then there must be a sequence $\{x_n\} \subset A$ converging to $x$. Thus, in a topological space $X$, if $A \subset X$ and $x \in \overline{A}$ absence of any sequence in $A$ converging to $x$ would assure the non-metrizability of $X$. With this we consider :

**Example 1.** ***If X is any uncountable set $\mathbf{R}^X$ with its product topology is not metrizable.***

**Proof :** Let us denote by $\hat{o}$ the point in $\mathbf{R}^X$ each of whose coordinate is zero; let

$$A = \{(x_\alpha) \in \mathbf{R}^X : x_\alpha = 0 \text{ for finitely many values of } \alpha \text{ and } x_\alpha = 1 \text{ for all other values of } \alpha\}$$

We first show that $\hat{o} \in \overline{A}$; let $\Pi V_\alpha$ be a basis element containing $\hat{o}$. Then $V_\alpha \neq R$ for only finitely many values of $\alpha$, say for $\alpha = \alpha_1, ..., \alpha_n$. Let $(x_\alpha)$ be the point of $A$ defined by letting $x_\alpha = 0$ for $\alpha = \alpha_1, ..., \alpha_n$ and $x_\alpha = 1$ for all other values of $\alpha$; then $(x_\alpha) \in A \cap \Pi V_\alpha$, as desired.

We now show that no sequence in $A$ converges to $\hat{o}$ thereby assuring the non-metrizability of $\mathbf{R}^X$ with the help of Note 13.

Let $\hat{a}_n$ be a sequence of points in A. Each point $\hat{a}_n \in \mathbf{R}^X$ has only finitely many coordinates equal to zero. Given $n$, let $X_n$ denote the subset of $X$ consisting of those indices $\alpha$ for which the $\alpha$th coordinate of $\hat{a}_n$ is zero. Obviously each $X_n$ is finite and consequently $\bigcup_{n=1}^{\infty} X_n$ is countable. But $X$ being uncountable, $X \setminus \bigcup_{n=1}^{\infty} X_n$ is non-null and thereby, there exists an index, say $\beta$ belonging to $X \setminus \bigcup_{n=1}^{\infty} X_n$. Hence, for each of the points $\hat{a}_n$, the $\beta$th coordinate equals 1.

Let $U_\beta = (-1, 1)$ in $\mathbf{R}$ and let $U = p_\beta^{-1}(U_\beta) \in \mathbf{R}^X$ where $p_\beta$ is the usual projection. Obviously $\hat{o} \in U$ but $U$ contains none of the points $\hat{a}_n$ and consequently $\hat{a}_n$ cannot converge to $\hat{o}$.

**Example 2.** ***Every metrizable space is normal.***

**Proof :** Let $X$ be a metrizable space with metric $d$. Let $A$ and $B$ be disjoint closed subsets of $X$. For each $a \in A$, there exists an open sphere $S(a, \varepsilon_a)$ such that $S(a, \varepsilon_a) \cap B = \phi$; similarly, for each $b \in B$, there exists an open sphere $S(b, \varepsilon_b)$ such that $S(b, \varepsilon_p) \cap A = \phi$. Define

$$U = \bigcup_{a \in A} S(a, \varepsilon_{a/2}), \text{ and } V = \bigcup_{b \in B} S(b, \varepsilon_{b/2})$$

Obviously, $U$ and $V$ are both open sets in $X$ and $A \subset U$, $B \subset V$; we are done if we can prove that $U \cap V = \phi$. Let, if possible, $z \in U \cap V$; then, for some $a \in A$ and $b \in B$, $d(a, z) < \varepsilon_{a/2}$ and $d(b, z) < \varepsilon_{b/2}$. Hence

$$d(a, b) \le d(a, z) + d(z, b) < \frac{\varepsilon_a + \varepsilon_b}{2}$$

If $\varepsilon_a \le \varepsilon_b$, then $d(a, b) < \varepsilon_b$ *i.e.*, $a \in S(b, \varepsilon_b)$—a contradiction; similarly, if $\varepsilon_b \le \varepsilon_a$ we again arrive at contradiction. Hence $U \cap V = \phi$.

## Art. 6 COUNTABILITY AND COVERING PROPERTIES IN METRIC SPACES

**Theorem 6 :** *In metric spaces, the concepts of second countability, separability and Lindeloff are all equivalent.*

**Proof :** Separability $\Rightarrow$ second countability : Let $\{x_i\}$ be a countable dense set in a metric space $(X, d)$; the family

$$\mathcal{B} = \{S(x_i, r) \mid r - \text{rational}, i = 1, 2, \ldots\}$$

is countable.

We now apply Theorem 1, Chapter 1 to prove that the family $\mathcal{B}$ is a base for the metric topology $T(d)$ : in fact, let $U$ be open in $X$ and let $x \in U$; there exists $\varepsilon > 0$ such that $x \in S(x, \varepsilon) \subset U$; since $\{x_i\}$ is dense in $X$, there exists a $k \in \mathbf{N}$ such that $d(x, x_k) < \varepsilon/2$. We claim that $x \in S(x_k, \varepsilon/2) \subset U$ and we are done if we can prove our claim.

Let $y \in S(x_k, \varepsilon/2)$, then

$$d(x, y) \leq d(x, x_k) + d(x_k, y) < \varepsilon/2 + \varepsilon/2 = \varepsilon$$

*i.e.*, $y \in S(x, \varepsilon) \subset U$; that $x \in S(x_k, \varepsilon/2)$ is obvious.

Hence $\mathcal{B}$ is a countable base for $T(d)$ and $(X, T(d))$ is second countable.

By Theorem 1 of Chapter 5, second countability always implies Lindeloff property.

We now show that Lindeloff property implies separability; we first show that for each $\varepsilon > 0$, there exists a countable set $A_\varepsilon \subset X$ such that $d(x, A_\varepsilon) < \varepsilon$ for every $x \in X$ : in fact, from the covering $\{S(x, \varepsilon) \mid x \in X\}$, we extract a countable covering $\{S(x_i, \varepsilon) \mid i \in \mathbf{N}\}$ and let $A_\varepsilon = \{x_i \mid i \in N\}$.

Let $D = \bigcup_{n=1}^{\infty} A_{1/n}$; obviously, each $A_{1/n}$ is a countable set and so is $D$; that $\overline{D} = X$ is evident. Hence $(X, T(d))$ is separable.

In Chapter 6 we have defined four types of compactness; in fact, in addition to compactness there are three others: countable compactness, sequential compactness and Frechet compactness. We first show that :

**Theorem 7 :** *A metric space $(X, d)$ is sequentially compact iff it is Frechet compact.*

**Proof :** From Theorem 16 (*a*), Chapter 6, we know that every sequentially compact space is countably compact and from Theorem 15 (*a*), Chapter 6, we know that every countably compact space is Frechet compact. Thus, one side of the theorem is obvious.

We prove the other part. Let $X$ be Frechet compact *i.e.*, we assume that every infinite subset of $X$ has a limit point; let $\{x_n\}$ be an arbitrary sequence in $X$. If $\{x_n\}$ has a point which is repeated infinitely, then it has a constant subsequence which is evidently convergent and thus, we are done.

If no point of $\{x_n\}$ is infinitely repeated, then the set $A$ of points of this sequence is infinite. By hypothesis it has a limit point $x$; since for every $\varepsilon > 0$, we have, for an infinite number of indices, $d(x_n, x) < \varepsilon$, we have, in particular, for a suitable $n = k_1$, $d(x_{k_1}, x) < 1$; for a suitable $n = k_2 > k_1$, we have similarly, $d(x_{k_2}, x) < 1/2$, and in general, for a suitable $n = k_\gamma > k_{\gamma-1}$ $d(x_{k_\gamma}, x) < 1/\gamma$ $(\gamma = 2, 3, \ldots)$. For the subsequence $\{x_n'\} \equiv \{x_{k_n}\}$ thus picked out, we have $x_n' \to x$; hence $X$ is sequentially compact.

**Theorem 8 :** *A metric space* $(X, d)$ *is Frechet compact iff it is countably compact.*

**Proof :** From Theorem 15 (*a*), any countably compact space is Frechet compact and by Theorem 15 any Frechet compact $T_1$-space is countably compact. As observed in Example 6, $(X, T(d))$ is a $T_2$-space and thus every Frechet compact metric space is countably compact.

**Note 14 :** To show that a countably compact metric space is compact we require the following definition :

**Definition 4 :** *In a metric space* $(X, d)$, *let* $F \subset X$ *be a finite set;* $F$ *is said to be an* $\varepsilon$*-net for* $X$ *iff, given any point* $p \in X$, *there exists a point* $q$ *of* $F$ *such that* $d(p, q) < \varepsilon$.

**Lemma 3 :** *Let* $(X, d)$ *be a countably compact metric space; then* $X$ *has an* $\varepsilon$*-net for every* $\varepsilon > 0$.

**Proof :** If possible, let there be an $\varepsilon > 0$ such that $X$ has no $\varepsilon$-net; for $p_1 \in X$, obviously there exists at least one point $p_2$ such that $d(p_1, p_2) \geq \varepsilon$ since the subset of $X$ consisting of the single point $p_1$ is not an $\varepsilon$-net for $X$. Since $\{p_1, p_2\}$ is not an $\varepsilon$-net for $X$, there must exist a point $p_3 \in X$ which is not in $U_1 \cup U_2$ where $U_1 = S(p_1, \varepsilon)$ and $U_2 = S(p_2, \varepsilon)$; also $d(p_i, p_j) \geq \varepsilon$ for $i \neq j$, $i = 1, 2, 3$; $j = 1, 2, 3$; we continue this process; suppose that we have found $n$ points $p_1, p_2, \ldots, p_n$ in $X$ such that $d(p_i, p_j) \geq \varepsilon$ for $i \neq j$, $i = 1, 2, \ldots, n$; $j = 1, 2, \ldots, n$ and define $U_i = S(p_i, \varepsilon)$, for $i = 1, 2, \ldots, n$; if $F = \{p_1, \ldots, p_n\}$ then $F$ is not an $\varepsilon$-net for $X$ enabling the existence of a point $p_{n+1}$ not in $\bigcup_{i=1}^{n} U_i$; at once, we thus have $d(p_i, p_j) \geq \varepsilon$ for $i \neq j$, $i = 1, 2, \ldots, (n+1)$; $j = 1, 2, \ldots, (n+1)$.

By mathematical induction, we now have a sequence $\{p_n\}$ of distinct points of the countably compact metric space $(X, d)$ such that $d(p_i, p_j) \geq \varepsilon$ for $i \neq j$. If $K = \bigcup_{n=1}^{\infty} \{p_n\}$, $K$ is an infinite set in the countably compact space $X$ and so, by Theorem 8 has a limit point $q$ in $X$. Thus $S(q, \varepsilon/3)$

must contain infinitely many elements of $K$ contradicting the fact that $d(p_i, p_j) \geq \varepsilon$ for $i \neq j$. Thus $X$ must have an $\varepsilon$-net.

**Lemma 4 :** *Every countably compact metric space $(X, d)$ is separable.*

**Proof :** By Lemma 3, for every $n \in \mathbf{N}$, $X$ must have an $\frac{1}{n}$-net $F_n$. If $F = \bigcup_1^\infty F_n$, then $F$, being a countably union of finite sets, must be countable; let $F = \{p_n\}\ n \in \mathbf{N}$. We show that $X = \overline{F}$, thereby proving separability.

For this, let $p \in X$; for an arbitrary real number $\delta > 0$, let $m \in N$ be such that $\frac{1}{m} < \delta$; since $F_m$ is an $\frac{1}{m}$-net, there exists a point $\omega \in F_m$ such that $d(p, \omega) < \frac{1}{m} < \delta$. Since $\omega \in F$, $\omega = p_k$ for same $k \in \mathbf{N}$, we have thus proved that, for every $\delta > 0$, $S(p, \delta)$ contains a point of $F$ and thus $\overline{F} = X$.

**Note 15 :** By Theorem 6, and Lemma 4 we do have that every countably compact metric space is Lindeloff and thus, by Theorem 14 (*b*) in Chapter 6, it is compact as well. By Theorem 14 (*a*) in Chapter 6, every compact space is countably compact and hence :

**Theorem 9 (*a*) :** *A subset A in a metric space $(X, d)$ is compact iff it is countably compact.*

**Theorem 9 (*b*) :** *A subset A in metric space $(X, d)$ is compact iff it is sequentially compact.*

## Art. 7 RELATIVELY COMPACT AND TOTALLY BOUND SETS IN A METRIC SPACE (*X*, *d*); THE IDEA OF COMPLETENESS

**Definition 5 :** *A set A in $(X, d)$ is called relatively compact if $\overline{A}$ is compact in X.*

**Note 16 :** From the discussion in Art. 6, it can easily be shown that:

**Theorem 10 :** *A set A in a metric space is relatively compact iff every sequence $\{x_n\}$ whose elements belong to A has a convergent subsequence.*

**Definition 6 :** *A set A in a metric space $(X, d)$ is called pre-compact or totally bounded if for each $\varepsilon > 0$, there is a finite set of points (say) $x_1, x_2, ..., x_n$ in X such that A is contained in the union of the open spheres $S(x_i, \varepsilon)$, $i = 1, 2, ..., n$.*

**Note 17 :** Obviously, $(X, d)$ itself is totally bounded iff, for every $\varepsilon > 0$, $X$ has an $\varepsilon$-net.

**Note 18 :** The empty set $\phi$ is clearly totally bounded.

**Note 19 :** If $A$ is a non empty totally bounded set in $(X, d)$, it is easy to see that $A$ is contained in the union of a finite number of open spheres, with centres in $A$, each sphere having radius $\varepsilon$.

**Note 20 :** Evidently, the definition of total boundedness can be rewritten as follows :

$A \subset (X, d)$ is pre-compact, if corresponding to any $\varepsilon > 0$, there exists a finite number of sets $A_1, A_2, ..., A_n$, each having diameter not exceeding $\varepsilon$ such that $A \subset A_1 \cup ... \cup A_n$.

**Theorem 11 :** *A relatively compact set A in a metric space (X, d) is totally bounded.*

**Proof :** If possible, let $A$ be not totally bounded; then, there exists an $\varepsilon > 0$ such that $A$ is not contained in the union of any finite number of open spheres each with radius $\varepsilon$. For $x_1$ in $A$, choose $x_2$ in $A$ such that $d(x_1, x_2) \geq \varepsilon$. Then choose $x_3$ in $A$ so that $d(x_1, x_3) \geq \varepsilon$ and $d(x_2, x_3) \geq \varepsilon$. When $x_1, x_2, ..., x_n$ have been chosen, there is some point $x_{n+1}$ of $A$ not in the set $\bigcup_{i=1}^{n} S(x_i, \varepsilon)$. We have thus chosen a sequence $\{x_n\}_{n=1}^{\infty}$ of distinct points in $A$. Since $A$ is relatively compact, by Theorem 10, $\{x_n\}$ has a convergent subsequence and thus the set $F = \bigcup_{n=1}^{\infty} \{x_n\}$ has a limit point, say $x$; hence, $S(x, \varepsilon/3)$ must contain infinitely many elements of $\{x_n\}$ contradicting the definition of $\{x_n\}$; hence $A$ must be totally bounded.

**Note 20 :** It is interesting to observe that the converse of Theorem 11 need not be true; for this, we have the following example :

**Example 1 :** ***If Q denotes the set of all rationals in R we can consider Q as a metric space with the metric $d^* = d/Q \times Q$ where d is the usual metric in R; it is obvious that $T(d^*)$ is the subspace topology in Q.***

Consider $A = \{x \in \mathbf{Q} : 0 \leq x \leq 1\}$ obviously, $A$ is totally bounded but it is not relatively compact in $(\mathbf{Q}, T(d^*))$.

**Note 21 :** Thus the following question arises; when is a totally bounded subset of a metric space relatively compact? For the answer we require the following concept of completeness.

**Definition 7 :** *Let (X, d) be a metric space.*

(*a*) *If* $\{x_n\}$ *is a sequence in* $X$ *such that* $d(x_m, x_n) < \varepsilon$ *whenever* $n \geq k$, $m \geq k$ *for some* $k \in N$, *for a preassigned* $\varepsilon > 0$, *then we say that* $\{x_n\}$ *is a Cauchy sequence.*

(*b*) *If every, Cauchy sequence is convergent in a metric space* $(X, d)$ *we say that* $(X, d)$ *is complete.*

**Example 2 :** ***Consider the open interval (0, 1) with the usual metric inherited from the real line R; then*** $\left\{\frac{1}{n}\right\}_{n=1}^{\infty}$ ***is a Cauchy sequence in (0, 1) which is not convergent in (0, 1); thus (0, 1) is not complete.***

**Note 22 :** In Example 22 of Misc. exercises I we have observed that (0, 1) and $R$ are homeomorphic. By definition of completeness, $R$ is complete, where as (0, 1) is not complete. Thus completeness is not a topological invariant.

**Note 23 :** It is easy to observe that every convergent sequence is a Cauchy sequence.

**Theorem 12 :** *If* $(X, d)$ *is a complete metric space, a totally bounded set* $A$ *in* $X$ *is relatively compact.*

**Proof :** By definition of completeness and by Theorem 10, is suffices to show that every sequence in $A$ has a Cauchy subsequence.

We start with a sequence $\{x_n\}$ in $A$. Since $A$ is totally bound $A$ is contained in the finite union of open spheres, each of radius 1. At least one of these spheres must contain a subsequence from $\{x_n\}$. Let such a subsequence be denoted by $x_{11}, x_{12}, \ldots$; next, we consider a finite number of spheres of radius $\frac{1}{2}$ whose union would contain $A$ by its property of total boundedness and find a sphere of radius $\frac{1}{2}$ which contains a subsequence of $\{x_{1n} : n = 1, 2, \ldots\}$, say $\{x_{2n} : n = 1, 2, \ldots\}$; we proceed by induction, obtaining sequences $\{x_{ki} : i = 1, 2, \ldots\}$ $(k = 1, 2, \ldots)$, each sequence a subsequence of the predecessor, and the $k$th sequence contained in a sphere of radius $\frac{1}{k}$; thus, if $m > k$, $d(x_{mm}, x_{kk}) < \frac{1}{m}$ and thus $d(x_{mm}, x_{kk}) \to 0$ as $m \to \infty$ and $k \to \infty$ giving $\{x_{nn} : n = 1, 2, \ldots\}$ to be a Cauchy subsequence of $\{x_n\}$.

**Note 24 :** It is obvious from Theorem 12 that if a complete metric space $(X, d)$ is totally bounded then it must be compact. To prove the converse, it suffices to show that a compact metric space is complete.

**Lemma 5 :** *A Cauchy sequence is convergent if it has a convergent subsequence.*

**Proof :** If a Cauchy sequence is itself convergent, then it is a convergent subsequence of itself.

Conversely, let a Cauchy sequence $\{x_n\}_\infty$ in a metric space $(X, d)$, have a convergent subsequence $\{x_{n_k}\}_{k=1}^{\infty}$; then, there exists a $x \in X$ such that, corresponding to a preassigned $\varepsilon > 0$ there exists $p_1 \in N$ with $d(x_{n_k}, x) < \varepsilon/2$ whenever $k > p_1$; since $\{x_n\}$ is a Cauchy sequence, there exists $p_2 \in N$ such that $d(x_n, x_m) < \varepsilon/2$ whenever $n, m > p_2$; choose $k$ sufficiently large and greater than $p_1$ such that $n_k > p_2$, then $d(x_{n_k}, x_m) < \varepsilon/2$ whenever $m > p_2$. Then $d(x_m, x) \le d(x_m, x_{n_k}) + d(x_{n_k}, k) < \varepsilon/2 + \varepsilon/2 = \varepsilon$, whenever. $m > p_2$.

**Theorem 13 :** *A compact metric space* $(X, d)$ *is complete.*

**Proof :** Let $\{x_n\}$ be a Cauchy sequence in $(X, d)$. By Theorem 9 $(b)$, $(X, d)$ is sequentially compact and hence $\{x_n\}$ has a convergent subsequence; by Lemma 5, $(X, d)$ is complete.

**Theorem 14 :** *A metric space is compact iff it is complete and totally bounded.*

**Proof :** Follows from Note 24 and Theorem 13.

## Art. 8 COMPLETENESS AND COMPLETION

With the concept of completeness, as in definition 7 $(b)$, we can call those metrics spaces, which are not complete, as incomplete. When we study the real line **R** or $\mathbf{R}^n$ $(n \ge 2)$ we always have a complete metric space; but there are metric spaces which are not complete as we have seen already.

But we can get rid of this limitation of incompleteness by sheerly enlarging an incomplete metric space so as to yield a complete metric space and that too in a minimal way by sheer adjunction of those and only those points which are absolutely necessary to obtain a completeness space.

Actually what we would do is as follows :

(*i*) we shall construct a complete metric space $Y$ of such a character that:

(*ii*) there exists an isometric mapping $f : X \to Y$ such that :

(*iii*) $f(X)$ is dense in $Y$; finally,

(*iv*) we would identify $X$ and $f(X)$.

For this we need the definition of isometric mapping at the very outset.

**Definition 8 :** *Let* $(X_1, d_1)$ *and* $(X_2, d_2)$ *be metric spaces with metrics* $d_1$ *and* $d_2$ *respectively; if there exists a mapping* $f: X_1 \to X_2$ *such that* $d_2 [f(x_1), f(x_2)] = d_1 (x_1, x_2)$ *for each pair* $x_1, x_2$ *in X, we call f an isometry or an isometric mapping.*

**Note 25 :** It is easy to see that :

(*i*) $f(x_1) = f(x_2)$ implies $x_1 = x_2$ so that $f^{-1}$ exists;

(*ii*) $f$ is continuous and so is $f^{-1} : f(X) \to X$.

If $f(X) = Y$ we say that $X$ and $Y$ are in isometric correspondence.

Further, we shall call $Y$ in (*i*) above the completion of $X$. To assert the existence of the completion of an incomplete metric space $(X, d)$ we go through a series of Lemmas :

**Lemma 6 :** *If* $\{x_n\}$ *and* $\{y_n\}$ *are Cauchy sequences in* $(X, d)$, *let us define a relation '~' as follows :* $\{x_n\} \sim \{y_n\}$ *iff* $\lim_{n \to \infty} d(x_n, y_n) = 0$. *Then '~' is an equivalence relation.*

Proof follows from the definition of a metric '$d$'.

**Note 26 :** The collection of all equivalence classes of cauchy sequences in $(X, d)$ with respect to the relation '~' shall be denoted by $X^*$. Any two Cauchy sequences, belonging to the same equivalence class are called equivalent.

**Lemma 7 :** *For* $x^*$ *and* $y^*$ *belonging to* $X^*$, *let us define*

$$d^* (x^*, y^*) = \lim_{n \to \infty} d(x_n, y_n) \qquad ...(1)$$

*where* $\{x_n\} \in x^*$ *and* $\{y_n\} \in y^*$; *then :*

(*i*) *limit (1) exists;*

(*ii*) *the limit does nor depend on which particular member of the equivalence class is chosen.*

**Proof :** (*i*) Let $d(x_n, y_n) = s_n$ for every $n$; then

$$|d(x_n, y_n) - d(x_m, y_m)| \le |d(x_n, y_n) - d(x_n, y_m)| + |d(x_n, y_m) - d(x_m, y_m)| \le d(y_n, y_m) + d(x_n, x_m) \qquad ...(2)$$

where the last inequality follows from the fact that, if $d$ is any metric of a metric space $(X, d)$ and if $x, y, z \in X$, then

$$|d(x, y) - d(x, z)| \le d(y, z) \qquad ...(3)$$

From (2), it follows that $\{s_n\}$ is a Cauchy sequence since $\{x_n\}$ and $\{y_n\}$ are so; since the set of real numbers is complete the limit $\lim_{n\to\infty} s_n$ exists.

(*ii*) suppose $\{x_n\}, \{x_n'\} \in x^*$ and $\{y_n\}, \{y_n'\} \in y^*$ then

$$|d(x_n, y_n) - d(x_n', y_n')| \leq |d(x_n, y_n) - d(x_n', y_n)| + |d(x_n', y_n) - d(x_n', y_n')| \leq d(x_n, x_n') + d(y_n, y_n') \quad \text{(from (3))};$$

since $\{x_n\} \sim \{x_n'\}$ and $\{y_n\} \sim \{y_n'\}$, the r.h.s. of (4) tends to zero, by definition of equivalence, as $n$ tends to infinity; hence

$$\lim_{n\to\infty} d(x_n, y_n) = \lim_{n\to\infty} d(x_n', y_n').$$

**Note 27 :** From (*i*) and (*ii*) above we can assert that $d^*$ is well-defined.

**Lemma 8 :** *$(X^*, d^*)$ is a metric space.*

**Proof :** It is clear that $d^*(x^*, y^*) \geq 0$. We must also show that $d^*(x^*, y^*) = 0$ iff $x^* = y^*$. Let $d^*(x^*, y^*) = 0$; then $\lim_{n\to\infty} d(x_n, y_n) = 0$, which implies $\{x_n\} \sim \{y_n\}$ *i.e.*, $x^* = y^*$. The other side is obvious.

The symmetry of $d^*$ is evident because $d(x_n, y_n) = d(y_n, x_n)$ for every $n$.

To verify the triangle inequality, let $x^*, y^*, z^* \in X^*$ and suppose $\{x_n\} \in x^*$, $\{y_n\} \in y^*$ and $\{z_n\} \in z^*$ respectively. Now the obvious inequality $d(x_n, z_n) \leq d(x_n, y_n) + d(y_n, z_n)$ holds for every $n$ and is true in the limit as well giving the triangle inequality $d(x^*, z^*) \leq d(x^*, y^*) + d(y^*, z^*)$.

**Lemma 9 :** *$X^*$ contains a subspace $X_0$ isometric to $X$.*

**Proof :** Let $X_0$ denote the collection of all equivalence classes $x'$ which contain the sequence $(x, x, \ldots, x, \ldots)$.

Let $$\left.\begin{array}{c} f : X \to X_0 \\ x \to x' \end{array}\right\}$$

We show that $f$ determines an isometric correspondence between $X$ and $X_0$.

It is obvious that $\{y\} \notin x'$ if $x \neq y$ and thus the correspondence is one-one; it is a surjection as well; finally, for any two points $x', y'$ in $X_0$, $d^*(x', y') = \lim_{n\to\infty} d(x, y) = d(x, y)$ and we are done.

**Lemma 10 :** $\overline{X}_0 = X^*$ *where the closure is taken with respect to the topology* $T(d^*)$.

**Proof :** Let $x^* \in X^*$ and let $\{x_n\} \in x^*$. Since $\{x_n\}$ is a Cauchy sequence; there exists $k \in \mathbf{N}$ such that for all $n, m > k$, $d(x_n, x_m) < \varepsilon/2$ for an $\varepsilon > 0$. Now $d^*(x^*, x_n') = \lim_{m \to \infty} d(x_m, x_n) \leq \varepsilon/2$ (since $(x_n, x_n, x_n, ...) \in x_n'$); thus, for any $\varepsilon > 0$ $S(x^*, \varepsilon)$ contains a member of $X_0$ giving $\overline{X}_0 = X^*$.

**Lemma 11 :** *If* $x_1', x_2', ..., x_i', ...$ *is a Cauchy sequence in* $X^*$, *then it is convergent.*

**Proof :** We first observe, that by definition $(x_i, x_i, ..., x_i, ...) \in x_i'$ for any $i = 1, 2, ...$; we consider $f$ as in Lemma 9 and observe that $f(x_i) = x_i'$ for every $i$; obviously, $d^*(x_m', x_n') = d(x_m, x_n)$ since $f(x_m) = x_m'$ and $f(x_n) = x_n'$ and $f$ is an isometry; since $\{x_m'\}$ is a Cauchy sequence in $X^*$, so is $\{x_m\}$ in $X$; evidently $\{x_m\} \in x^*$ for some $x^* \in X^*$ and

$$\lim_{n \to \infty} d^*(x_n', x^*) = \lim_{n \to \infty} \lim_{m \to \infty} d(x_n, x_m) = 0$$

since $\{x_n\}$ is a Cauchy sequence in $X$; thus $\{x_n'\}$ is convergent to $x^*$.

**Lemma 12 :** $(X^*, d^*)$ *is complete.*

**Proof :** Consider an arbitrary Cauchy sequence $\{x_n^*\}$ in $X^*$; since $\overline{X}_0 = X^*$, there exist points in $X_0$, $x_1', x_2', ..., x_n', ...$ such that $d^*(x_n^*, x_n') < 1/n$ for every $n$; obviously,

$$d^*(x_n', x_m') \leq d^*(x_n', x_n^*) + d^*(x_n^*, x_m^*) + d^*(x_m^*, x_m') < \frac{1}{n} + \varepsilon + \frac{1}{m}$$

for $m$ and $n$ large enough. Therefore $\{x_n'\}$ is a Cauchy sequence. As noted above, then, $\{x_n'\}$ must converge to some $y^* \in X^*$ by Lemma 11; now, $d^*(y^*, x_n^*) \leq d^*(y^*, x_n') + d^*(x_n', x_n^*)$ where, as previously noted, each term on the r.h.s. goes to zero as $n$ tends to $\infty$. Hence $\{x_n^*\}$ must converge to $y^*$ as well, ascertaining the completeness of $X^*$.

**Theorem 15 :** *Let $(X, d)$ be a metric space. Then there exists a completion $(X^*, d^*)$ of $(X, d)$. Proof follows from the previous Lemmas.*

**Note 28 :** In Theorem 15 we have established the fact that every metric space $(X, d)$ has a completion $(X^*, d^*)$. One very pertinent question now arises : if there exists another completion $(X^{**}, d^{**})$ of $X$ then what would be the relation between these two completions?

Before answering the question we need some new definitions and results.

## Art. 9 COMPLETENESS AND UNIFORM CONTINUITY

We have already mentioned that every metric space is a topological space but not every topological space is metrizable; consequently, there are some properties which are valid for metric spaces but not for topological spaces, we have discussed some of them in 6, 9 (*a*) and (*b*). We refer to another such here now :

**Lemma 13 :** *Let $(X, d)$ be a metric space and $p \in X$; then there exists a decreasing sequence $Q_1 \supset Q_2 \supset Q_3 \supset \dots$ of nbds of $p$ such that if $U$ is any nbd of $p$, then there exists $Q_n$ for some $n$ such that $p \in Q_n \subset U$.*

**Proof :** Since $(X, T(d))$ is first countable, there exists a countable fundamental system of *nbds* $U_1, U_2, \dots, U_n, \dots$ such that if $V$ is any *nbd* of $p$ then there exists an integer $m$ such that $p \in U_m \subset V$; set $Q_1 = U_1$; $Q_2 = U_2 \cap Q_1$, $Q_3 = U_3 \cap Q_2$, $\dots$ $Q_n = U_n \cap Q_{n-1}$, $\dots$ and so on. Obviously, $\{Q_i\}_{i=1}^{\infty}$ is a decreasing sequence of *nbds* of $p$ and further, let $V$ be any *nbd* of $p$; for some $m \in \mathbf{N}$, $p \in U_m \subset V$; but $p \in Q_m \subset U_m \subset V$ and hence we get the desired result.

**Theorem 16 :** *Let $(X, d)$ be a metric space and $M \subset X$; if $p$ is a limit point of $M$, then $M$ contains an infinite sequence of distinct points converging to $p$.*

**Proof :** To prove this, let $\{Q_i\}_{i=1}^{\infty}$ be as in the lemma; since $p$ is a limit point of $M$, for each $n$, $Q_n$ contains a point $p_n$ of $M$ distinct from $p$. For any $n$, there exists a *nbd* $U$ of $p$ not containing $p_n$ and an $m$, such that $Q_m \subset U$. Hence $p_n \neq p_{m+i}$ $(i = 0, 1, 2, \dots)$. It may so happen that $p_n = p_{n+1} = \dots = p_{m-1}$ and thus any given point $p_n$ occurs in the sequence $\{p_n\}$ only a finite number of times, which is why there are infinitely many distinct points $p_n$. Any *nbd* of $p$ contain all but finite number of the sets

$Q_1, Q_2, \ldots$ and consequently, all but finitely many members of the sequence $\{p_n\}$; thus $p_n \to p$ and we are done.

**Note 29 :** The result in Theorem 16 may not be true in an arbitrary topological space $(X, T)$. Consider $T$ to be the co-countable topology, in **R**; if $A \subset R$ and $x \in \mathbf{R} \setminus A$ where $A$ is an infinite set, then, for any sequence $\{x_n\}$ of infinitely many distinct points in $A$, the set $\mathbf{R} \setminus \bigcup_{n=1}^{\infty} \{x_n\}$ is an open set containing $x$ and as such, $\{x_n\}$ cannot converge to $x$.

**Definition 9 :** *(i) Let $(X_1, d_1)$ and $(X_2, d_2)$ be metric spaces and $f: A \to X_2$ where $A \subset X_1$; $f$ is said to be uniformly continuous on A provided that for any $\varepsilon > 0$, there exists a $\delta > 0$ such that if x and y are any two points of A with $d_1(x, y) < \delta$, then $d_2(f(x), f(y)) < \varepsilon$.*

*(ii) If $f: (X_1, d_1) \to (X_2, d_2)$, then we say that f is continuous provided $f: (X_1, T(d_1)) \to (X_2, T(d_2))$ is continuous.*

**Example 1.** ***If $f: (X_1\ d_1) \to (X_2, d_2)$ is an isometric correspondence then, obviously, f is uniformly continuous; further $f^{-1}$ is uniformly continuous as well.***

**Note 30 :** It is interesting to note that an isometry $f$, as above, is a homeomorphism between $(X_1, T(d_1))$ and $(X_2, T(d_2))$ but a homeomorphism between two metric spaces need not be an isometry. We know that there exists a homeomorphismf between $R$ and $(0, 1)$ [with their usual metrics] but $f$ is not an isometry.

**Theorem 17 :** *If A in Definition 9 is compact and $f: A \to X_2$ is continuous, then f is uniformly continuous.*

**Proof :** If possible, let $f$ be continuous, but not uniformly continuous on $A$; then, for some $\varepsilon > 0$ there exist two sequences of points $(x_n)$ and $(y_n)$ in $A$ such that, for each $n$, $d_1(x_n, y_n) < \frac{1}{n}$ but $d_2(f(x_n), f(y_n)) \geq \varepsilon$; since $A$ is compact we can suppose them so chosen that $x_n \to x$ where $x \in A$; the condition $d_1(x_n, y_n) < \frac{1}{n}$ for each $n$ requires that $y_n \to x$ and thus, by continuity $f(x_n) \to f(x)$ and $f(y_n) \to f(x)$ [Theorem 49, Chapter 2], so that, for sufficiently large $m$, $d_2(f(x_n), f(y_n) < \varepsilon$, contrary to the definition of $(x_n)$ and $(y_n)$.

**Theorem 18 :** *Let $(X_1, d_1)$ and $(X_2, d_2)$ be two metric spaces where $(X_2, d_2)$ is complete; let $D \subset X_1$ and $f : D \to X_2$ be uniformly continuous. Then f can be extended to all limit points of D in one and only one way so that the extended function is uniformly continuous on $\overline{D}$ and is identical with f on D.*

**Proof :** Let $x \in \overline{D}$; by Theorem 16, there exists a sequence $(x_n) \subset D$ converging to $x$, agreeing that if $x \in D$, $(x_n)$ is simply the sequence

$$\{x, x, \dots x, \dots\} \qquad \text{...(1)}$$

By uniform continuity of $f$, corresponding to an $\varepsilon > 0$, there exists $\delta > 0$ such that $d_1(x_m, x_n) < \delta$ implies $d_2(f(x_m), f(x_n)) < \varepsilon$. Now, since $(x_m)$ is a convergent sequence, it is a Cauchy sequence as well and hence there exists $k \in N$ such that for all $m, n > k$ $d_1(x_m, x_n) < \delta$; hence, for all $m, n > k$, $d_2(f(x_m), f(x_n)) < \varepsilon$. Accordingly, $(f(x_n))$ is a Cauchy sequence in $(X_2, d_2)$; by completeness of $(X_2, d_2)$, there exists $y \in X_2$ such that $f(x_n) \to y$ as $n \in \infty$. Put $y = \hat{f}(x)$. By (1), obviously, if $x \in D$, $\hat{f}(x) = f(x)$. To show that the definition of $\hat{f}(x)$ is independent of the sequence converging to $x$ we proceed as follows : Let $(y_n)$ be another sequence converging to $x$; obviously the sequence $x_1, y_1, x_2, y_2, x_3, y_3, \dots$ converges to $x$ and in the same way, it follows that

$$f(x_1), f(y_1), f(x_2), f(y_2) \dots \qquad \text{...(2)}$$

converges in $X_2$. But $f(x_1), f(x_2), \dots$ is a subsequence of this sequence converging to $y$ (as defined) and hence (2) also converges to $y$; hence $(f(y_n))$, another subsequence of (1) also converges to $y$. Thus, $\hat{f}(x)$ is well-defined.

Now to prove uniform continuity of $\hat{f}$ on $\overline{D}$, let $\varepsilon > 0$ and let $\delta > 0$ be given by the uniform continuity of $f$ on $D$. Let $x, y \in \overline{D}$ with $d_1(x, y) < \delta/3$ and let $(x_i)$, $(y_i)$ be the sequences in $D$ converging to $x$ and $y$ respectively. Since $d_1(x_n, y_n) \le d_1(x_n, x) + d_1(x, y) + d_1(y_1, y_n)$ we can choose $k \in N$ such that $d_1(x_n, x) < \delta/3$ and $d_1(y_1, y_n) < \delta/3$ for all $n > k$. Then $d_1(x_n, y_n) < \delta$ giving $d_2(f(x_n), f(x_m)) \le \varepsilon$. Taking limit, as $n \to \infty$, we can assure, by the continuity of $d_2$, that $d_2(\hat{f}(x), \hat{f}(y)) \le \varepsilon$ giving the uniform continuity of $\hat{f}$ on $\overline{D}$.

To show that $\hat{f}$ is unique, as an extension, if possible, let there be another extension of $f$ on $D$, say, $\hat{g}$ on $\overline{D}$ which is uniformly continuous on $\overline{D}$. Obviously, for any $x \in D$, $\hat{f}(x) = \hat{g}(x) = f(x)$; hence, if $\hat{f}$ and $\hat{g}$ are different, there must be same point $x \in \overline{D} \setminus D$, such that $\hat{f}(x) \neq \hat{g}(x)$.

Let $(x_n)$ be a sequence in $D$ converging to $x$. By the continuity of $\hat{f}$, $\hat{f}(x_n)$ converges to $\hat{f}(x)$ and since $\hat{f}(x_n) = \hat{g}(x_n)$ for every $n$, $\hat{g}(x_n)$ also converges to $\hat{f}(x)$; but, by the continuity of $\hat{g}$, $\hat{g}(x_n)$ converges to $\hat{g}(x)$. [Theorem 49, Chapter 2]. But since $(X_2, d_2)$ is a Hausdorff space, $\hat{g}(x_n)$ cannot converge to two limits; hence $\hat{f}(x) = \hat{g}(x)$—a contradiction.

Thus the extension $\hat{f}$ of $f$ is unique.

**Note 31 :** (*i*) It follows from the proof above that if $f : (X_1, d_1) \to (X_2, d_2)$ be uniformly continuous and if $(x_n)$ is a Cauchy sequence in $(X_1, d_1)$, then $(f(x_n))$ is a Cauchy sequence in $(X_2, d_2)$;

(*ii*) If $D \subset X$, $f : (D, T_D) \to (Y, T')$, where $T$ is the topology on $X$, be continuous and if $\overline{D} = X$, then continuous extension of $f$ to $X$, if there be any, is unique when $Y$ is Hausdorff.

**Theorem 19 :** *Let $f : D \to f(D)$ be an isometry where $D \subset X_1$, $f(D) \subset X_2$, $(X_1, d_1)$ and $(X_2, d_2)$ are complete metric spaces, $\overline{D} = X_1$ and $\overline{f(D)} = X_2$. Then $f$ has a unique extension $\hat{f} : X_1 \to X_2$ such that $\hat{f}$ establishes an isometric correspondence between $X_1$ and $X_2$.*

**Proof :** Since $f$ is an isometry, $f : D \to X_2$ is uniformly continuous as well. Let $y \in X_2$; since $\overline{f(D)} = X_2$, there exists $(x_n)$ in $D$ such that $f(x_n)$ converges to $y$; by the property of isometry, $(x_n)$ is Cauchy in $X_1$ and by completeness of $X_1$ converges to some $x \in X_1$. As in Theorem 18, since the extension $\hat{f}$ of $f$ is unique $\hat{f}(x) = y$ and thus $\hat{f}$ is a surjection on $X_2$.

For any two points $x$, $y$ in $X_1$, there exist sequences $(x_n)$ and $(y_n)$ in $D$ such that $x_n \to x$ and $y_n \to y$. Since $\overline{D} = X_1$; by the continuity of $d_1$, $\lim_{n \to \infty} d_1(x_n, y_n) = d_1(x, y)$; but $d_1(x_n, y_n) = d_2(f(x_n), f(y_n))$ and $\lim_{n \to \infty} d_1(x_n, y_n) = d_1(x, y) = \lim_{n \to \infty} d_2(f(x_n), f(y_n)) = d_2(\hat{f}(x), \hat{f}(y))$ by the continuity of $d_2$ and the definition of $\hat{f}(x)$ and $\hat{f}(y)$. Hence $\hat{f}$ is an isometric correspondence between $X_1$ and $X_2$.

**Theorem 20 :** *If $(X, d)$ be an incomplete metric space; if possible, let there be two completions $(X^*, d^*)$ and $(X^{**}, d^{**})$ of $(X, d)$. Then, there exists an isometric correspondence between $X^*$ and $X^{**}$.*

**Proof :** By Theorem 15, we can draw the diagram, where $\overline{X_0^*} = X^*$, $\overline{X_0^{**}} = X^{**}$, $\phi_1$ and $\phi_2$ are isometrics; obviously, $f = \phi_{2_0}\ \phi_1^{-1}$ establishes an isometric correspondence between $X_0^*$ and $X_0^{**}$. By Theorem 19 we can find an $\hat{f}$ which establishes an isometric correspondence between $X^*$ and $X^{**}$.

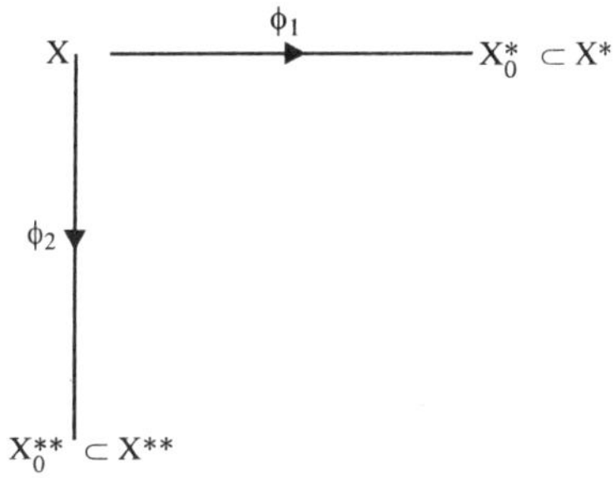


**Note 32 :** Hence, we can say that the completion of an incomplete space is unique upto isometry.

## Art. 10 SOME MORE FACTS ABOUT COMPLETE METRIC SPACES

**Theorem 21 :** *Let $Y$ be a subset of a metric space $(X, d)$. Then, $Y$ is complete if it is closed. If $(X, d)$ is complete, then, if $Y$ is closed, then it is complete.*

**Proof :** Let $Y$ be complete as a subspace of $X$; we show that $Y$ is closed. Let $y$ be a point of accumulation of $Y$; then, for each $n \in N$, $S(y, 1/n)$ contains a point $y_n$ in $Y$. Obviously, for $m > n$, $S(y, 1/m) \subset S(y, 1/n)$ and consequently, $S(y, 1/n)$ contains all but finitely many elements of $y_n$ which is why $(y_n)$ converges to $y$ in $X$; but $(y_n)$ is Cauchy in $Y$ and $Y$ being complete, it converges in $Y$ as well. Thus $y \in Y$ and $Y$ is closed.

Conversely, let $Y$ be closed and let $(y_n)$ be a Cauchy sequence in $Y$. It is also a Cauchy sequence in $X$ and so, converges to a limit $y \in X$. We show that $y$ is in $Y$. If $(y_n)$ has only finitely many distinct points, then $y$ is that point which is infinitely repeated and thus is in $Y$. On the other hand, if $\{y_n\}$ has infinitely many distinct points, then $y$ is obviously a limit point of the set of points of the sequence; it is therefore also a limit point of $Y$; $y \in Y$ since $Y$ is closed. Thus, $Y$ is complete.

**Theorem 22 :** *A metric space $(X, d)$ is complete iff every decreasing countable sequence $\{F_i \because i \in N\}$ of non-empty closed sets in $(X, d)$, for* *which diam $F_i \leq \varepsilon_i$ with $\varepsilon_i \to 0$ as $i \in \infty$, has a non-empty intersection consisting of exactly one point.*

**Proof :** At the very outset, we observe that the assumption diam $F_i \leq \varepsilon_i$ with $\varepsilon_i \to 0$ implies that $F = \cap \{F_i : i \in \mathbf{N}\}$ cannot contain more than one point; thus, it suffices to show that $F$ is non-empty. Let $x_n \in F_n$; since diam $(F_n) \to 0$, $(x_n)$ is a Cauchy sequence : in fact $d(x_m, x_n) \leq$ diam $F_m$ if $n > m$ and thus $d(x_m, x_n) \to 0$ as $n, m \to \infty$. Since $X$ is complete $(x_n)$ converges to a point $x$ in $X$. We claim that $x \in F$. If $\{x_n\}$ has only finitely many distinct points, then $x$ is that point which is infinitely repeated and thus $x \in F_{n_0}$ for some $n_0$ and $x \in F_n$ for all $n \geq n_0$. Since $\{F_n\}$ is a decreasing sequence $x \in F$. It $(x_n)$ has infinitely many distinct points, then $x$ is a limit point of the set of points of the sequence and also a limit point of the subset $P = \underset{n \geq n_0}{\cup} \{x_n\}$ of the set of points of the sequence for any $n_0 \in N$ and thus, a fortiori, is a limit point of $F_{n_0}$; since $F_{n_0}$ is closed, $x \in F_{n_0}$. Thus $x \in F$.

To prove the converse we shall show that if the space $(X, d)$ is not complete *i.e.*, if there exists a Cauchy sequence in $X$ which does not have a limit, then it is possible to construct a decreasing sequence $\{F_n\}$ of closed sets such that diam $(F_n) \to 0$ as $n \to \infty$ but $\overset{\infty}{\underset{n=1}{\cap}} F_n$ is empty.

For this let $\bar{S}(x, y) = \{y \in X : d(x, y) \leq r\}$; let $n_1$ be such that $d(x_{n_1}, x_m) < 1/2$ for all $m > n_1$; let $S_1 = \bar{S}(x_{n_1}, 1)$; further, let $n_2 > n_1$ be such that $d(x_{n_2}, x_m) < \frac{1}{4}$ for all $m > n_2$. Let $S_2 = \bar{S}(x_{n_2}, 1/2)$. If $z \in S_2$, then $d(x_{n_1}, z) \leq d(x_{n_1}, x_{n_2}) + d(x_{n_2}, z) < \frac{1}{2} + \frac{1}{2} = 1$ and thus $z \in S_1$; hence $S_2 \subset S_1$. Now let $n_3 > n_2$ be such that $d(x_{n_3}, x_m) < \frac{1}{8}$ for all $m > n_3$ and let $S_3 = \bar{S}(x_{n_3}, 1/4)$, and so forth. If we continue this process of construction we obtain a decreasing sequence of closed sets $\{S_n\}$ where $S_n$ has radius $(1/2)^{n-1}$ *i.e.*, diam $S_n \to 0$ as $n \to \infty$.

We claim that this sequence of closed sets has a void intersection: in fact, if possible, let $x \in \underset{k}{\cap} S_k$. Now $S_k$ contains all points $x_n$ beginning with $x_{n_k}$ and consequently $d(x, x_n) < (1/2)^{k-1}$ for all $n > n_k$. But, by assumption, the sequence $(x_n)$ does not have a limit. Therefore $\bigcap_{n=1}^{\infty} S_n = \phi$.

**Theorem 23 :** *Let $(X, d)$ be a complete metric space; then $X$ cannot be expressed as a countable union $\bigcup_{n=1}^{\infty} A_n$ where each $A_n$ is nowhere dense.*

**Proof :** We start with $A_1$ which is nowhere dense. By Theorem 33, Chapter 2, there is an open sphere $S_1$ of radius less than 1 such that $S_1 \cap A_1 = \phi$.

In this Theorem we repeatedly apply Theorem 33 of Chapter 2. Let $F_1$ be a closed sphere, concentric with $S_1$, whose radius is one-half of that of $S_1$ and consider its interior; since $A_2$ is nowhere dense, Int $(F_1)$ contains an open sphere $S_2$ of radius less than $1/2$ which is disjoint with $A_2$. Let $F_2$ be a closed sphere, concentric with $S_2$, whose radius is one-half of that of $S_2$. Obviously, Int $(F_2)$ contains an open sphere $S_3$ of radius less than $\frac{1}{2^2}$ which is disjoint from $A_3$ which is nowhere dense. We proceed in the same fashion only to get a decreasing sequence $\{F_n\}$ of closed sets such that diam $(F_n) \to 0$ as $n \to \infty$. By Theorem 22, $\bigcap_{n=1}^{\infty} F_n$ consists of a single point '$x$' since $X$ is complete. But $x \notin A_n$ for any $n$ whatsoever. Hence $X \neq \bigcup_{n=1}^{\infty} A_n$.

**Corollary 2 :** *Any complete metric space is a set of the second category.*

## Art. 11 CONNECTEDNESS IN METRIC SPACES

**Definition 10 :** *A set $K$ is said to be non-degenerate iff $K$ contains more than one point.*

**Theorem 24 :** *Let $A$ be a non-degenerate connected subset of a metric space $(X, d)$. Then $A$ is uncountable.*

**Proof :** Let $a$ and $b$ be two distinct points of $A$.

Let $\left.\begin{array}{l} f: A \in R \\ x \to d(x, a) \end{array}\right\}$. Now that $f$ is a continuous function and the continuous image of a connected set is connected, we have $f(A)$ to be connected in $R$. Obviously, $f(a)=0$ and $f(b)=d(a, b)=c$ (say) both belonging to $R$. Obviously, $f(A)$ must contain the interval $(0, c)$ of $\mathbf{R}$. Obviously $(0, c)$ is uncountable and hence, so is $A$.

**Definition 11 :** *Let $(X, d)$ be a metric space; '$d$' is called an M-metric if $S_d(p, r)$ for every $p \in X$ and every $r > 0$ is a connected set.*

**Theorem 25 :** *Let $(X, T)$ be a metrizable space such that $T = T(d)$ where $d$ is an M-metric. If $(Y, T')$ be homeomorphic to $(X, T)$, then the topology $T' = T'(e)$ where $e$ is also an M-metric.*

**Proof :** Proof is easy and is left as an exercise.

**Theorem 26 :** *Let $(X, T)$ be a metrizable space. Then, there is an M-metric for $X$ giving the topology $T$ iff $(X, T)$ is connected and locally connected.*

**Proof :** Let '$d$' be an $M$-metric on $X$ such that $T(d) = T$. By definition of $M$-metric, $X$ is locally connected.

Now, let $x_0 \in X$; obviously, $X = \bigcup_{n \in N} S(x_0, n)$. Thus $X$ is the union of a collection of connected sets each of which contains $x_0$. By Theorem 5, Chapter 7, $X$ is connected; conversely, let $X$ be a metrizable connected and locally connected space. By Note 9, the topology of $X$ is given by a $B$-metric $e$. Let $d : X \times X \to \mathbf{R}$ :

$$d(x, y) = g.l.b.\ \text{diam}_e(K) \text{ for all connected subsets } K \subset X$$
$$\text{suchthat } K \text{ contains both } x \text{ and } y;$$

we show that $d$ is the desired $M$-metric giving the topology of $X$.

We first observe that for any two points $x$ and $y$ there is at least one connected set, bounded with respect to $e$ containing both $x$ and $y$. Thus $d$ is well-defined for any two points $x$ and $y$ in $X$ and obviously $d(x, y) \geq 0$; also, $d(x, y) = d(y, x)$ by definition. We also note that, for any connected set $K$ containing $x$ and $y$, $\text{diam}_e(K) \geq e(x, y)$. Consequently,

$$d(x, y) \geq e(x, y) \qquad \ldots(1)$$

by definition. Hence $d(x, y) = 0 \Rightarrow e(x, y) = 0 \Rightarrow x = y$. Also, if $x = y$, $d(x, y) = 0$ by definition

For triangle inequality, let $x, y, z \in X$ and let $\varepsilon > 0$. By definition of $g.l.b.$ and further, by definition of '$d$' there exist connected subsets $K_1$ and $K_2$ of $X$ such that $K_1$ contains $\{x, y\}$, $K_2$ contains $\{y, z\}$ and $\text{diam}_e(K_1) \leq d(x, y) + \varepsilon$ and diam $(K_2) \leq d(y, z) + \varepsilon$. Hence

$$d(x, z) \leq \text{diam}_e(K_1 \cup K_2) \leq \text{diam}_e(K_1) + \text{diam}_e(K_2) + d(K_1, K_2)$$

[by Note 11]

But $K_1 \cap K_2 \neq \phi \Rightarrow d(K_1, K_2) = 0$ and thus $d(x, z) \leq \text{diam}_e (K_1) + \text{diam}_e (K_2) \leq d(x, y) + d(y, z) + 2\varepsilon$; since this holds for every $\varepsilon > 0$, it follows that $d(x, z) \leq d(x, y) + d(y, z)$. Thus '$d$' is a metric on $X$.

Now we show that $T(d) = T(e)$; we have already observed in (1) that $d(x, y) \geq e(x, y)$; consequently, for any $r > 0$,

$$S_d(x, r) \subset S_e(x, r) \qquad \text{...(2)}$$

for any $x \in X$. On the other hand, let $S_d(x, r)$ be an arbitrary open sphere with centre $x$. Since $(X, T(e))$ is locally connected we have a connected open set $V$ (belonging to $T(e)$) containing $x$ and contained in $S_e(x, r/3)$. Thus $\text{diam}_e(V) < r$ giving

$$V \subset S_d(x, r) \qquad \text{...(3)}$$

In fact, if $y \in V$, both of $x$ and $y$ are contained in the connected set $V$ and by definition $d(x, y) \leq \text{diam}_e(V) < r$.

By Equations (2) and (3), applying Theorem 4, Chapter 2, we can say that $T(d) = T(e)$.

The only thing left is now to show that '$d$' is an $M$-metric on $X$. Let $S_d(x, r)$ be an arbitrary open sphere containing $x$. Let $d(x, y) = \alpha$ where $y \in S_d(x, r)$. Then $d(x, y) = \alpha < r$; it follows that there exists a connected subset $K(y)$ of $X$ containing $x$ and $y$ and of diameter (w.r.to $e$) less than $\frac{(\alpha + r)}{2}$ (by definition of $d(x, y)$). Then $S_d(x, r) = \bigcup_{y \in S_d(x, r)} K(y)$ : in fact, if $z \in K(y)$ both $x$ and $z$ are contained in the connected set $K(y)$ and by definition $d(x, z) \leq \text{diam}_e K(y) < \frac{\alpha + r}{2} < r$ *i.e.*, $z \in S_d(x, r)$. Now that for every $y \in S_d(x, r)$, $x \in K(y)$ we have $S_d(x, r)$ is connected establishing the fact that $d$ is an $M$-metric on $X$.

## Art. 12 MISCELLANEOUS EXERCISES

**Example 1.** ***Let $X = N \cup \{b\}$ where $N$ is the set of natural numbers and $b$ is an object not belonging to $N$.***

Define
$$d(x, y) = 1 \text{ if } x, y \in \mathbf{N},\ x \neq y$$
$$d(b, x) = d(x, b) = 1 + \frac{1}{x} \text{ if } x \in N$$
$$d(x, y) = 0 \text{ if } x = y$$

show that : (*i*) '$d$' is a metric for $X$;

(*ii*) if $A = \mathbf{N}$, $B = \{b\}$, find $d(A, B)$;

(*iii*) show that $(X, d)$ is a complete metric space;

(*iv*) do there exist a pair of points $a \in A$, $b \in B$ for which
$$d(A, B) = d(a, b)\ ?$$

**Example 2.** Let $(X_1, d_1)$ and $(X_2, d_2)$ be metric spaces consider

$$d[(x, y), (w, z)] = \sqrt{d_1^2(x, w) + d_2^2(y, z)}$$

where $(x, y)$ and $(w, z)$ both belong to $X_1 \times X_2$;

(*i*) is $d$ a metric for $X_1 \times X_2$ ?

(*ii*) if $T(d_1) \times T(d_2)$ denotes the product topology or $X_1 \times X_2$, is it true that $T(d_1) \times T(d_2) = T(d)$ ?

(*iii*) if $(X_1, d_1)$ and $(X_2, d_2)$ are both complete, is it true that $(X_1 \times X_2, d)$ is complete as well?

**Example 3.** ***Show that every metric space is completely normal.***

**Example 4.** ***Let $(X_1, d_1)$ and $(X_2, d_2)$ be two metric spaces. Let $f: (X_1, d_1) \to (X_2, d_2)$; following Definition 9 (ii), show that f is continuous iff, given any real number $\varepsilon > 0$ and any point $x_0$ of $X_1$, there exists a real number $\delta > 0$ such that, if $x \in X_1$ and $d_1(x, x_0) < \delta$ then***

$$d_2(f(x), f(x_0)) < \varepsilon.$$

**Example 5.** ***Let $(X_1, d_1)$ and $(X_2, d_2)$ be two metric spaces. Let $f: (X_1, d_1) \to (X_2, d_2)$. Following Definition 9 (ii), show that f is continuous iff, given any real number $\varepsilon > 0$ and any point $x_0 \in X_1$, there exists a real number $\alpha > 0$ such that, if U is any nbd of $x_0$ with diam (U) $< \alpha$, then diam f (U) $< \varepsilon$.***

**Example 6.** ***Let $(X, d)$ be a metric space and let $p \in X$. Prove that there exists a monotone decreasing sequence $\{U_n\}$ of nbds of p such that***

***$U_n \supset \overline{U}_{n+1}$ for every n, and $\{p\} = \bigcap_{n=1}^{\infty} \overline{U}_n = \cap U_n$.***

**Example 7.** ***Let K be a subset of a metric space $(X, d)$ and let $r > 0$. Define $S_r(K)$ to consist of all points $x \in S$ such that $d(x, y) < r$ for at least one point y of K.***

(*a*) Prove that $S_r(K)$ is open.

(*b*) If $K_1$ and $K_2$ are disjoint closed subsets of $X$ and $K_1$ is countably compact, prove that there exists a real number $r > 0$, such that $S_r(K_1)$ and $S_r(K_2)$ are disjoint.

**Example 8.** *Let $(X, T)$ be a topological space, $(Y, d)$ be a metric space and $g : A \to Y$ be a mapping where $A \subset X$. Let $A \subset B \subset X$. Suppose that $f : B \to Y$ is an extension of the mapping from $A$ to $B$. Then*

(*a*) The mapping $g$ is continuous at a point $p$ of $A$ iff the oscillation of $g$, at the point $p$, defined as *g.l.b.* diam $(g(U \cap A))$ is zero. [$\mathbf{N}_p$, as usual, denotes the *nbd* filter of the point $p$ and $U \in \mathbf{N}_p$

(*b*) If $w_f(p)$ and $w_g(p)$ denote, respectively, the oscillations of $f$ and $g$ at a point $p$ of $X$, then show that $w_f(p) \geq w_g(p)$.

(*c*) If there exists a point $p \in B$ for which $w_g(p) > 0$, then show that there exists no continuous extension of $g$ from $A$ to $B$.

**Example 9.** *Let $(X_1, d_1)$ and $(X_2, d_2)$ be metric spaces, $A \subset X$, and $f : A \to Y$ a mapping. If $w_f(p) = 0$, and $\{x_n\}$ and $\{y_n\}$ are sequence of points of A, each converging to p, then show that $\{f(x_n)\}$ and $\{f(y_n)\}$ are equivalent Cauchy sequences [as in Note 26].*

**Example 10.** *Let $\{x_n\}$ and $\{y_n\}$ be two Cauchy sequences of points of a metric space $(X, d)$. Prove that, $\{x_n\}$ and $\{y_n\}$ are equivalent Cauchy sequences iff the sequence $\{z_n\}$ of points of X, where $z_{2n} = x_n$ for every n and $z_{2n-1} = y_n$ for every n, is a Cauchy sequence.*

**Example 11.** *Let $C[0, 1]$ denote the set of all continuous mappings $f : [0, 1] \to \mathbf{R}$. Show that the mapping $d : C[0, 1] \times C[0, 1] \to R$ defined by $d(f, g) = \underset{x \in [0, 1]}{l.u.b.} |f(x) - g(x)|$ is a metric for the set $C[0, 1]$.*

*Further show that $C[0, 1]$ is complete.*

**Example 12.** *Let $(X, T)$ be a topological space and $(Y, d)$ a metric space. A sequence of mappings $\{f_n\}$ of X into Y is said to converge uniformly to a mapping $f : X \to Y$ iff, given any $\varepsilon > 0$, there exists an integer N such that $n > N$ implies $d(f_n(x), f(x)) < \varepsilon$ for all $x \to X$. If $\{f_n\}$ is a sequence of continuous mappings of X into Y which converge uniformly to a mapping fof X into Y, then show that f is continuous.*

**Example 13.** *Let $(X_1, d_1)$ and $(X_2, d_2)$ be metric space; let $K \subset X_1$. If $f : K_1 \to X_2$ is uniformly continuous, then show that the oscillation $w(p)$ of f is zero at every point p of $X_1$.*

**Solution :** For each point $p$ of $K$ we have $w(p) = 0$ since $f$ is continuous at $p$ (Example 8 (*a*)). For each point $p \in X_1 - \overline{K}$, we must have $0 \le w(p) \le \text{diam}\,[f(X_1 - \overline{K}) \cap K)] = 0$, since $X_1 - \overline{K}$ is a *nbd* of $p$ having no intersection with $K$.

It remains to consider the case $p \in \overline{K} - K$. Let $\varepsilon > 0$ be preassigned. By uniform continuity of $f$, there exists $\delta > 0$ such that whenever $d_1(r, q) < \delta$ we have

$$d_2(f(r), f(q)) < \varepsilon/2; \qquad ...(1)$$

we claim that $$0 \le w(p) \le \text{diam}\,[f(S(p, \delta/2) \cap K)] < \varepsilon \qquad ...(2)$$

In fact, let $q, r \in S(p, \delta/2) \cap K$. Then $d_1(q, r) \le d_1(q, p) + d_1(p, r) < \delta/2 + \delta/2 = \delta$; by (1), $d_2(f(q), f(r)) < \varepsilon/2$ giving

$$\text{diam}\,[f(S(p, \delta/2) \cap K)] \le \varepsilon/2 < \varepsilon.$$

Thus, from (1) $0 \le w(p) < \varepsilon$ for every $\varepsilon > 0$; it follows that $w(p) = 0$ for every $p \in \overline{K} - K$. But $X_1 = (X_1 - \overline{K}) \cup (\overline{K} - K) \cup K$. Thus $w(p) = 0$ for every point $p$ of $X_1$.

**Example 14.** ***Use Note 31 to prove the following :***

***Let*** $\left.\begin{array}{l} f : (0, 1] \to R \\ x \to \dfrac{1}{x} \end{array}\right\}$ ***with the usual metric in (0, 1]; show that f is not uniformly continuous.***

**Example 15.** ***Let*** $T_n$ ***be the terminal sequence*** $\{i \in N\}\, i \ge n\}$, ***(X, d) be a metric space and let*** $\left.\begin{array}{l} \phi : N \to X \\ \phi(n) = y_n \end{array}\right\}$ ***be a sequence. Show that*** $\phi$ ***is Cauchy iff for all*** $\varepsilon > 0$, ***there exists*** $n \in N$ ***such that diam*** $[\phi(T_n)] < \varepsilon$.

**Example 16.** ***Let (X, d) be a metric space and let*** $\phi$ ***be a sequence as in Example 15;*** $\phi$ ***is said to accumulate at*** $y_0$ ***iff*** $y_0 \in \bigcap_1^\infty \overline{\phi(T_n)}$; ***if*** $\phi$ ***accumulate at*** $y_0$, ***show that*** $\phi$ ***converges to*** $y_0$.

**Example 17.** ***Let (X, d) be a metric space; let, for each*** $x \in X$, ***there exist an*** $\varepsilon > 0$ ***such that*** $\overline{S(x, \varepsilon)}$ ***is compact; show that (X, d) is complete.***

**Solution :** Let $\phi$ be a Cauchy sequence in $(X, d)$. By Example 15, corresponding to $\varepsilon > 0$, then exists $n \in N$, sufficiently large such that diam

$[\phi(T_n)] < \varepsilon/2$; then $\phi(T_n) \subset \overline{S[\phi(n), \varepsilon]}$; but $\overline{S[\phi(n), \varepsilon]}$ being compact, $\phi(T_n)$ must accumulate at some $x$ [Verify]. By Example 16, it follows that $\phi \to x$.

**Example 18.** *Let $(X^*, d^*)$ be a completion of a metric space $(X, d)$. If $(X, d)$ is totally bounded, show that, so is $(X^*, d^*)$.*

**Example 19.** *If we replace the axiom '1' in the definition 1 as follows: $d(x, y) \geq 0$ and is equal to zero when $x = y$, then $d$ is called a pseudo-metric. If $(X, d)$ be a pseudo-metric space, $A \neq \phi$ be closed, $B \neq \phi$ be countably compact, then show that $d(A, B) > 0$ if $A \cap B = \phi$.*

**Example 20.** *Let $(X, d)$ be a pseudo-metric space and let $x \in X$. Show that $\{\overline{x}\} = \{y : d(x, y) = 0\}$.*

**Example 21.** *For every pseudo-metric space $(X, d)$, let $Y = \{\{\overline{x}\} : x \in X\}$; define $e(\{\overline{x}\}, \{\overline{y}\}) = d(x, y)$, show that : (i) $(Y, e)$ is a metric space; (ii) the function $\left.\begin{array}{c} f : x \to \{\overline{x}\} \\ (X, d) \to (Y, e) \end{array}\right\}$ is a distance preserving surjection.*

**Example 22.** *Let $(X, d)$ be a pseudometric space. Show that every closed set in $(X, T(d))$ is $G_\delta$.*

**Solution :** Let $S_A(\varepsilon) = \cup \{S(x, \varepsilon) : x \in A\}$ where $A$ is a closed set in $X$; obviously, $S_A(\varepsilon)$ is an open set containing $A$. We claim that

$$A = \bigcap_{n \in N} S_A(1/n);$$

the claim is justified and $A$ becomes a $G_\delta$-set.

**Example 23.** *A topological space $(X, T)$ is called perfectly normal iff it is normal and every closed set in $(X, T)$ is $G_\delta$. Show that every metric space $(X, T)$ is perfectly normal.*

**Example 24.** *Let $f : (X, d) \to (Y, e)$ be a continuous function and let $(X, d)$ be a countably compact space. Show that $f$ is uniformly continuous.*

**Example 25.** *Let $l_2$ denote the set of all real-valued sequences $\{x_n\}$ such that $\sum x_n^2$ is convergent. If $x = \{x_n\}$, $y = \{y_n\}$ then, show that $d(x, y) = \left(\sum_{n=1}^{\infty} |x_n - y_n|^2\right)^{1/2}$ always exists and is a metric in $l_2$. Show that $(l_2, d)$ is complete and separable as well.*

**Example 26.** ***Show that the derived set of a countably compact set in a metric space is countably compact.***

**Example 27.** ***Let (X, d) and (Y, e) be two metric space and let*** $f: X \to Y$ ***be uniformly continuous. Let*** $A, B \subset X$ ***be such that*** $d(A, B) = 0$***; show that*** $e(f(A), f(B)) = 0$.

**Example 28.** ***Let*** $X = \{x, y, z, w\}$***; let*** $D: X \times X \to R$ ***be defined as follows :*** $d(x, y) = d(x, z) = d(y, z) = z$; $d(x, w) = d(y, w) = d(z, w) = 1$. ***Show that (X, d) is a metric space with some added conditions.***

**Example 29.** ***Let C, as usual denote the set of complex numbers. Define*** $d(x, y) = |x| + |y|$ ***if*** $\arg x \neq \arg y$, $d(x, y) = |x - y|$ ***if*** $\arg x = \arg y$ ***for any two points x and y in C. Show that (C, d) is a metric space; is the metric space (C, d) separable? Justify your answer.***

**Example 30.** ***Show that a quotient of a locally connected space is locally connected.***

**Solution :** We apply Theorem 12 of Chapter 7 for the solution *i.e.*, for any open set $V$ in $Y$ where $f: X \to Y$ is a quotient map we show that the component $C$ of $V$ is open.

$f(X) = Y$ since $f$ is a surjection; it will be enough to show that $f^{-1}(C)$ is open in $X$ since, in that case $C$ will be open in $Y$ by the definition of quotient map.

Let $x \in f^{-1}(C)$ : then $x \in f^{-1}(V) = U$ where $U$ is open in $X$ by the continuity of $f$. Since $X$ is locally connected, the component $K$ of $U$ containing $x$ is an open set. $f$ being continuous, $f(K)$ is connected in $Y$ and $f(x) \in f(K) \subset U$. Since $f(x) \in C$, $f(K) \cap C \neq \phi$. Hence $f(K) \cup C$ is connected in $Y$ and contained in $V$; since $C$ is a component in $V$, $f(K) \subset C$. Hence $x \in K \subset f^{-1}(C)$ *i.e.*, $f^{-1}(C)$ is an open set in $X$ and we are done.

**Example 31.** ***Show that a Hausdorff space X, which is the image of a closed interval [a, b] under a continuous mapping, is connected and locally connected.***

**Solution :** Since an interval is connected so is $X$ which is its continuous image. Further, since $f$ is continuous, for any closed subset $K$ contained in $[a, b]$, which is compact as well, so is $f(K)$ in $X$ which is Hausdorff. Thus $f(K)$ is closed in $X$ and $f$ is a closed mapping. By Theorem 67, Chapter 2, $f$ is a quotient mapping. By Example 30, $X$ is locally connected since $[a, b]$ is so.

**Example 32.** ***Let X be a metrizable space. Show that X is connected and locally connected iff X has a metric which is both a B-metric and a M-metric.***

*Chapter 9*

# HOMOTOPY AND FUNDAMENTAL GROUPS

## INTRODUCTION

Homotopy Theory plays an important role in topology. For any topological space $X$ and any point $x_0 \in X$, we define a group $\pi_1(X, x_0)$, consisting of homotopy classes of closed paths called loops in $X$ based at $x_0$, which is called the *fundamental group* of $X$ based at $x_0$. This group is an algebraic invariant in the sense that the fundamental groups $\pi_1(X, x_0)$ and $\pi_1(Y, y_0)$ of two homeomorphic ($\approx$) pointed topological spaces $(X, x_0)$ and $(Y, y_0)$ are isomorphic. But an isomorphism between two fundamental groups $\pi_1(X, x_0)$ and $\pi_1(Y, y_0)$ of two pointed-topological spaces $(X, x_0)$ and $(Y, y_0)$ does not in general imply that the pointed spaces $(X, x_0)$ and $(Y, y_0)$ are homeomorphic. By using the fundamental groups, topological problems about topological spaces and continuous functions may be sometimes reduced to algebraic problems about groups and homomorphisms for solutions.

## Art. 1 HOMOTOPY : INTRODUCTORY CONCEPTS

Let $\mathbf{I} = [0, 1]$ be the closed unit interval with the topology induced by usual topology on the real line.

**Definition 1 :** *A path $f$ in a topological space $X$ is a continuous map $f : \mathbf{I} \to X$. $f(0)$ is called the initial point and $f(1)$ is called the terminal point of the path $f$.*

**Definition 2 :** *A path $f : \mathbf{I} \to X$ is called a loop in $X$ based at $x_0 \in X$ if $f(0) = f(1) = x_0$. In particular, the constant map $c : \mathbf{I} \to x_0 \in X$ is called a constant path or null path in $X$ at $x_0$.*

**Definition 3 :** *Let $f$ and $g$ be two paths in $X$ such that*

$$f(0) = g(0) = x_0 \in X \text{ and } f(1) = g(1) = x_1 \in X$$

*i.e., $f$ and $g$ are two paths in $X$ having the same initial point $x_0$ and the same terminal point $x_1$. Then $f$ is said to be path homotopic to $g$ written*

*$f \underset{p}{\simeq} g$ if there exists a continuous function $F: \boldsymbol{I} \times \boldsymbol{I} \to X$ such that $F(t, 0) = f(t)$, $F(t, 1) = g(t)$, $\forall\, t \in \boldsymbol{I}$; $F(0, s) = x_0$ and $F(1, s) = x_1$, $\forall\, s \in \boldsymbol{I}$.*

If $f$ and $g$ are paths homotopic under $F$, we represent it by $F: f \underset{p}{\simeq} g$ and $F$ is called a path homotopy between $f$ and $g$.

$F$ represents a continuous deformation from the path $f$ to the path $g$. Moreover, for each $s \in \mathbf{I}$, the path $t \to F(t, s)$ is a path in $X$ from $x_0$ to $x_1$ by keeping the end points of the paths $f$ and $g$ fixed during the deformation. Thus intuitively, we say that two paths having the same initial point and the same terminal point are homotopic if one can be continuously deformed into another by keeping the end points of the paths fixed.

We now state and prove 'Pasting Lemma' which will be used in our subsequent development.

**Lemma 1 :** (Pasting Lemma) : Let $X$ be a topological space and $A$, $B$ be closed subsets of $X$ such that $X = A \cup B$. Let $Y$ be a topological space and $f: A \to Y$ and- $g: B \to Y$ be continuous. If $f(x) = g(x)$, $\forall\, x \in A \cap B$, then the function $h: X \to Y$ defined by

$$h(x) = \begin{cases} f(x), & \forall\, x \in A \\ g(x), & \forall\, x \in B \end{cases}$$

is continuous.

**Proof :** Let $C$ be a closed subset of $Y$. Then

$$h^{-1}(C) = f^{-1}(C) \cup g^{-1}(C)$$

(see Theorem 1.17 of Book [1]).

$f$ is continuous $\Rightarrow f^{-1}(C)$ is closed in $A \Rightarrow f^{-1}(C)$ is closed in $X$. Similarly, $g$ is continuous $\Rightarrow g^{-1}(C)$ is closed in $B \Rightarrow g^{-1}(C)$ is closed in $X$.

Hence $h^{-1}(C)$ is closed in $X$. Consequently, $h$ is continuous.

**Theorem 1 :** *Let $P(X)$ be the set of all paths in $X$ having the same end points $x_0$ and $x_1 \in X$. Then the path homotopy relation $\underset{p}{\simeq}$ is an equivalence relation on $P(X)$.*

**Proof :** Let $f, g, h \in P(X)$. Then $f(0) = g(0) = h(0) = x_0$ and

$$f(1) = g(1) = h(1) = x_1.$$

Let a map $F: \mathbf{I} \times \mathbf{I} \to X$ be defined by $F(t, s) = f(t)$, $\forall\, t, s \in \mathbf{I}$.

Then $F$ is continuous because it is the composite of the projection map onto the first factor and the continuous map $f$. Thus $F$ is a continuous map such that $F(t, 0) = f(t)$, $F(t, 1) = f(t)$, $\forall\, t \in \mathbf{I}$ and $F(0, s) = f(0) = x_0$, $F(1, s) = f(1) = x_1$, $\forall\, s \in \mathbf{I}$. Hence $f \underset{p}{\simeq} f$, $\forall\, f \in P(X) \Rightarrow \underset{p}{\simeq}$ is reflexive.

Next, let $f \underset{p}{\simeq} g$ and $F; f \underset{p}{\simeq} g$. Then $F : \mathbf{I} \times \mathbf{I} \to X$ is a continuous map such that $F(t, 0) = f(t)$, $F(t, 1) = g(t)$, $\forall\, t \in I$ and $F(0, s) = x_0$, $F(1, s) = x_1$, $\forall\, s \in \mathbf{I}$.

Let $G : \mathbf{I} \times \mathbf{I} \to X$ be the map defined by $G(t, s) = F(t, 1 - s)$. Since the maps $\mathbf{I} \to \mathbf{I}$, $s \to 1 - s$ and $t \to t$ are both continuous, $G$ is continuous.

Now $G(t, 0) = F(t, 1) = g(t)$, $G(t, 1) = F(t, 0) = f(t)$, $\forall\, t \in I$ and $G(0, s) = F(0, 1 - s) = x_0$, $G(1, s) = F(1, 1 - s) = x_1$.

Hence $G : g \underset{p}{\simeq} f \Rightarrow \underset{p}{\simeq}$ is symmetric.

Finally, let $f \underset{p}{\simeq} g$ and $g \underset{p}{\simeq} h$. Then there exist continuous maps $F$, $G : \mathbf{I} \times \mathbf{I} \to X$ such that $F : f \underset{p}{\simeq} g$ and $G : g \underset{p}{\simeq} h$. Consequently, $F(t, 0) = f(t)$, $F(t, 1) = g(t)$, $F(0, s) = x_0$, $F(1, s) = x_1$, $G(t, 0) = g(t)$, $G(t, 1) = h(t)$, $G(0, s) = x_0$ and $G(1, s) = x_1$.

We construct a map $H : \mathbf{I} \times \mathbf{I} \to X$ by the rule

$$H(t, s) = \begin{cases} F(t, 2s), & 0 \le s \le 1/2 \\ G(t, 2s - 1), & 1/2 \le s \le 1 \end{cases}$$

At $s = \frac{1}{2}$, $F(t, 2s) = F(t, 1) = g(t)$ and $G(t, 2s - 1) = G(t, 0) = g(t)$, $\forall\, t \in I$.

Hence, by Pasting Lemma, $H$ is continuous.

Moreover, $H(t, 0) = F(t, 0) = f(t)$

$$H(t, 1) = G(t, 1) = h(t)\ \forall\, t \in I,$$

$$H(0, s) = \begin{cases} F(0, 2s), & 0 \le s \le 1/2 \\ G(0, 2s - 1), & 1/2 \le s \le 1 \end{cases}$$

$$= x_0, \ \forall\, s \in \mathbf{I}$$

and

$$H(1, s) = \begin{cases} F(1, 2s), & 0 \le s \le 1/2 \\ G(1, 2s - 1), & 1/2 \le s \le 1 \end{cases}$$

$= x_1, \ \forall \ s \in I$

Hence $H; f \underset{p}{\simeq} h \Rightarrow \underset{p}{\simeq}$ is transitive.

Consequently, $\underset{p}{\simeq}$ is an equivalence relation on $P(X)$.

**Corollary 1 :** *Let $\Omega(X, x_0)$ be the set of all loops in X based at $x_0$. Then the path homotopy relation $\underset{p}{\simeq}$ on $\Omega(X, x_0)$ is an equivalence relation.*

**Definition 4 :** *The quotient set $\Omega(X, x_0)/\underset{p}{\simeq}$ denoted by $\pi_1(X, x_0)$ consists of homotopy classes of loops in X based at $x_0$ and admits a group structure (see Theorem 6) and is called the fundamental group of X based at $x_0$.*

In Section 2, we shall study the fundamental group $\pi_1(X, x_0)$.

Instead of studying path homotopy, we now consider homotopy of arbitrary continuous maps.

**Definition 5 :** *Let X and Y be two topological spaces and $f, g :$ $X \to Y$ be two continuous maps. Then f is said to be homotopic to g denoted by $f \simeq g$ if there exists a continuous map $F : X \times \mathbf{I} \to Y$ such that*

$$F(x, 0) = f(x) \text{ and } F(x, 1) = g(x), \ \forall \ x \in X$$

We write $F : f \simeq g$ to represent a homotopy from $f$ to $g$.

Thus two continuous maps $f, g : X \to Y$ are said to be homotopic if there exists a continuous family of maps $F_t : X \to Y, \ \forall \ t \in \mathbf{I}$ such that $F_0 = f, \ F_1 = g$, where $F_t : X \to Y$ is defined by the rule

$$F_t(x) = F(x, t), \ \forall \ x \in X \text{ and } \forall \ t \in \mathbf{I}.$$

By saying that the maps $F_t$ form a continuous family of maps, we mean that $F$ is continuous with respect to both $x$ and $t$ *i.e.*, $F$ is continuous as a function of the product space $X \times \mathbf{I}$ to $Y$.

## Interpretation of $F_t$

Consider a continuous map $F_t : X \to Y$ such that

$$F_0(x) = f(x), \ F_1(x) = g(x), \ \forall \ x \in X$$

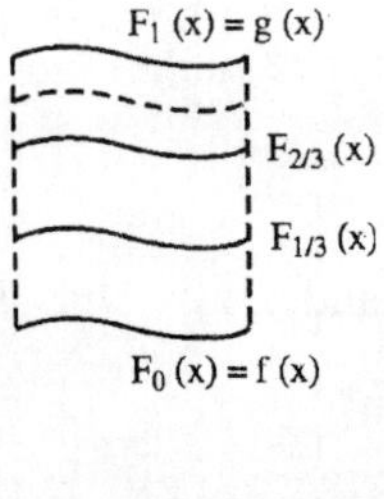


Then $F_t : f \simeq g$.

Think of parameter $t$ as representing time, then at $t = 0$, the map $F_0$ coincides with $f$. As $t$ varies from 0 to 1, the map $F_t : X \to Y$ varies continuously so that at time $t = 1$, $F_1$ coincides with $g$ as shown in the diagram. For this reason, the homotopy $F : f \simeq g$ is often called a *continuous deformation* of the continuous map $f$ into the continuous map $g$ and $f$ is said to be continuously deformed into $g$.

**Definition 6 :** *A topological space X is said to be contractible if the identity map $1_X : X \to X$ is homotopic to a constant map.*

**Example 1.** (*i*) **I** and the real line **R** are contractible spaces.

(*ii*) Let $X = Y = \mathbf{R}^n$ and $f, g : X \to Y$ be defined by $f(x) = x$ and $g(x) = 0 = (0, 0, ..., 0) \in \mathbf{R}^n$. Then $f$ is the identity map $1_X$ and $g$ is the constant map of $\mathbf{R}^n$ to its origin 0.

Then $F : \mathbf{R}^n \times \mathbf{I} \to \mathbf{R}^n$ defined by $F(x, t) = (1 - t)x$ is a continuous map such that $F : f \simeq g$.

Hence $\mathbf{R}^n$ is a contractible space.

**Remark :** *Contractible spaces are characterised by the Theorem 5, page 229.*

(*iii*) Let $X$ be a topological space and $Y$ be a subspace of $\mathbf{R}^n$ and let $f, g : X \to Y$ be two continuous maps. If for each $x \in X$, $f(x)$ and $g(x)$ can be joined by a line segment in $Y$, then $f \simeq g$.

Define a map $F : X \times \mathbf{I} \to Y$ by the rule

$$F(x, t) = (1 - t) f(x) + t g(x), \ \forall\, x \in X \text{ and } \forall\, t \in \mathbf{I}$$

Since $f(x)$ and $g(x)$ can be joined by a line segment in $Y$ by hypothesis, $F(x, t) \in Y$, $\forall\, x \in X$ and $\forall\, t \in \mathbf{I}$ and hence $F$ is well defined.

To show the continuity $F$, take $x, u \in X$ and $t, s \in \mathbf{I}$.

Then $F(u, s) = (1 - s)\, f(u) + s\ g(u)$.

Now $F(u, s) - F(x, t) = (s - t)\ (g(u) - f(u)) + (1 - t)\ (f(u) - f(x))$
$+ t\,(g(u) - g(x))$.

Let $\in\, > 0$ be an arbitrary small positive number. Then

$$\|F(u, s) - F(x, t)\| \le |(s - t)|\ \|g(u) - f(u)\| + |(1 - t)|\ \|f(u) - f(x)\| + |t|\ \|g(u) - g(x)\| \quad ...(1)$$

$f$ and $g$ being continuous, there exist open neighbourhood $U_1$ and $U_2$ of $x$ in $X$ such that

$$u \in U_1 \Rightarrow \|f(u) - f(x)\| < \in/3,$$

$$u \in U_2 \Rightarrow \|g(u) - g(x)\| < \in/3$$

Thus if $u \in U_1 \cap U_2$, then

$$\|g(u) - f(u)\| \le \|g(u) - g(x)\| + \|g(x) - f(x)\| + \|f(x) - f(u)\| < c,$$

where $c$ is the positive constant $\in/3 + \|g(x) - f(x)\| + \in/3$.

Thus if $|s - t| < \in/3c$, then from (1) it follows that

$$\|F(u, s) - F(x, t)\| < \in/3c \cdot c + \in/3 + \in/3 = \in \qquad ...(2)$$

Since the set $(U_1 \cap U_2) \times (t - \in/3c,\ t + \in/3c)$ is open in $X \times \mathbf{I}$, $F$ is continuous.

Finally, $F(x, 0) = f(x)$ and $F(x, 1) = g(x)$, $\forall\ x \in X$.

Hence $F : f \simeq g$.

(*iv*) Let $f, g : X \to \mathbf{R}^n$ be two continuous maps.

Define $F : X \times \mathbf{I} \to \mathbf{R}^n$ by $F(x, t) = (1 - t) f(x) + t g(x)$

Then $F : f \simeq g$.

$F$ is called a *straight line homotopy*, because $F$ shifts the point $f(x)$ to the point $g(x)$ along the straight-line segment joining $f(x)$ and $g(x)$ for every $x \in X$.

(*v*) Let $X = \mathbf{R}^2 - 0$, be the punctured Euclidean plane, where $0 = (0, 0) \in \mathbf{R}^2$. Consider the paths $f, g, h : \mathbf{I} \to X$ defined by

$$f(t) = (\cos \pi t, \sin \pi t);\ g(t) = (\cos \pi t, 2 \sin \pi t)$$

and $$h(t) = (\cos \pi t, -\sin \pi t).$$

Then $f \underset{p}{\simeq} g$ but $f \not\simeq h$ under a straight line homotopy.

Construct a map $F : \mathbf{I} \times \mathbf{I} \to X$ by

$$F(t, s) = (1 - s) f(t) + s g(t)$$

Then $F$ is well defined and continuous and is such that

$$F(t, 0) = f(t),\ F(t, 1) = g(t),\ \forall\ t \in \mathbf{I}$$

Hence $F : f \underset{p}{\simeq} g$.

To the contrary, at $t = \frac{1}{2}$, $f(t) = f\left(\frac{1}{2}\right) = (0, 1)$ and

$$h(t) = h\left(\frac{1}{2}\right) = (0, -1)$$

and hence the line segment joining $f\left(\frac{1}{2}\right)$ and $h\left(\frac{1}{2}\right)$ must pass through the point $(0, 0) \notin X$. In other words, there exists no straight line homotopy between $f$ and $h$.

Intutively, we cannot deform the path $f$ to the path $h$ because of the presence of the hole at $(0, 0) \in \mathbf{R}^2$.

(*vi*) Let $S^n$ be the unit $n$-sphere defined by $S^n = \{x \in \mathbf{R}^{n+1} : \|x\| = 1\}$ and $X$ be any topological space. If $f, g : X \to S^n$ are two continuous maps such that $f(x) \neq -g(x)$ for any $x \in X$, then $f \simeq g$.

By hypothesis, $f(x) \neq -g(x)$ for any $x \in X \Rightarrow f(x)$ and $g(x)$ are not diametrically opposite points of $S^n$ and hence the line segment joining the points $f(x)$ and $g(x)$ cannot pass through the origin

$$0 = (0, 0, ..., 0) \in \mathbf{R}^{n+1}.$$

Thus we may consider the maps $f, g$ as $f, g : X \to \mathbf{R}^{n+1} - 0$.

Then by (*iii*), there exists a homotopy $F : X \times \mathbf{I} \to \mathbf{R}^{n+1} - 0$ defined by $F(x, t) = (1-t) f(x) + t g(x)$, $\forall x \in X$, $\forall t \in \mathbf{I}$. Clearly, $F : f \simeq g$.

Again consider the map $\psi : \mathbf{R}^{n+1} - 0 \to S^n$ defined by $\psi(x) = \frac{x}{\|x\|}\cdot$ Consider the composite map $G : X \times \mathbf{I} \xrightarrow{F} \mathbf{R}^{n+1} - 0 \xrightarrow{\psi} S^n$ *i.e.*,

$$G = \psi \circ F : X \times \mathbf{I} \to S^n$$

defined by $$G(x, t) = \frac{(1-t) f(x) + t g(x)}{\|(1-t) f(x) + t g(x)\|}$$

Then $G$ is a well defined and continuous map such that

$$G(x, 0) = \frac{f(x)}{\|f(x)\|} = f(x), \text{ since } f(x) \in S^n;$$

$$G(x, 1) = \frac{g(x)}{\|g(x)\|} = g(x), \text{ since } g(x) \in S^n$$

Thus $G : f \simeq g$.

(*vii*) If $f : X \to S^n$ is a continuous non-surjective maps, then $f$ is homotopic to a constant map from $X$ to $S^n$. $f$ is non-surjective $\Rightarrow f(X) \neq S^n$

$\Rightarrow \exists$ a point $s_0 \in S^n$ such that $s_0 \notin f(X)$. Define $g : X \to S^n$ by $g(x) = -s_0$.

Apply the result of (*vi*) to show that $f \simeq g$.

We now extend the Definitions 3 and 5 by defining homotopy of maps between pairs of spaces.

A *topological pair* $(X, A)$ consists of a topological space $X$ and a subspace $A$ of $X$. If $A$ is a empty set denoted by $A = \phi$, we shall not distinguish between the pair $(X, \phi)$ and the space $X$. Given a pair $(X, A)$ we denote by $(X, A) \times \mathbf{I}$ the pair $(X \times \mathbf{I}, A \times \mathbf{I})$.

A continuous map $f : (X, A) \to (Y, B)$ between pairs is a continuous function $f : X \to Y$ such that $f(A) \subset B$.

**Definition 7 :** *Given pairs $(X, A)$ and $(Y, B)$, two continuous maps $f, g : (X, A) \to (Y, B)$ such that $f|_A = g|_A$ are said to be homotopic relative to $A$ written as $f \simeq g$ rel $A$, if there exists a continuous map*

$$F : (X \times \mathbf{I}, A \times \mathbf{I}) \to (Y, B)$$

*such that $F(x, 0) = f(x)$, $F(x, 1) = g(x)$, $\forall x \in X$ and $F(a, t) = f(a) = g(a)$, $\forall a \in A$ and $\forall t \in \mathbf{I}$.*

If $F$ is a homotopy from $f$ to $g$ relative to $A$, then we write

$$F : f \simeq g \text{ rel } A.$$

**Example 2.** (*i*) Let $X = Y = \mathbf{R}^n$ and $f, g : X \to Y$ be defined by $f(x) = x$ and $g(x) = 0$, $\forall x \in \mathbf{R}^n$. Then $f \simeq g$ rel 0.

Define $F : X \times \mathbf{I} \to Y$ by $F(x, t) = (1 - t)x$.

Then $F(x, 0) = x = f(x)$; $F(x, 1) = 0 = g(x)$, $\forall x \in X$.

and $F(0, t) = 0 = f(0) = g(0)$, $\forall t \in \mathbf{I}$

Hence $F : f \simeq g$ rel 0.

(*ii*) Let $X$ be any topological space and $Y$ be a convex subspace of $\mathbf{R}^n$. If $f, g : X \to Y$ are two continuous maps such that $f | A = g | A$ for any subspace $A$ of $X$, then $f \simeq g$ rel $A$.

Define $F : X \times \mathbf{I} \to Y$ by

$$F(x, t) = (1 - t) f(x) + t g(x), \ \forall x \in X, \ \forall t \in \mathbf{I}$$

Then $F(x, 0) = f(x)$, $F(x, 1) = g(x)$, $\forall x \in X$ and

$F(a, t) = (1 - t) f(a) + t g(a) = f(a) = g(a)$, $\forall a \in A$ and $\forall t \in \mathbf{I}$,

since $f | A = g | A$. Hence $F : f \simeq g$ rel $A$.

**Theorem 2 :** *Let $X$ and $Y$ be topological spaces and $C(X, Y)$ be the set of all continuous maps from $X$ to $Y$. Then the homotopy relation $\simeq$ is an equivalence relation on $C(X, Y)$.*

**Proof :** Let $f \in C(X, Y)$. Define $H : X \times \mathbf{I} \to Y$ by the rule

$$H(x, t) = f(x), \ \forall\, x \in X \text{ and } \forall\, t \in I.$$

Then $\quad H(x, 0) = f(x), \ H(x, 1) = f(x), \ \forall\, x \in X$

Hence $\quad H : f \simeq f, \ \forall\, f \in C(X, Y) \Rightarrow \simeq$ is reflexive.

Let $f, g \in C(X, Y)$ be such that $f \simeq g$.

Then there exists a continuous map $H : X \times \mathbf{I} \to Y$ such that $H : f \simeq g$. Now $H(x, 0) = f(x)$ and $H(x, 1) = g(x), \ \forall\, x \in X$.

Define $G : X \times \mathbf{I} \to Y$ by the rule

$$G(x, t) = H(x, 1 - t), \ \forall\, x \in X \text{ and } \forall\, t \in \mathbf{I}$$

Then $\quad G(x, 0) = H(x, 1) = g(x), \ \forall\, x \in X$

and $\quad G(x, 1) = H(x, 0) = f(x), \ \forall\, x \in X$

Hence $\quad G : g \simeq h \Rightarrow \simeq$ is symmetric.

Finally, let $f, g, h \in C(X, Y)$ be such that $f \simeq g$ and $g \simeq h$. Then there exist continuous maps $F, G : X \times \mathbf{I} \to Y$ such that

$$F : f \simeq g \text{ and } G : g \simeq h$$

Thus $F(x, 0) = f(x)$, $F(x, 1) = g(x)$, $G(x, 0) = g(x)$ and $G(x, 1) = h(x), \ \forall\, x \in X$.

Define $H : X \times \mathbf{I} \to Y$ by the rule

$$H(x, t) = \begin{cases} F(x, 2t), & 0 \le t \le 1/2 \\ G(x, 2t - 1), & 1/2 \le t \le 1 \end{cases}$$

Since $H$ is continuous on each of the closed subsets $X \times \left[0, \frac{1}{2}\right]$ and $X \times \left[\frac{1}{2}, 1\right]$, by Pasting Lemma $H$ is continuous on $X \times \mathbf{I}$.

Now, $\quad H(x, 0) = F(x, 0) = f(x), \ \forall\, x \in X$

and $\quad H(x, 1) = G(x, 1) = h(x), \ \forall\, x \in X$

Thus $\quad H : f \simeq h \Rightarrow \simeq$ is transitive

Consequently $\simeq$ is an equivalence relation on $C(X, Y)$.

**Definition 8 :** *The equivalence classes of $C(X, Y)$ under the homotopy relation $\simeq$ are called the homotopy classes of $C(X, Y)$ denoted by $[X, Y]$ and the equivalence class of $f \in C(X, Y)$ is denoted by $[f]$.*

**Theorem 3 :** *Composites of homotopic maps are homotopic.*

**Proof : Case 1 :** Let $f, g : X \to Y$ and $h : Y \to Z$ be continuous maps such that $f \simeq g$. We prove that $hof \simeq hog$. Now, $f \simeq g \Rightarrow$ there exists a continuous map $F : X \times \mathbf{I} \to Y$ such that $F : f \simeq g$. Then $F(x, 0) = f(x)$ and $F(x, 1) = g(x)$, $\forall\, x \in X$.

Consider the composite map $X \times \mathbf{I} \xrightarrow{F} Y \xrightarrow{h} Z$ determined by the rule $(hoF)(x, t) = h(F(x, t))$, $\forall\, x \in X$, $\forall\, t \in \mathbf{I}$.

Then $hoF$ is a continuous map such that

$$(hoF)(x, 0) = h(F(x, 0)) = h(f(x)) = (hof)(x)$$

and $$(hoF)(x, 1) = h(F(x, 1)) = h(g(x)) = (hog)(x), \ \forall\, x \in X$$

$\Rightarrow$ $$hoF : hof \simeq hog.$$

**Case 2 :** Let $f, g : X \to Y$ and $h : Z \to X$ be continuous maps such that $f \simeq g$. We prove that $foh \simeq goh$.

Consider the composite map $G$:

$$Z \times I \xrightarrow{h \times \text{identity } id} X \times I \xrightarrow{F} Y$$

defined by

$$G(z, t) = (F\, o\, (h \times 1d))(z, t) = F(h(z), t), \ \forall\, z \in Z \text{ and } \forall\, t \in I.$$

Then $G$ is a continuous map such that

$$G(z, 0) = F(h(z), 0) = f(h(z)) = (foh)(z)$$

and $$G(z, 1) = F(h(z), 1) = g(h(z)) = (goh)(z), \ \forall\, z \in Z$$

$\Rightarrow$ $$G : foh \simeq goh.$$

We now utilize the Theorem 3 to prove the following Theorem :

**Theorem 4 :** *A topological space X is contractible iff for any topological space Y, any two continuous maps $f, g : Y \to X$ are homotopic.*

**Proof :** Taking in particular, $Y = X$, $f = 1_X$ and $g = c : X \to x_0 \in X$, we find that $X$ is contractible.

Conversely, let $X$ be contractible. Then $1_X \simeq c$, where $c$ is a constant map from $X$ to itself. Let $f, g : Y \to X$ be two arbitrary continuous maps. Then $1_X \simeq c \Rightarrow f = 1_X of \simeq cof$ and $g = 1_X og \simeq cog$ by Theorem 3.

Hence $cof = cog \Rightarrow f \simeq g$.

**Definition 9 :** *A map $f \in C(X, Y)$ is said to be a homotopy equivalence if there exists a map $g \in C(Y, X)$ such that $gof \simeq 1_X$ and $fog = 1_Y$. The map g is called the homotopy inverse of f.*

**Definition 10 :** *Two topological spaces X and Y are said to be homotopy equivalent if there exists a homotopy equivalence $f \in C(X, Y)$ i.e., X and*

*$Y$ are said to be homotopy equivalent denoted by $X \simeq Y$ if there exist continuous maps $f: X \to Y$ and $g: Y \to X$ such that $gof \simeq 1_X$ and $fog \simeq 1_Y$.*

**Remark :** A space $X$ is said to be of the same homotopy type as $Y$ if $X \simeq Y$.

Homeomorphic spaces are always homotopically equivalent. Its converse is not true as the following Theorem shows (also see W.O. Example 2).

**Theorem 5 :** *A topological space X is contractible iff X is homotopically equivalent to a one-point space.*

**Proof :** Let $X$ be contractible. Then $1_X : X \to X$ is homotopic to the constant map $c(x) = x_0 \in X$. Let $Y = \{x_0\}$ and $j : Y \to X$ be the inclusion map. Then $coj = 1_Y$ (identity map on $Y$) and $joc = c$ is homotopic to $1_X$. Thus $coj = 1_Y$ and $joc \simeq 1_X \Rightarrow j : Y \to X$ is a homotopy equivalence

$$\Rightarrow \qquad X \simeq Y = \{x_0\}$$

Conversely, let $X \simeq Y = \{x_0\}$. Then there exists a homotopy equivalence $f: X \to Y$ with homotopy inverse $g: Y \to X$. Then $gof \simeq 1_X$ and $fog \simeq 1_Y$. Now $gof: X \to X$ is a constant map $c$ (say) such that $c \simeq 1_X$. Then $1_X \simeq c \Rightarrow X$ is contractible.

We now study inclusion maps. Let $X$ be a topological space and $A$ be a subspace of $X$ and $i: A \to X$ be the inclusion map defined by $i(a) = a, \ \forall \ a \in A$.

**Definition 11 :** *A subspace A of X is called a retract of X if there exists a continuous map $r: X \to A$ such that $roi = 1_A$ i.e., $r \mid A = 1_A$; r is called a retraction.*

**Definition 12 :** *A subspace A of X is called a weak retract of X if there exists a continuous map $r: X \to A$ such that $roi \simeq 1_A$. If $F: roi \simeq 1_A$, then F is called a weak retraction.*

**Definition 13 :** *A subspace A of X is called a weak deformation retract of X if the inclusion map $i: A \to X$ is a homotopy equivalence.*

**Definition 14 :** *A subspace A of X is called a strong deformation retract of X if there exists a retraction $r: X \to A$ such that $1_X \simeq ior$ rel A. If $F: 1_X \simeq ior$ rel A, then F is called a strong deformation retraction of X to A.*

**Definition 15 :** *A subspace A of X is called a deformation retract of X if there exists a retraction $r: X \to A$ such that $1_X \simeq ior$.*

If $F: 1_X \simeq ior$, then $F$ is called a *deformation retraction* of $X$ to $A$.

## EXAMPLES

1. Let $X$ and $Y$ be two homeomorphic spaces. Then $X$ and $Y$ are necessarily homotopy equivalent spaces.

**Solution :** $X \approx Y \Rightarrow \exists$ a homeomorphism $f: X \to Y$ with a continuous map $g: Y \to X$ such that $gof = 1_X$ and $fog = 1_Y$. Thus $gof = 1_X \simeq 1_X \Rightarrow gof \simeq 1_X$ and $fog = 1_Y \simeq 1_Y \Rightarrow fog \simeq 1_Y$.

Consequently, $X \simeq Y$. Thus $X \approx Y \Rightarrow X \simeq Y$.

The following example 2 shows that $X \simeq Y \not\Rightarrow X \approx Y$.

2. Let $X$ be the unit circle $S^1$ in $\mathbf{R}^2$ and $Y$ be the unit circle $S^1$, together with the line segment $I_1$ joining the point (1, 0) and (2, 0) *i.e.*, $I_1 = \{(r, 0) : r \in \mathbf{R}$ and $1 \le r \le 2\}$. Then $X \simeq Y$ but $X \not\approx Y$.

**Solution :** The removal of the point (1, 0) from $Y$ makes $Y$ disconnected but the removal of any point from $X$ leaves $X$ connected. Hence $X \not\approx Y$.

To show that $X \simeq Y$, we define $f: X \to Y$ by $f(x) = x$, $\forall\, x \in X$ and $g: Y \to X$ by

$$g(y) = \begin{cases} y, & \forall\, y \in S^1 = X \\ (1, 0), & \forall\, y \in I_1 \end{cases}$$

Then $g$ is a continuous map by Pasting Lemma.

Now $fog$, $1_Y: Y \to Y$ be two continuous maps such that

$$(fog)(y) = f(g(y)) = g(y), \quad \forall\, y \in Y$$

and
$$1_Y(y) = \begin{cases} y, & \forall\, y \in X \\ (r, 0), & \text{if } (r, 0) \in I_1 \end{cases}$$

Then $f$, $g$, $1_X$ and $1_Y$ are continuous maps such that $gof = 1_X$ and also $fog \simeq 1_Y$ by Example 1 (*iii*). In other words $X \simeq Y$. Thus $X \not\approx Y$ but $X \simeq Y$.

3. Let $A$ be a subspace of a topological space $X$. If $A$ is a deformation retract of $X$, then $A \simeq X$.

**Solution :** Let $i: A \to X$ be the inclusion map and $A$ be a deformation retract of $X$. Then there exists a retraction $r: X \to A$ such that $ior \simeq 1_X$. Again $r: X \to A$ is a retraction $\Rightarrow roi = 1_A \simeq 1_A$. Hence $ior \simeq 1_X$ and $roi \simeq 1_A \Rightarrow A \simeq X$.

**4.** The subspace $X$ of the space $Y$ defined in Example 2 is a strong deformation retract of $Y$.

The map $g: Y \to X$ defined in Example 2 is a retraction such that $iog \simeq 1_Y \operatorname{rel} X$, where $i: X \to Y$ is the inclusion map.

Hence $S^1 = X$ is a strong deformation retract of $Y$.

## EXERCISES

**1.** Let $[A, X]$ be the set of all homotopy classes of maps from a fixed space $A$ to an arbitrary space $X$.

Then a continuous map $f: X \to Y$ induces a function :

$f_*: [A, X] \to [A, Y]$ with the following properties :

(*i*) For $f, g \in C(X, Y)$, $f \simeq g \Rightarrow f_* = g_*: [A, X] \to [A, Y]$.

(*ii*) $1_X: X \to X$ is the identity map $\Rightarrow 1_X^*: [A, X] \to [A, X]$ is the identity function.

(*iii*) $f \in C(X, Y)$, $g \in C(Y, Z) \Rightarrow (gof)_* = g_* o f_*: [A, X] \to [A, Z]$.

(*iv*) $X \approx Y \Rightarrow$ the sets $[A, X]$ and $[A, Y]$ are equipotent.

(*v*) $X \simeq Y \Rightarrow$ the sets $[A, X]$ and $[A, Y]$ are equipotent.

Determine the corresponding results (called dual results) for the set $[X, A]$ when $X$ varies and $A$ is kept fixed.

[**Hint :** Define a function $f_*: [A, X] \to [A, Y]$ by the rule

$$f_*([\alpha]) = [f o \alpha].$$

To show that $f_*$ is well defined, take $\beta \in [\alpha]$. Then $\alpha \simeq \beta$. Hence $f o \alpha \simeq f o \beta$ by Theorem 3 $\Rightarrow f_*([\alpha]) = f_*([\beta]) \Rightarrow f_*$ is well defined.

Verify the properties (*i*)-(*v*).

For dual results, define $f^*: [Y, A] \to [X, A]$ by the rule

$$f^*([\alpha]) = [\alpha o f].$$

Verify that $f^*$ is well defined and then proceed].

**2.** Any two continuous maps of an arbitrary space to a contractible space are homotopic.

**3.** Constant maps $f, g: X \to Y$ for arbitrary spaces $X$ and $Y$ need not be homotopic.

**4.** Any continuous map homotopic to a homotopy equivalence is a homotopy equivalence.

**5.** Two contractible spaces have the same homotopy type and any continuous map between contractible spaces is a homotopy equivalence.

**6.** A space $X$ is contractible iff $X$ is deformable into one of its points.

**7.** A subspace $A$ of $X$ is a deformation retract of $X$ if there is a homotopy $F : X \times I \to X$ such that $F(x, 0) = x$ and $F(x, 1) \in A$, $\forall x \in X$.

**8.** A subspace $A$ of $X$ is a strong deformation retract of $X$ iff there is a homotopy $F : X \times \mathbf{I} \to X$ such that $F(x, 0) = x$, $\forall x \in X$, $F(x, 1) \in A$ $\forall x \in X$, $F(x, 1) = x$, $\forall x \in A$ and $F(a, t) = a$, $\forall a \in A$ and $\forall t \in \mathbf{I}$.

**9.** If $A$ is a strong deformation retract of $X$, then $A$ is a deformation retract of $X$.

**10.** Consider $D^n = \{x \in \mathbf{R}^n : \|x\| \leq 1\}$. Then $D^n$ is a retract of $\mathbf{R}^n$.

[**Hint :** Define $r : \mathbf{R}^n \to D^n$ by $r(x) = \begin{cases} x/\|x\|, & \text{if } \|x\| > 1 \\ x, & \text{other} \end{cases}$]

**11.** $S^n = \{x \in \mathbf{R}^n : \|x\| = 1\}$ is a strong deformation retract of punctured Euclidean space $\mathbf{R}^{n+1} - 0$.

[**Hint :** Define $H : (\mathbf{R}^{n+1} - 0) \times \mathbf{I} \to \mathbf{R}^{n+1} - 0$ by

$$H(x, t) = (1 - t)\, x + \frac{tx}{\|x\|}, \quad \forall\, x \in \mathbf{R}^{n+1} - 0, \ \forall\, t \in \mathbf{I}.$$

Then $H$ is a strong deformation retraction].

**12.** Let $D^2 = \{z \in \mathbf{C} : z = re^{i\theta},\ r \in \mathbf{R} \text{ and } 0 \leq r \leq 1\}$ and

$$S^1 = \{z \in \mathbf{C} : z = e^{i\theta}\}.$$

Define $f : (D^2, S^1) \to (D^2, S^1)$ to be the identity map and

$$g : (D^2, S^1) \to (D^2, S^1)$$

to be the reflection map about the origin defined by

$$g(re^{i\theta}) = re^{i(\theta + \pi)}$$

Then $$f \simeq g \text{ rel } 0.$$

**13.** The relation of being homotopy equivalent on any set of topological spaces is an equivalence relation.

**14.** The relation of being homotopy equivalence on any set of topological spaces is an equivalence relation.

**15.** Every contractible space is pathwise connected.

**16.** Every retract of a contractible space is contractible.

**17.** A continuous map $f: S^n \to X$ can be continuously extendable over $D^{n+1}$ iff $f$ is homotopic to a constant map.

**18.** Any continuous map from $S^n$ to a contractible space has a continuous extension over $D^{n+1}$.

## Art. 2 THE FUNDAMENTAL GROUP

In this section we study an algebraic invariant of a topological space $X$, namely $\pi_1(X, x_0)$ defined in Definition 4. Recall that $\pi_1(X, x_0)$ is the set of homotopy classes of continuous maps $f: (\mathbf{I}, \dot{\mathbf{I}}) \to (X, x_0)$ where $\dot{\mathbf{I}} = \{0, 1\}$. We prove in Theorem 6 that $\pi_1(X, x_0)$ admits the structure of a group in a natural way and this group is homotopy type invariant in the sense that the for homotopy equivalent spaces $(X, x_0)$ and $(Y, y_0)$, the groups $\pi_1(X, x_0)$ and $\pi_1(Y, y_0)$ are isomorphic. We develop sufficient machinery to prove the Brouwer Fixed-point Theorem (Theorem 12).

**Definition 16 :** *Let $u, v: I \to X$ be two loops in $X$ based at $x_0$. Then $u$ and $v$ are said to be homotopic relative to $\dot{I} = \{0, 1\}$ denoted by $u \simeq v$ rel $(0, 1)$ if there exists a continuous map $F: I \times I \to X$ such that $F(t, 0) = u(t)$; $F(t, 1) = v(t)$, $\forall\, t \in I$ and $F(0, s) = f(1, s) = x_0$, $\forall\, s \in I$.*

**Remark :** For loops $u$ and $v$ in $X$ based at $x_0$, $u \simeq v$ rel $(0, 1) \Rightarrow$ the continuous maps $u, v: (\mathbf{I}, \dot{\mathbf{I}}) \to (X, x_0)$ are homotopic relative to the subspace $\dot{\mathbf{I}}$ of $\mathbf{I}$.

**Definition 17 :** *Given loops $u, v: I \to X$ based at $x_0$, their product $u * v: (I, \dot{I}) \to (X, x_0)$ is defined by*

$$(u * v)(t) = \begin{cases} u(2t), & 0 \le t \le 1/2 \\ v(2t-1), & 1/2 \le t \le 1 \end{cases}$$

*Then $u * v$ is well defined. Moreover $u * v$ is continuous by Pasting Lemma.*

Finally, $(u * v)(0) = u(0) = x_0 = v(0) = (u * v)(1) \Rightarrow u * v$ is a loop in $X$ based at $x_0$.

We now extend this definition for the product of three loops.

**Definition 18 :** *Given loops $u, v, w : \mathbf{I} \to X$ in X based at $x_0$, their product $u * v * w : \mathbf{I} \to X$ is defined by*

$$(u * v * w)(t) = \begin{cases} u(3t), & 0 \le t \le 1/3 \\ v(3t-1), & 1/3 \le t \le 2/3 \\ w(3t-2), & 2/3 \le t \le 1 \end{cases}$$

Then $u * v * w$ is a loop in $X$ based at $x_0$.

**Definition 19 :** *The loop $c : \mathbf{I} \to X$ defined by $c(t) = x_0 \in X$, $\forall\, t \in \mathbf{I}$ is called a null loop or constant loop in X based at $x_0$.*

**Definition 20 :** *Let $u : \mathbf{I} \to X$ be a loop in X based at $x_0$. Then its inverse loop $u^{-1} : \mathbf{I} \to X$ defined by $u^{-1}(t) = u(1-t)$, $\forall\, t \in \mathbf{I}$ is also a loop in X based at $x_0$.*

**Remark :** The loops $u$ and $u^{-1}$ give the same set of points of $X$ but their directions are opposite.

**Proposition 1 :** *If $u_1$, $u_2$ and $v_1$, $v_2$ are loops in X based at $x_0$ such that $u_1 \simeq u_2$ rel (0, 1) and $v_1 \simeq v_2$ rel (0, 1), then*

$$u_1 * v_1 \simeq u_2 * v_2 \text{ rel } (0,\ 1).$$

**Proof :** $u_1 \simeq u_2$ rel (0, 1) $\Rightarrow$ there exists a homotopy

$$F : u_1 \simeq u_2 \text{ rel } (0,\ 1).$$

Then $\quad F(t, 0) = u_1(t),\ F(t, 1) = u_2(t),\ \forall\, t \in \mathbf{I}$

and $\quad F(0, s) = F(1, s) = x_0,\ \forall\, s \in \mathbf{I}.$

Similarly, $v_1 \simeq v_2$ rel (0, 1) $\Rightarrow$ there exists a homotopy

$$G : v_1 \simeq v_2 \text{ rel } (0,\ 1).$$

Then $\quad G(t, 0) = v_1(t),\ G(t, 1) = v_2(t)$

and $\quad G(0, s) = G(1, s) = x_0.$

Define a mapping $H : \mathbf{I} \times \mathbf{I} \to X$ by

$$H(t, s) = \begin{cases} F(2t, s), & 0 \le t \le 1/2 \\ G(2t-1, s), & 1/2 \le t \le 1 \end{cases}$$

Then $H$ is well defined. Moreover, $H$ is continuous by Pasting Lemma.

Now $\quad H(t, 0) = \begin{cases} F(2t, 0), & 0 \le t \le 1/2 \\ G(2t-1, 0), & 1/2 \le t \le 1 \end{cases}$

$$= \begin{cases} u_1(2t), & 0 \le t \le 1/2 \\ v_1(2t-1), & 1/2 \le t \le 1 \end{cases}$$

$$= (u_1 * v_1)(t), \ \forall\, t \in \mathbf{I}$$

Similarly, $H(t, 1) = (u_2 * v_2)(t), \ \forall\, t \in \mathbf{I}$

Moreover, $H(0, s) = F(0, s) = x_0, \ \forall\, s \in \mathbf{I}$

and $H(1, s) = G(1, s) = x_0, \ \forall\, s \in \mathbf{I}$

Hence $H : u_1 * v_1 \simeq u_2 * v_2 \text{ rel } (0, 1)$.

**Proposition 2 :** *If $u, v : \boldsymbol{I} \to X$ are loops in $X$ based at $x_0$ such that $u \simeq v$ rel $(0, 1)$, then their inverse loops $u^{-1}$, $v^{-1} : \boldsymbol{I} \to X$ are such that $u^{-1} \simeq v^{-1}$ rel $(0, 1)$.*

**Proof :** $u \simeq v$ rel $(0, 1) \Rightarrow$ there exists a homotopy $F : \mathbf{I} \times \mathbf{I} \to X$ such that

$$F(t, 0) = u(t), \ F(t, 1) = v(t), \ \forall\, t \in \mathbf{I}$$

and $F(0, s) = F(1, s) = x_0$

Define $F^{-1} : \mathbf{I} \times \mathbf{I} \to X$ by

$$F^{-1}(t, s) = F(1 - t, s).$$

Then $F^{-1}$ is well defined and continuous.

Moreover, $F^{-1}(t, 0) = F(1-t, 0) = u(1-t) = u^{-1}(t)$,

$$F^{-1}(t, 1) = F(1-t, 1) = v(1-t) = v^{-1}(t), \ \forall\, t \in \mathbf{I}$$

and $F^{-1}(0, s) = F(1, s) = x_0, \ F^{-1}(1, s) = F(0, s) = x_0, \ \forall\, s \in \mathbf{I}$

Hence $F^{-1} : u^{-1} \simeq v^{-1} \text{ rel } (0, 1)$.

**Proposition 3 :** *If $u, v, w : \boldsymbol{I} \to X$ are loops in $X$ based at $x_0$, then $u * (v * w) \simeq (u * v) * w$ rel $(0, 1)$.*

**Proof :** $$(u * (v * w))(t) = \begin{cases} u(2t), & 0 \le t \le 1/2 \\ (v * w)(2t-1), & 1/2 \le t \le 1 \end{cases}$$

$$= \begin{cases} u(2t), & 0 \le t \le 1/2 \\ v(4t-2), & 1/2 \le t \le 3/4 \\ w(4t-3), & 3/4 \le t \le 1 \end{cases}$$

Then $u * (v * w)$ is well defined and continuous and a loop in $X$ based at $x_0$.

On the other hand $$((u * v) * w)(t) \begin{cases} (u * v)(2t), & 0 \le t \le 1/2 \\ w(2t-1), & 1/2 \le t \le 1 \end{cases}$$

$$= \begin{cases} u\,(4t), & 0 \le t \le 1/4 \\ v\,(4t-1), & 1/4 \le t \le 1/2 \\ w\,(2t-1), & 1/2 \le t \le 1 \end{cases}$$

Then $(u * v) * w$ is a loop in $X$ based at $x_0$.

Define a map $F : \mathbf{I} \times \mathbf{I} \to X$ by the rule

$$F\,(t,\ s) = \begin{cases} u\left(\dfrac{4t}{1+s}\right), & 0 \le t \le \dfrac{1+s}{4} \\ v\,(4t-1-s), & \dfrac{1+s}{4} \le t \le \dfrac{2+s}{4} \\ w\left(1 - \dfrac{4\,(1-t)}{2-s}\right), & \dfrac{2+s}{4} \le t \le 1 \end{cases}$$

$F$ is well defined and continuous. Moreover,

$$F : (u * v) * w \simeq u * (v * w) \text{ rel } (0,\ 1)$$

**Proposition 4 :** *If* $u : \boldsymbol{I} \to X$ *is a loop is based at* $x_0 \in X$ *and* $c : I \to X$ *is the constant loop at* $x_0$ *defined by* $c\,(t) = x_0,\ \forall\ t \in \boldsymbol{I}$, *then* $u * c \simeq u$ *rel* $(0,\ 1)$.

**Proof :** $u * c : \mathbf{I} \to X$ is defined by

$$(u * c)\ (t) = \begin{cases} u\,(2t), & 0 \le t \le 1/2 \\ c\,(2t-1), & 1/2 \le t \le 1 \end{cases}$$

$$= \begin{cases} u\,(2t), & 0 \le t \le 1/2 \\ x_0, & 1/2 \le t \le 1 \end{cases}$$

Then $u * c$ is a loop in $X$ based at $x_0$.

Define a map $F : \mathbf{I} \times \mathbf{I} \to X$ by

$$F\,(t,\ s) = \begin{cases} u\left(\dfrac{2t}{1+s}\right), & 0 \le t \le \dfrac{1+s}{2} \\ x_0, & \dfrac{1+s}{2} \le t \le 1 \end{cases}$$

Then $\quad F : u * c \simeq u \text{ rel } (0,\ 1)$.

Similarly, $\quad c * u \simeq u \text{ rel } (0,\ 1)$.

**Proposition 5 :** *If $u : \mathbf{I} \to X$ is a loop in $X$ based at $x_0$, then $u * u^{-1} \simeq c$ rel (0, 1) and $u^{-1} * u \simeq c$ rel (0, 1) where $c : \mathbf{I} \to x_0 \in X$ is a constant loop.*

**Proof :** $u * u^{-1} : \mathbf{I} \to X$ is defined by

$$(u * u^{-1})\ (t) = \begin{cases} u\,(2t), & 0 \le t \le 1/2 \\ u^{-1}\,(2t-1), & 1/2 \le t \le 1 \end{cases}$$

$$= \begin{cases} u\,(2t), & 0 \le t \le 1/2 \\ u\,(1-\overline{2t-1}), & 1/2 \le t \le 1 \end{cases}$$

$$= \begin{cases} u\,(2t), & 0 \le t \le 1/2 \\ u\,(2-2t), & 1/2 \le t \le 1 \end{cases}$$

Then $u * u^{-1}$ is a loop in $X$ based at $x_0$.

Define $F : \mathbf{I} \times \mathbf{I} \to X$ by

$$F\,(t,\ s) = \begin{cases} u\,(2t\,(1-s)), & 0 \le t \le 1/2 \\ u\,((2-2t)\ (1-s)), & 1/2 \le t \le 1 \end{cases}$$

Then $F$ is well defined and continuous. Moreover,

$$F\,(t,\ 0) = \begin{cases} u\,(2t), & 0 \le t \le 1/2 \\ u\,(2-2t), & 1/2 \le t \le 1 \end{cases}$$

$$= (u * u^{-1})\ (t),\ \forall\ t \in \mathbf{I},$$

$$F\,(t,\ 1) = u\,(0) = x_0,\ \ \forall\ t \in \mathbf{I},$$

$$F\,(0,\ s) = u\,(0) = x_0 = F\,(1,\ s),\ \ \forall\ s \in \mathbf{I}$$

Hence $\quad F : u * u^{-1} \simeq c$ rel (0, 1)

Similarly, $\quad u^{-1} * u \simeq c$ rel (0, 1).

**Theorem 6 :** *$\pi_1\,(X,\ x_0)$ is a group. (This group is called the fundamental group of $X$ based at $x_0$).*

**Proof :** Let $[u],\ [v] \in \pi_1\,(X,\ x_0)$. Define $u * v$ by Definition 17. This law of composition '$*$' carries over to homotopy classes to give the composition '$o$' by the rule $[u]\ o\ [v] = [u * v]$. The composition '$o$' is independent of the choice of the representatives of the classes by Proposition 1 and hence it is well defined. This composition '$o$' is associative by Proposition 3, $[c]$ is the identity element with respect to this composition by Proposition 4 and any element $[u] \in \pi_1(X, x_0)$ has an inverse $[u^{-1}] \in \pi_1(X, x_0)$ by Proposition 5.

Hence $\pi_1(X, x_0)$ is a group.

**Theorem 7 :** *If X is a path connected space, then for any pair of points $x_0$ and $x_1$ in X, the fundamental groups $\pi_1(X, x_0)$ and $\pi_1(X, x_1)$ are isomorphic.*

**Proof :** As $X$ is path connected and $x_0, x_1 \in X$, then there exists a path $f : \mathbf{I} \to X$ in $X$ from $x_0$ to $x_1$. Then its inverse path $f^{-1} = \bar{f} : \mathbf{I} \to X$ is defined by $\bar{f}(t) = f(1-t)$ and $\bar{f}$ is a path in $X$ from $x_1$ to $x_0$.

For each loop $u$ in $X$ based at $x_0$, we define a loop $\bar{f} * u * f$ in $X$ based at $x_1$. This induces a mapping :

$$\tilde{f} : \pi_1(X, x_0) \to \pi_1(X, x_1) \text{ given by}$$
$$f([u]) = [\bar{f} * u * f], \ \forall \ [u] \in \pi_1(X, x_0).$$

Then $f$ is well defined. We claim that $\tilde{f}$ is a group isomorphism. Now for $[u], [v] \in \pi_1(X, x_0)$,

$$\begin{aligned} \tilde{f}([u] \, o \, [v]) &= \tilde{f}([u * v]) = [\bar{f} * (u * v) * f] \\ &= [\bar{f} * u * (f * \bar{f}) * v * f] \\ &= [(\bar{f} * u * f) * (\bar{f} * v * f)] = [\bar{f} * u * f] \, o \, [\bar{f} * v * f] \\ &= \tilde{f}([u]) \, o \, \tilde{f}([v]) \end{aligned}$$

Hence $\tilde{f}$ is a group homomorphism.

Denote $\bar{f}$ by $g$. Then $\bar{g} = f$ and $g$ induces a homomorphism

$$\tilde{g} : \pi_1(X, x_1) \to \pi_1(X, x_0) \text{ given by}$$
$$\tilde{g}([u]) = [\bar{g} * u * g] = [f * u * \bar{f}], \ \forall \ [u] \in \pi_1(X, x_1)$$

$$\begin{aligned} \text{Now } (\tilde{g} o \tilde{f})([u]) &= \tilde{g}(\tilde{f}([u])) = \tilde{g}([\bar{f} * u * f]) \\ &= [\bar{g} * (\bar{f} * u * f) * g] = [\bar{g} * (\bar{f} * u * f) * g] \\ &= [f * (\bar{f} * u * f) * \bar{f}] \\ &= [(f * \bar{f}) * u * (f * \bar{f})] = [u], \ \forall \ [u] \in \pi_1(X, x_0). \end{aligned}$$

$$\Rightarrow \qquad \tilde{g} o \tilde{f} = \text{Identity homomorphism on } \pi_1(X, x_0)$$

Similarly, $\tilde{f} o \tilde{g}$ = Identity homomorphism on $\pi_1(X, x_1)$

Consequently, $\tilde{f} : \pi_1(X, x_0) \to \pi_1(X, x_1)$ is an isomorphism of groups.

**Corollary 2 :** *If $f : \mathbf{I} \to X$ is a path in X from $x_0$ to $t_1$, then f induces an isomorphism. $\tilde{f} : \pi_1(X, x_0) \to \pi_1(X, x_1)$ defined by $\tilde{f}([u]) = [\bar{\bar{f}} * u * f]$, where $\bar{f}$ is the inverse path of f in X.*

**Theorem 8 :** *Let $\mathbf{R}^n$ be the Euclidean n-space and*

$$D^n = \{x \in \mathbf{R}^n : \|x\| \leq 1\}$$

*be the n-disk. If $x_0 \in \mathbf{R}^n$ and $d_0 \in D^n$, then*

(*a*) $\pi_1(\mathbf{R}^n, x_0) = \{0\}$ *(the trivial group)*

(*b*) $\pi_1(D^n, x_0) = \{0\}$.

**Proof :** Let $u$ be a loop in $\mathbf{R}^n$ based at $x_0$ and $c : \mathbf{I} \to X$ be the constant loop in $\mathbf{R}^n$ based at $x_0$ defined by $c(t) = x_0, \ \forall\, t \in \mathbf{I}$.

Define a map $F : \mathbf{I} \times \mathbf{I} \to \mathbf{R}^n$ by

$$F(t, s) = sx_0 + (1 - s)\, u(t)$$

As $\mathbf{R}^n$ is a convex set, $F$ is well defined and continuous.

Now $\quad F(t, 0) = u(t),\ F(t, 1) = x_0 = c(t),\ \forall\, t \in I,$

$$F(0, s) = sx_0 + (1 - s)\, u(0) = x_0 = u(0) = c(0)$$

and $\quad F(1, s) = sx_0 + (1 - s)\, u(1) = x_0 = u(1) = c(1),\ \forall\, s \in I$

Hence $\quad F : u \simeq c \text{ rel } (0, 1) \Rightarrow [u] = [c] = 0$

Consequently, $\pi_1(\mathbf{R}^n, x_0) = \{0\}$.

(*b*) Proceeds as in (*a*).

## Fundamental Group of the Circle $S^1$

We now find the fundamental group of the circle $S^1$ endowed with topology induced by the usual topology on $\mathbf{R}^2$. Since $S^1$ is a path connected space, all fundamental groups $\pi_1(S^1, s)$ are isomorphic, $\forall\, s \in S^1$. So, to compute $\pi_1(S^1, s)$, it is sufficient to take a fixed point $s_0$ on $S^1$. We describe an intutive approach to find $\pi_1(S^1, s_0)$. Analytical approach is available in any text book of algebraic topology including the book [2 or 11].

A loop in $S^1$ based at $s_0 \in S^1$ is a closed path which begins and ends at $s_0$. Then $u$ is either a null path or $u$ is given by one or more complete description of the circle.

Let $u$, $v$ be two loops in $S^1$ based at $s_0$ such that $u$ describes $S^1$ $m$ times and $v$ describes $S^1$ $n$ times. If $m > n$, then $u \not\simeq v$ rel (0, 1), because, $u * v^{-1}$ is a path describing $S^1$ $(m - n)$ times, and such a path is not homotopic to a null path. Thus the homotopy classes of loops in $S^1$ based at $s_0$ are in bijective correspondence with the set of integers **Z**. Hence there exists a bijection $\psi : \pi_1 (X, x_0) \to \mathbf{Z}$. But $(\mathbf{Z}, +)$ is an infinite cyclic group and hence $\pi_1 (S^1, s_0)$ is an infinite cyclic group. Thus we have the following Theorem :

**Theorem 9 :** $\pi_1 (S^1, s_0) \cong \mathbf{Z}$ *(isomorphic).*

**Theorem 10 :** *Every continuous map* $f : (X, x_0) \to (Y, y_0)$ *induces a homomorphism* $f_* : \pi_1 (X, x_0) \to \pi_1 (Y, y_0)$.

**Proof :** Define $f_* : \pi_1 (X, x_0) \to \pi_1 (Y, y_0)$ by

$f_* ([u]) = [fou]$ where $fou$ is the composite of maps :

$$I \xrightarrow{\quad u \quad} X \xrightarrow{\quad f \quad} Y.$$

Since $fou$ is a loop in $Y$ based at $y_0$ and for $v \in [u]$,

$$u \simeq v \text{ rel } (0, 1) \Rightarrow fou \simeq fov \text{ rel } (0, 1)$$

$\Rightarrow f_*$ is well defined. We now show that $f_*$ is a homomorphism.

Now for $[u]$, $[v] \in \pi_1 (X, x_0)$,

$$f_* ([u] \, o \, [v]) = f_* ([u * v]) = [f o (u * v)]$$

Again, $u * v : \mathbf{I} \to X$ is a loop given by

$$(u * v) \ (t) = \begin{cases} u (2t), & 0 \le t \le 1/2 \\ v (2t - 1), & 1/2 \le t \le 1 \end{cases}$$

This shows that $fo (u * v)$ is a loop $I \to Y$ given by

$$(fo (u * v)) \ (t) = \begin{cases} f (u (2t)), & 0 \le t \le 1/2 \\ f (v (2t - 1)), & 1/2 \le t \le 1 \end{cases}$$

$$= \begin{cases} (fou) \ (2t), & 0 \le t \le 1/2 \\ (fov) \ (2t - 1), & 1/2 \le t \le 1 \end{cases}$$

$$= ((fou) * (fov))\ (t),\ \forall\ t \in \mathbf{I}$$

$$\Rightarrow \qquad fo\,(u * v) = (fou) * (fov)$$

Hence $[fo\,(u * v)] = [fou]\, o\, [fov],\ \forall\ [u],\ [v] \in \pi_1\,(X,\ x_0)$

$$\Rightarrow \qquad f_*\,([u]\, o\, [v]) = f_*\,([u])\, o\, f_*\,([v]),\ \forall\ [u],\ [v] \in \pi_1\,(X,\ x_0)$$

$\Rightarrow f_*$ is a group homomorphism.

$f_* : \pi_1\,(X,\ x_0) \to \pi_1\,(Y,\ y_0)$ is called the *homomorphism induced* by the continuous map $f: (X,\ x_0) \to (Y,\ y_0)$.

**Corollary 3 :** *If* $f: (X,\ x_0) \to (Y,\ y_0)$ *and* $g: (Y,\ y_0) \to (Z,\ z_0)$ *are continuous maps, then the continuous map*

$gof: (X,\ x_0) \to (Z,\ z_0)$ *induces a homomorphism*

$(gof)_* : \pi_1\,(X,\ x_0) \to \pi_1\,(Z,\ z_0)$

such that $(gof)_* = g_*\, of_*$.

**Proof :** It follows from the Definition of induced homomorphism.

**Corollary 4 :** *If* $I_X : (X,\ x_0) \to (X,\ x_0)$ *is the identity map, then its induced homomorphism* $I_{X*} : \pi_1\,(X,\ x_0) \to \pi_1\,(X,\ x_0)$ *is the identity automorphism.*

**Proof :** It follows from the definition of $I_X^*$.

**Theorem 11 :** $S^1$ *is not a retract of* $D^2$.

**Proof :** If possible $S^1$ is a retract of $D^2$, then for $S^1 \overset{i}{\subset} D^2$, there exists a retraction.

$r: D^2 \to S^1$ such that $roi = {}^1S^1$ (identity map on $S^1$).

Hence the composite homomorphism

$$\pi_1\,(S^1,\ s_0) \xrightarrow{\ i_*\ } \pi_1\,(D^2,\ d_0) \xrightarrow{\ r_*\ } \pi_1\,(S^1,\ s_0)$$

must be identity automorphism on $\pi_1\,(S^1,\ s_0) = \mathbf{Z}$. But $\pi_1\,(D^2,\ d_0) = 0$. Hence $r_*\, oi_* \neq$ identity homomorphism.

Because, the composite homomorphism $\mathbf{Z} \xrightarrow{\ i_*\ } O \xrightarrow{\ r_*\ } \mathbf{Z}$ can never be the identity homomorphism.

In other words, $S^1$ is not a retract of $D^2$.

**Corollary 5 :** *The identity map* $1_{S^1} : S^1 \to S^1$ *can not continuously extended over* $D^2$.

We now in a position to prove Brouwer Fixed Point Theorem for $D^2$.

**Theorem 12 :** Every continuous map $f: D^2 \longrightarrow D^2$ has a fixed point.

**Proof :** If possible $f$ has no fixed point. Then $f(x) \neq x$ for any $x \in D^2$.

For each $x \in D^2$, define $r(x)$ to be the point where the line from $r(x)$ through $x$ meets the circle $S^1$. Then $r : D^2 \longrightarrow S^1$ is a continuous map such that $r \mid S^1 = 1_{S^1}$. In other words, $S^1$ is a retract of $D^2$ (equivalently, $1_{S^1}$ has a continuous extension over $D^2$). But this is a contradiction (see Theorem 11 or its Corollary). Consequently, $f$ has a fixed point.

**Remark :** The higher-dimensional analogue of Theorem 12 holds *i.e.*, every continuous map $f : D^n \longrightarrow D^n$ $(n \geq 2)$ has a fixed point.

The proof is beyond the scope of this book. (For proof see Example 5 p. 238 of [1]).

## EXERCISES

1. Let $f, g : (X, x_0) \to (Y, y_0)$ be two continuous maps such that $f \simeq g$. Then they will induce the same homomorphism (see Theorem 10) *i.e.*, $f_* = g_* : \pi_1(X, x_0) \longrightarrow \pi_1(Y, y_0)$.

2. If $(X, x_0)$ and $(Y, y_0)$ are homeomorphic, then the fundamental groups $\pi_1(X, x_0)$ and $\pi_1(Y, y_0)$ are isomorphic.

   [**Hint :** If $(X, x_0)$ and $(Y, y_0)$ are homeomorphic, then there exist continuous maps $f : (X, x_0) \longrightarrow (Y, y_0)$ and $g : (Y, y_0) \longrightarrow (X, x_0)$ such that $gof = I_X$ and $fog = I_Y$. Hence $(gof)_* = (I_X)_*$ and $(fog)_* = (I_Y)_*$ by the Theorem 9 $\Rightarrow g_* o f_* = 1d$ and $f_* o g_* = id$ (Identity homomorphism) by Corollaries 3 and 4 $\Rightarrow f_* : \pi_1(X, x_0) \longrightarrow \pi_1(Y, y_0)$ is an isomorphism].

3. If $(X, x_0)$ and $(Y, y_0)$ are homotopy equivalent spaces, then the fundamental groups $\pi_1(X, x_0)$ and $\pi_1(Y, y_0)$ are isomorphic.

   [**Hint :** By hypothesis, there exist continuous maps $f : (X, x_0) \to (Y, y_0)$ and $g : (Y, y_0) \to (X, x_0)$ such that $gof \simeq 1_X$ and $fog \simeq 1_Y$. Then

$(gof)_* = 1_{X^*}$ and $(fog)_* = 1_{Y^*}$ by Exericse 1. Now proceed as in Exercise 2 to show that $f_*$ is an isomorphism].

**4.** If $f: (X, x_0) \longrightarrow (Y, y_0)$ is a homotopy equivalence, then the induced homomorphism $f_* : \pi_1 (X, x_0) \longrightarrow \pi_1 (Y, y_0)$ is an isomorphism. [**Hint :** Proceed as in Exercise 3].

**5.** Let $f$ and $g$ be two paths in $X$ from $x_0$ to $x_1$, then their induced isomorphisms $\tilde{f}$ and $\tilde{g}$ (see Corollary 2) are equal *i.e.*, $\tilde{f} = \tilde{g}$ iff the fundamental group $\pi_1 (X, x_0)$ is abelian.

**6.** A space $X$ is called simply connected if it is path connected and $\pi_1 (X, x) = 0$ for some $x_0 \in X$ and hence for every $x_0 \in X$. Let $X$ be a simply connected space. Then any two paths having the same initial point and the same terminal point are homotopic.

**7.** Let $A$ be a subspace of $X$ and $x_0 \in A$ and $r : (X, x_0) \longrightarrow (A, x_0)$ be a retraction *i.e.*, $r \mid A = 1_A$ *i.e.*, $roi = 1_A$, where $i : (A, x_0) \longrightarrow (X, x_0)$ is the inclusion map. Then the induced homomorphism $i_* : \pi_1 (X, x_0) \longrightarrow \pi_1 (X, x_0)$ is a monomorphism and the induced homomorphism $r_* : \pi_1 (X, x_0) \longrightarrow \pi_1 (A, x_0)$ is an eqimorphism.

[**Hint :** $roi = I_A \Rightarrow (roi)_* = I_A^* \Rightarrow r_* \, oi_* = 1d$ (identity) $\Rightarrow r_*$ is an epimorphism and $i_*$ is a monomorphism (see Theorem 1.1.9 of [1]].

**8.** Let $A$ be a subspace of $\mathbf{R}^n$ and $f: (A, x_0) \longrightarrow (Y, y_0)$ be a continuous map. If $f$ is continuously extended over $\mathbf{D}^n$ into $Y$, then the induced homomorphism $f_* : \pi_1 (A, x_0) \longrightarrow \pi_1 (Y, y_0)$ is a zero homomorphsim.

[**Hint :** Let $f$ have a continuous extension

$$F : (\mathbf{R}^n, x_0) \longrightarrow (Y, y_0).$$

Then $Foi = f \Rightarrow F_* oi_* = f_* \Rightarrow$ the composite homomorphism.

$$\pi_1 (A, a_0) \xrightarrow{i_*} \pi_1 (\mathbf{R}^n, a_0) \xrightarrow{F_*} \pi_1 (Y, y_0) \text{ is } f_*.$$

But $(\mathbf{R}^n, a_0) = \{0\} \Rightarrow f_*$ is the zero homomorphism].

**9.** Let $f, g : I \rightarrow X$ be two paths in $X$ and $\bar{g}$ be the inverse path of $g$ in $X$ such that $f * \bar{g}$ exists and is a closed path in $X$. Then $f * \bar{g}$ is homotopic to the null path iff $f \simeq g$.

**10.** Let $X$ and $Y$ be two topological spaces. If $x_0 \in X$ and $y_0 \in Y$ then the groups $\pi_1 (X \times Y, (x_0, y_0))$ and $\pi_1 (X, x_0) \times \pi_1 (Y, y_0)$ are isomorphic.

Hence show that the fundamental group of the torus $T (= S^1 \times S^1)$ is $\mathbf{Z} \times \mathbf{Z}$.

# BIBLIOGRAPHY


1. Adhikari, M.R., Groups, Rings and Modules with Applications, Universities Press (India) Ltd., 1999.
2. Adhikari M.R. and Ganguly, S., A Basic Course in Algebraic Topology (To appear).
3. Alexandrov, P.S., Introduction to Set Theory and General Topology, Moscow, 1979.
4. Burger, D.C.J., Analytical Topology, Von Nostrand, Preinceton, 1966.
5. Bushaw, D., Elements of General Topology, John Wiley and Sons N.Y. 1963.
6. Copson, E.T., Metric spaces, Cambridge, Tracts in Math and Math Physics No. 57, Cambridge University Press, London, 1968.
7. Dugundji, J., Topology, Allyn and Bacon, Bosten, 1966.
8. Halmos, P.R., Naive Set Theory, Ven Nostrand Reinhold Co., N.Y., 1960.
9. Kelly, J.L., General Topology, Van Nostrand Reinhold Co., N.Y. 1955.
10. Maunder, C.R.F., Algebraic Topology, Van Nostrand Reinhold Co., London, 1970.
11. Munkers, J.R., Topology, A First Course, Prenctice-Hall of India Pvt. Ltd., New Delhi, 1978.
12. Patterson, E.M., Topology, Oliver and Boyd Ltd., 1959.
13. Spanier, E.H., Algebraic Topology, Tata Mc-Graw Publication Co. Ltd., New Delhi, 1976.
14. Stephen, Willard, General Topology, Addision-Wesley Pub. Co. Inc., 1970.

# INDEX